Tectonostratigraphic Terranes and Tectonic Evolution of Mexico

Richard L. Sedlock
Department of Geology
San Jose State University
San Jose, California 95192-0102

Fernando Ortega-Gutiérrez
Instituto de Geología
Universidad Nacional Autónoma de México, Apartado 70-296
México 20, D.F.

and

Robert C. Speed
Department of Geological Sciences
Northwestern University
Evanston, Illinois 60208

SPECIAL PAPER

278

1993

Published by The Geological Society of America, Inc.
3300 Penrose Place, P.O. Box 9140, Boulder, Colorado 80301

Printed in U.S.A.

GSA Books Science Editor Richard A. Hoppin

Library of Congress Cataloging-in-Publication Data
Sedlock, Richard L., 1958–
    Tectonostratigraphic terranes and tectonic evolution of Mexico /
Richard L. Sedlock, Fernando Ortega-Gutiérrez, Robert C. Speed.
        p.   cm. — (Special paper ; 278)
    Includes bibliographical references and index.
    ISBN 0-8137-2278-0
    1. Geology, Structural—Mexico.   2. Geology, Stratigraphic.
I. Ortega-Gutiérrez, Fernando.   II. Speed, Robert C.   III. Title.
IV. Series: Special papers (Geological Society of America) ; 278.
QE629.S43   1993
551.8'0972—dc20   93-3196
                                                          CIP

Cover photo: View to the southwest of Isla Santa Margarita,
Baja California Sur, México; Isla Magdalena in foreground. Both
islands are underlain by ophiolitic, arc, and blueschist terranes
whose history of amalgamation and displacement may be representative of the tectonic evolution of most of México.

10   9   8   7   6   5   4   3   2   1

Contents

Acknowledgments

This work benefited from discussions with or comments by Carlos Aiken, Emilio Almazán-Vázquez, Thomas Anderson, Suzanne Baldwin, David Bottjer, Richard Buffler, Thierry Calmus, Alejandro Carrillo-Chávez, Zoltan de Cserna, Ray Ethington, Gordon Gastil, Roger Griffith, James Handschy, Jonathan Hagstrum, Uwe Herrmann, Norris Jones, Keith Ketner, David Kimbrough, Jeff Lee, James McKee, John Minch, Pete Palmer, Barney Poole, Kevin Robinson, Jaime Roldán-Quintana, Jack Stewart, James Stitt, and George Viele.

Preprints and reprints were provided by Luis Delgado-Argote, Nick Donnelly, Hans-Jurgen Gursky, Christopher Henry, Gregory Horne, Hugh McLean, James McKee, Mark McMenamin, Dieter Michalzik, Douglas Smith, and James Wilson.

Unpublished data were provided by Alejandro Carrillo-Chávez, Luis Delgado-Argote, Nick Donnelly, Gregory Horne, Uwe Herrmann, Kevin Robinson, and Douglas Smith.

Thorough, helpful reviews were supplied by Christopher Henry, Jack Stewart, and an anonymous reviewer, who are not to blame for whatever speculations and mistakes persist in this version.

Kinematic studies in Part 2 were partly supported by National Aeronautics and Space Administration Grant NAG 5-1008. Sedlock thanks the faculty, students, and staff of the Department of Geological Sciences at Northwestern University and the Department of Geological Sciences at the University of Missouri–Columbia for computer and technical support.

Geological Society of America
Special Paper 278
1993

Tectonostratigraphic Terranes
and Tectonic Evolution of Mexico

ABSTRACT

Part 1 of this work (Sedlock, Ortega-Gutiérrez, and Speed) is a synthesis of geo-scientific data pertaining to México and northern Central America using the framework of a new division of these regions into tectonostratigraphic terranes. First, we review the morphotectonic provinces and the modern plate tectonic framework of the region. Next, we present data for 17 terranes that, except for North America, are named after indigenous cultures. Terrane descriptions are based on published and unpublished geophysical and geologic data of all types, utilizing a much more extensive data base than that used in previous terrane divisions. Each terrane description includes, if possible, an interpretive geologic and tectonic history focusing on distinctive features; a description of constituent rock units, with extended descriptions of especially significant or controversial units; a schematic tectonostratigraphic column, which in many cases shows geographic variation in the form of a structure section; and a compilation of radiometric data, including dates, system used, errors, and sources. Finally, we discuss the rationale for distinguishing individual terranes and summarize data concerning the orientation, nature, and kinematic history of terrane boundaries. An extended reference list is included.

Part 2 of this work (Sedlock, Speed, and Ortega-Gutiérrez) is a speculative model of the Late Precambrian to Cenozoic tectonic evolution of the terranes that comprise México and northern Central America. First, we discuss numerous formal premises on which the model is predicated, including Late Jurassic sinistral slip on the Mojave-Sonora Megashear and Late Cretaceous–Paleogene northward displacement of Baja California. Next, we review constraints imposed by plate motion models on the tectonic evolution of the region. Finally, we present a reconstruction of the tectonic evolution of the region that, while certainly not a unique solution, is an internally consistent solution that is testable in many respects.

The following are a few of the salient features of the reconstruction. (1) Grenville basement in eastern and southern México is regarded to be far-traveled with respect to the southern termination of the Grenville belt in North America. (2) The late Paleozoic Ouachitan suture that marks the collision of North America and Gondwana does not and did not extend into central México. (3) The Permo-Triassic continental arc on the western margin of Pangea affected only the far eastern edge and far northwestern corner of México; most of what is now México was a complex assemblage of arcs, continental blocks, and basins in the oceanic region west and south of the Pangean continental arc. (4) Continental México grew most markedly toward its present form during the Late Triassic and Jurassic as terranes were episodically accreted to its southern and western flanks. (5) Mesozoic southward and westward continental growth was accompanied by a southward and westward shift of the locus of arc magmatism. (6) The tectonically active southern and western margins of México were sites of large margin-parallel translations of terranes that accommodated the tangential component of oblique convergence of México with oceanic lithosphere to the west. Convergence and terrane translation were sinistral from the Triassic(?) until the Early Cretaceous, and

Sedlock, R. L., Ortega-Gutiérrez, F., and Speed, R. C., 1993, Tectonostratigraphic Terranes and Tectonic Evolution of Mexico: Boulder, Colorado, Geological Society of America Special Paper 278.

dextral in the mid-Cretaceous and Paleogene. (7) Jurassic stretching and rifting in the Gulf of Mexico was not kinematically related to sinistral faulting on the Mojave-Sonora Megashear; instead, slip on the megashear and on other, more outboard, fault systems was controlled by left-oblique convergence of México with plates in the Pacific basin. (8) Paleomagnetic data that indicate about 15° of northward latitudinal displacement of Baja in the Late Cretaceous and Paleogene can be reconciled with geologic correlations only by postulating an earlier episode of southward displacement during left-oblique convergence. (9) The Cretaceous reconstruction is consistent with postulated origins at Mexican latitudes of terranes in the western United States and Canada. (10) The Caribbean plate, including the Chortis block, has been translated 1,000 to 2,000 km eastward on strike-slip faults along the southern margin of México since about 45 Ma. (11) Basin and Range extension has affected most of México north of about 20°N, an area much larger than the Basin and Range province in the United States.

We have endeavored to ensure that Part 1 of this volume and the extended reference list serve as up-to-date storehouses of accurate information for anyone interested in the geology of México and northern Central America. We hope that Part 2—which clearly is but a first attempt to explain the diverse aspects of this complex region in tectonic terms—is sufficiently provocative to spur further study and subsequent modification of our ideas.

GOALS AND OVERVIEW

The subject of this two-part work is the geology and tectonics of México. Where appropriate, we also discuss aspects of the geology and tectonics of the southwestern United States, northern Central America, the Caribbean region, and northwestern South America.

In Part 1, we present a new division of México into tectonostratigraphic terranes using a geologic and geophysical data base that is much more extensive than that used in previous divisions. We hope that this compilation and synthesis of published and unpublished information serves as a useful reference for current and future workers in México and adjacent areas, and that it helps pinpoint problematic topics and areas that would benefit from further study. Part 1 begins with a very brief outline of the morphotectonic provinces of México, followed by a review of the modern plate tectonic framework of the region. The bulk of Part 1 is devoted to the 17 tectonostratigraphic terranes that we recognize in the region, including a review and synthesis of published and unpublished data, schematic tectonostratigraphic columns, and tabulated radiometric data. We also offer brief interpretations of the geologic history of each terrane and of the displacement on terrane-bounding faults.

In Part 2, we offer a speculative kinematic model of the Late Precambrian to Cenozoic tectonic evolution of these terranes. Although some aspects of this model undoubtedly will be modified or altered by future work, the model is useful as the first attempt to reconstruct in detail the evolution of all of México in a regional context. Part 2 also includes discussions of the numerous premises on which the model is predicated, and key controversial or poorly understood aspects of the model. Finally, we hope that the compilation of hundreds of entries in the bibliography serves as a useful resource for future workers.

As has long been noted, investigations of the geology and

tectonics of México and Central America are handicapped by several serious drawbacks. Thick tropical vegetation obscures bedrock in much of southeastern México and Central America, and access to the sparse outcrops is difficult. Much of northern and central México is covered by widespread late Mesozoic marine strata (Gulf of Mexico sequence) and Cenozoic volcanic rocks (Sierra Madre Occidental and Trans-Mexican Volcanic Belt), making it very difficult to establish the significance of isolated basement outcrops. Published geochronologic data from pre-Cretaceous rocks include K-Ar, Rb-Sr, and obsolete Pb-α determinations, but very few U-Pb and $^{40}Ar/^{39}Ar$ data have been published in refereed journals. Much of the country has been mapped at scales of 1:250,000, and 1:50,000 (see de Cserna, 1989), and an eight-sheet 1:1,000,000 geologic map was published in 1980 by Secretaría de Programación y Presupuesta (SPP), now called Instituto Nacional de Estadística Geografía e Información, or INEGI (INEGI, 1980), but most work of the last 15 yr has not yet been incorporated into these maps. Much of the available literature is published in Spanish language journals that have fairly limited circulation, and many studies in both English and Spanish are available only in abstract form. In light of these problems, we caution that this volume represents our best attempt to integrate data available at this time, and we anticipate future modifications based on new geologic and geophysical data.

PART 1: TECTONOSTRATIGRAPHIC TERRANES OF MÉXICO

Richard L. Sedlock, Fernando Ortega-Gutiérrez, and Robert C. Speed

MORPHOTECTONIC PROVINCES

Here we very briefly delineate and describe the major morphotectonic provinces of México (Fig. 1), based on the excellent summaries presented by de Cserna (1989) and Lugo-Hubp (1990).

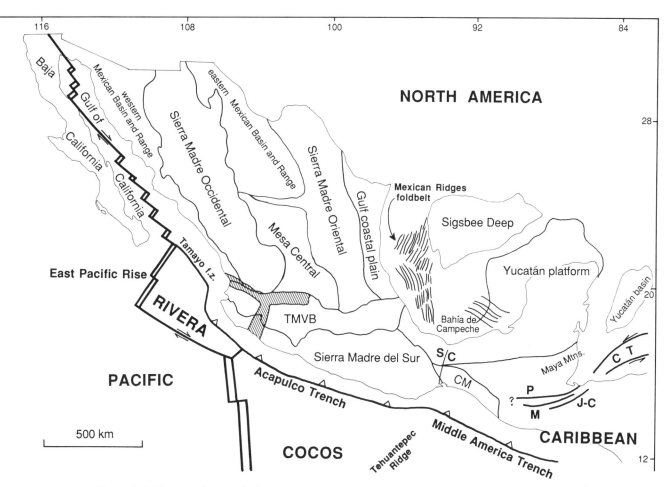

Figure 1. Major morphotectonic features and plate tectonic setting of México and vicinity. Major tectonic features identified by bold type; plate names given in all capital letters. Line segments in Gulf of Mexico are hinge traces of folds. Plate boundaries denoted by heavy lines; lighter lines bound labeled morphotectonic provinces. Lined region in southwestern México: Zacoalco (west), Colima (south), and Chapala (east) grabens. Abbreviations: CM, Chiapas Massif; CT, Cayman Trough; J-C, Jocotán-Chamelecón fault; M, Motagua fault; P, Polochic fault; SC, Salina Cruz fault; TMVB, Trans-Mexican Volcanic Belt.

The arid, rugged Baja California peninsula is separated from mainland México by the Gulf of California, in which most or all displacement on the Pacific–North America plate boundary currently is accommodated. The gulf is surrounded by an extensional province that includes the eastern margin of Baja California, the Laguna Salada–Salton Trough, coastal, northwestern, and central Sonora, and coastal Sinoloa and Nayarit. This extended province is bounded to the east by the Sierra Madre Occidental, a linear, north-northwest–elongate plateau of thick Tertiary volcanic rocks that, with few exceptions, completely obscure older rocks. East of the Sierra Madre Occidental is the Mexican Basin and Range province (Sierras y Valles or Sierras y Cuencas), another extended province characterized by north-northwest–trending basins and ranges that reaches from central México north into the southwestern United States. The margins of the Sierra Madre Occidental are affected by extension, but the plateau clearly has

undergone less net extension than provinces to the east and west. The extended provinces are the southern continuation of the Basin and Range province of the southwestern United States; we suggest that they be called the western and eastern Mexican Basin and Range provinces.

The eastern Mexican Basin and Range province partly overlaps, and is transitional with, the Sierra Madre Oriental to the east. The Sierra Madre Oriental consists of Mesozoic carbonates and clastic rocks that obscure underlying rocks and that were thrusted eastward and folded during Laramide orogenesis. The Sierra Madre Oriental extends as far south as the Trans-Mexican Volcanic Belt near 20°N; on the east it abuts the Gulf coastal plain, a low-lying region known mainly from exploratory drilling for hydrocarbons. The structural grain of the Sierra Madre Oriental generally is slightly west of north, but near 25°N the province includes a west-trending prong called the Monterrey-Torreón transverse system or Sierras Transversales. This transverse system separates the eastern Mexican Basin and Range province to the north from an elevated plateau to the south known as the Mesa Central, Meseta Central, or Altiplano. Irregular relief within the Mesa Central has been interpreted by some workers as due to block faulting related to the emplacement of Tertiary volcanic rocks in the Sierra Madre Occidental (Pasquaré and others, 1987) or, more probably, to late Cenozoic Basin and Range extension (Stewart, 1978).

The provinces of northern México outlined above are bounded on the south by the Trans-Mexican Volcanic Belt, a roughly east-west–trending belt of Miocene to Holocene volcanic rocks, stratovolcanoes, and active faults. South of this belt is the Sierra Madre del Sur, a rugged, geologically complex region that contains exposed basement rocks as old as Precambrian. The Chiapas Massif, or Sierra de Chiapas, intersects the Sierra Madre del Sur province in the vicinity of the narrow Isthmus of Tehuantepec and has geologic similarities with rocks in central Guatemala. The Yucatán platform is a broad, low-lying plateau underlain by subhorizontal Mesozoic strata and crystalline basement known only from wells.

MODERN PLATE TECTONIC FRAMEWORK

México is in the southwestern North American plate, with the exception of most of the Baja California peninsula, which is attached to the Pacific plate, and a small tract near Guatemala that probably moves partly or wholly with the Caribbean plate (Fig. 1). On its southern margin México abuts the Caribbean plate at a probable major sinistral strike-slip fault system. On its northwestern margin its contact with the Pacific plate is a system of transform faults and short segments of the East Pacific Rise. On its southwestern margin it is being underthrust by oceanic lithosphere of the Cocos and Rivera plates. Offshore to the southwest are the complex plate boundaries of the Cocos, Pacific, and Rivera oceanic plates.

Caribbean–North America plate boundary

In the Caribbean basin east of Central America, the Caribbean–North America plate boundary is the Cayman Trough (Fig. 1). In Central America, the plate boundary is widely interpreted as a sinistral fault zone consisting of the Motagua fault (site of a M_s = 7.5 left-slip earthquake in 1976), the Polochic (sometimes called Cuilco-Chixoy-Polochic) fault, and the Jocotán-Chamelecón fault (Malfait and Dinkelman, 1972; Muehlberger and Ritchie, 1975; Bowin, 1976; Plafker, 1976; Burkart, 1978, 1983; Schwartz and others, 1979; Sykes and others, 1982). All three faults have strong topographic expression and the Motagua and Polochic faults are interpreted to be active, but rates of displacement are unknown and demonstrable Quaternary sinistral offsets are minimal (Plafker, 1976; Schwartz and others, 1979; Burkart and others, 1987). The lack of surface expression of the Motagua fault in western Guatemala, the westward bifur-

cation and apparent termination of the Polochic fault (Case and Holcombe, 1980; Burkart and others, 1987), neotectonic field studies in southeastern México and northern Guatemala, and stability analysis of the boundary zone of the Caribbean, North American, and Cocos plates indicate that deformation in the diffuse Caribbean–North America–Cocos triple junction probably is accommodated by most or all of the following: sinistral slip on the Polochic and Motagua faults; sinistral slip on faults between Puerto Angel, Oaxaca and Macuspana, Tabasco; east-west extension in the vicinity of the Isthmus of Tehuantepec; and clockwise rotation of crust between the Puerto Angel-Macuspana and Polochic fault systems (Guzmán-Speziale and others, 1989; Delgado-Argote and Carballido-Sánchez, 1990).

Relative motion of the Caribbean plate with respect to the North American plate in the vicinity of Guatemala and the western Cayman Trough, as determined from the Euler vector of the NUVEL-1 global plate motion model (DeMets and others, 1990), is 12 ± 3 mm/yr, S75-80E. The magnitude previously had been estimated at 37 (Sykes and others, 1982), 20 (Macdonald and Holcombe, 1978), and 15 mm/yr (Stein and others, 1988). The Cayman Trough spreading rate, based on magnetic anomalies and subsidence, has been estimated to be about 15 mm/yr (Rosencrantz and others, 1988), but cumulative Caribbean–North America motion may be 20 mm/yr or more, based on tectonic interpretation of new SeaMARC mapping (Rosencrantz and Mann, 1991).

Pacific–North America plate boundary

In the Gulf of California, the Pacific–North America plate boundary consists of short segments of the northernmost East Pacific Rise spreading ridge separated by northwest-striking transform faults (Fig. 1). Normal oceanic crust, as defined by linear magnetic anomalies parallel to ridge segments, is present only between the Pescadero and Tamayo fracture zones at the mouth of the gulf. A thick blanket of sediment smooths basement topography in much of the gulf (van Andel, 1964; Curray, Moore, and others, 1982; Aguayo-C., 1984), but transform faults and as many as 16 ridge segments can be determined on the basis of bathymetry, sediment distribution, negative gravity anomalies, heat flow, and active seismicity (Dauphin and Ness, 1991; Couch and others, 1991; Ness and Lyle, 1991). Background heat flow values are about 200 mW/m² (von Herzen, 1963; Henyey and Bischoff, 1973), and maximum values are as high as 6,250 mW/m² in the vicinity of vigorous hydrothermal vents in the Guaymas basin near 27°N (Lonsdale and Becker, 1985; Becker and Fisher, 1991). Crustal thickness decreases from about 20 km in the Laguna Salada–Salton Trough north of the gulf to 13 km in the northern gulf to 8 km in the southernmost gulf, whereas low-density (3.1 to 3.15 g/cc) upper mantle is 90 km wide and 4 km thick beneath the northern gulf but about 230 km wide and 10 km thick beneath the southernmost gulf (Couch and others, 1991). The velocity structure of the mantle beneath the Gulf of California includes unusually low velocities to a depth of about 350 km (Walck, 1984).

Modern seismicity on the Pacific–North America plate boundary in México coincides fairly well with ridge segments and transform faults in the gulf but is more dispersed in the diffuse plate boundary zone in northern Baja California. Earthquakes in the gulf region are shallower and smaller than in the subduction regime of southern México; only two $M_s \geqslant 7.5$ earthquakes have occurred since 1900. On land, most earthquakes occur in swarms corresponding to mapped dextral strike-slip faults such as the Cerro Prieto, Imperial, Agua Blanca, and San Miguel–Vallecitos; normal faults that comprise the main gulf escarpment in northern Baja show little or no seismicity but may still be active (Brune and others, 1979; Frez and González-García, 1991a, b; Ness and Lyle, 1991; Suárez-Vidal and others, 1991).

The direction and magnitude of relative motion of the Pacific and North American plates can be measured only between the Pescadero and Tamayo transforms at the mouth of the Gulf of California. Marine magnetic anomalies on profiles normal to the Pescadero rift segment between these transforms are consistent with spreading rates of 49 (DeMets and others, 1987) or 66 mm/yr (Lyle and Ness, 1981). There is no statistical basis for preferring one rate over the other (Ness and others, 1991), but we favor the slow rate because it is predicted by the NUVEL-1 global plate motion model (DeMets and others, 1990) and because the fast rate requires about 11 mm/yr dextral slip in a zone in southern México that generally is interpreted to accommodate a small amount of sinistral slip (see below). Using the NUVEL-1 Euler pole, relative motion of the Pacific plate with respect to the North American plate ranges from 50 ± 1 mm/yr, N55W at latitude 23°N to 46.5 ± 1 mm/yr, N43W at latitude 31°N. Geologic and geodetic studies indicate only 35 mm/yr of dextral slip on the San Andreas fault system at the north end of the Gulf (Thatcher, 1979; Savage, 1983; Sieh and Jahns, 1984). Thus, Pacific-North America motion is not confined to the northern and central Gulf of California, and 11-15 mm/yr dextral slip must be taken up by faults within or west of Baja California and perhaps in mainland México (Sedlock and Hamilton, 1991). A laser trilateration geodetic study across the central Gulf of California (latitude 29°) found northwest-southeast relative motion of about 80 ± 30 mm/yr (Ortlieb and others, 1989). Preliminary determinations of relative motion determined by a Global Positioning System geodetic experiment across the southern Gulf of California are 44 ± 8, N53 $\pm 10°$W and 47 ± 7, N57 $\pm 6°$W, similar to the NUVEL-1 value within 1σ error (Dixon and others, 1991). However, this experiment, and any experiment of similar design, cannot measure displacement that may be occurring west of the southern Baja peninsula because no sites exist sufficiently close to the plate boundary on indisputible Pacific plate.

Pacific–North America relative motion includes a component of extension normal to the boundary, based on the orientation of the boundary and the NUVEL-1 Euler pole (DeMets and others, 1990). Extension currently is accommodated by rifting within the Gulf of California and by detachment faulting, high-angle normal faulting, and block tilting of continental crust along the east coast of the Baja California peninsula and, to a lesser

degree, western and central Sonora and western Sinaloa and Nayarit (e.g., Moore, 1973; Roldán-Quintana and González-León, 1979; Gastil and Fenby, 1991). Miocene extension and continental rifting that formed a proto-gulf probably were manifestations of Basin and Range extension (p. 117).

Cocos–North America and Cocos-Caribbean plate boundaries

The Cocos plate is being subducted beneath the North American and Caribbean plates at the Acapulco and Middle America trenches (Fig. 1). The NUVEL-1 plate motion model (DeMets and others, 1990) predicts that motion of the Cocos plate relative to North America is to the north-northeast, slightly counterclockwise from the normal to the trench, at rates of about 55 mm/yr near Colima, 60 near Acapulco, and 75 near the Chiapas-Guatemala border (see Fig. 2 for locations). Remote sensing, field, and plate motion studies support the inference that continental México east of the Colima graben and south of the Trans-Mexican Volcanic Belt is moving southeastward, parallel to the trench, along left-lateral faults at about 5 ± 5 mm/yr (Pasquaré and others, 1987, 1988; Johnson and Harrison, 1989; DeMets and Stein, 1990). Neotectonic studies recognize uplift of the Michoacán-Guerrero coast at rates as fast as 14 mm/yr (Corona-Esquivel and others, 1988).

Most large ($M_s > 7.5$) historical earthquakes recorded in southern México, including the 1985 Michoacán event (8.1) and aftershocks (J. Anderson and others, 1986), the 1907 (7.8) and 1957 (7.5) events near Acapulco (González-Ruiz and McNally, 1988), and the 1978 (7.7) Oaxaca event (Stewart and others, 1981), were associated with subduction of the Cocos plate beneath North America. Focal mechanisms indicate predominant shallow thrust events ascribed to underthrusting during subduction, and less common, deeper normal events that may represent tensional deformation of the subducted plate (Nixon, 1982; Burbach and others, 1984). In northern Central America and adjacent southeastern México west of 94°W, a well-defined Wadati-Benioff zone dips 15° to 20° to about 60 km and about 60° to maximum depths of 200 to 240 km (Burbach and others, 1984; LeFevre and McNally, 1985). In southern México east of 96°W, a less well-defined Wadati-Benioff zone dips 10° to 20° to a maximum depth of about 100 km (Havskov and others, 1982; Bevis and Isacks, 1984; Burbach and others, 1984; LeFevre and McNally, 1985; Valdes and others, 1986; Castrejón and others, 1988; Nava and others, 1988).

The continental arcs above the subduction zones also exhibit pronounced changes in the vicinity of 96–94°W (p. 7). One possible explanation for the different seismic and volcanic characteristics is the different buoyancy of lithosphere in the subducting plate on either side of the Tehuantepec Ridge, which intersects the trench at about 95°W (Fig. 1) (Bevis and Isacks, 1984). Another likely cause is differential plate motions: the transition at 96–94°W coincides with the diffuse Cocos–North America–Caribbean triple junction. The Cocos plate is more strongly coupled to the North American plate than to the Caribbean plate,

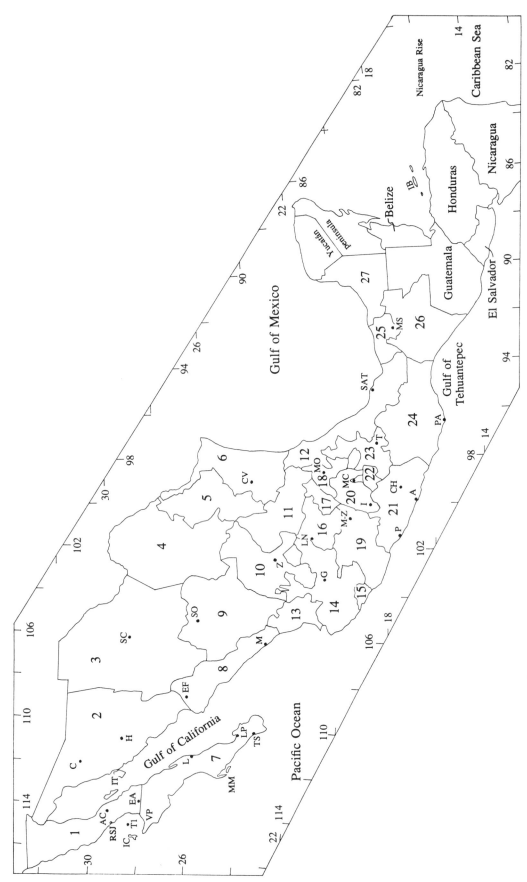

Figure 2. Geographic map of México and northern Central America, showing states (*estados*) and local-ities referred to in text. Numbered states: 1, Baja California; 2, Sonora; 3, Chihuahua; 4, Coahuila; 5, Nuevo León; 6, Tamaulipas; 7, Baja California Sur; 8, Sinaloa; 9, Durango; 10, Zacatecas; 11, San Luis Potosí; 12, Veracruz; 13, Nayarit; 14, Jalisco; 15, Colima; 16, Guanajato; 17, Querétaro; 18, Hidalgo; 19, Michoacán; 20, México; 21, Guerrero; 22, Morelos; 23, Puebla; 24, Oaxaca; 25, Tabasco; 26, Chiapas; 27, Campeche. Abbreviations: A, Acapulco; AC, Arroyo Calamajué; C, Caborca; CH, Chilpancingo; CV, Ciudad Victoria; EA, El Arco; EF, El Fuerte; G, Guadalajara; H, Hermosillo; I, Ixtapán de la Sal; IB, Islas de la Bahía (Bay Islands); IC, Isla Cedros; IT, Isla Tiburón; L, Loreto; LN, city of León; LP, La Paz; M, Mazatlan; MC, México City (Distrito Federal); MM, Magdalena-Margarita region; MO, Molango; MS, Macuspana; M-Z, Morelia-Zitácuaro region; P, Petatlán; PA, Puerto Angel; RSJ, Rancho San José; SAT, San Andres Tuxtla; SC, Sierra del Cuervo; SO, Santa María del Oro; T, Tehuacán; T1, Totoaba-1 (PEMEX well); TS, Todos Santos; VP, Vizcaíno peninsula; Z, Zacatecas city.

as indicated by backarc extension in Central America and back-arc contraction in southern México (Buffler and others, 1979; Weyl, 1980; McNally and Minster, 1981; de Cserna, 1989). The Caribbean plate appears to be motionless with respect to the subducted slab, resulting in a well-developed Wadati-Benioff zone and narrow arc, whereas motion of North America toward the slab has produced a broader, less continuous arc and increased the arc-trench distance (Burbach and others, 1984).

The alignment of terrestrial volcanic centers has been used to infer a segmented subducted slab beneath México and Central America (Stoiber and Carr, 1973; Carr and others, 1974; Nixon, 1982). Seismologic data have been interpreted to indicate that the subducted Cocos plate lithosphere is either smoothly curving and laterally continuous (Burbach and others, 1984) or segmented near 99°W where the O'Gorman Fracture Zone intersects the Middle America Trench (Singh and Mortera, 1991).

Rivera–North America plate boundary

The Rivera–North America plate boundary is the Acapulco Trench south of 20°N and probably the Tamayo Fracture Zone and Tres Marias escarpment north of 20°30′N (Fig. 1). Larson (1972) and Menard (1978) inferred that the Acapulco Trench was inactive and that the Rivera plate currently is part of the North American plate, but the lithosphere above the trench is seismically active, although to a lesser extent than the Middle America Trench along the Cocos–North America boundary (Ness and Lyle, 1991). The 1932 (8.1) Jalisco event clearly indicates shallow thrusting to the northeast, as would be expected if the Rivera plate underthrusts México (Nixon, 1982; Eissler and McNally, 1984). When compared with the NUVEL-1 plate motion model, spreading rates and slip vectors deduced from the boundaries of the Rivera plate indicate that the Rivera plate is kinematically distinct from both the North American and Cocos plates, and that the Rivera plate moves roughly orthogonally to the Rivera–North America plate boundary at rates that increase from about 6 mm/yr at the Tamayo transform fault to about 20 or even 30 at the Acapulco Trench (DeMets and others, 1990; DeMets and Stein, 1990). These results contrast sharply with earlier interpretations of a large dextral component of Rivera–North America relative motion (Minster and Jordan, 1979; Eissler and McNally, 1984). The nature of the Rivera–North America plate boundary north of 20°N is poorly understood, but the precipitous gravity gradient there may reflect a steeply dipping (transform?) fault boundary between continental and oceanic crust (Couch and others, 1991).

The Colima, Zacoalco (also called Tepic-Chapala), and Chapala grabens in southwestern México (Fig. 1) are characterized by active alkalic volcanism and about 1 mm/yr extension. Net extension across the rifts has been estimated at 4 to 8% (Barrier and others, 1990). It has been suggested that the Rivera triple junction (Pacific–North America–Rivera) is in the initial stages of jumping from the mouth of the Gulf of California to the junction of these grabens, and that the "Jalisco block," a fragment of North American continental crust in western Jalisco,

is being transferred to the Pacific plate (Luhr and others, 1985; Allan, 1986; Barrier and others, 1990; Allan and others, 1991). Offshore seismicity and bathymetric data are consistent with an incipient spreading ridge offshore southern Colima in a structurally complex zone of diffuse seismicity that probably encloses the indistinct Rivera–Cocos–North America and Pacific-Rivera-Cocos triple junctions (Bourgois and others, 1988; Dauphin and Ness, 1991; Ness and Lyle, 1991). However, plate motion kinematics imply that the Colima graben is a passive pull-apart basin and that the Chapala graben is the locus of transtensional displacement at the northwestern and northeastern boundaries, respectively, of a coastal sliver undergoing southeastward transport during oblique convergence (DeMets and Stein, 1990). Also, heat flow values offshore Colima are not elevated compared to other parts of the southern continental slope of México, suggesting that active rifting is not occurring (M. Khutorskoy and others, unpublished manuscript).

Trans-Mexican Volcanic Belt (TMVB)

The Trans-Mexican Volcanic Belt (TMVB) consists of late Miocene (11 Ma), chiefly andesitic to dacitic volcanic rocks and active volcanoes that extend across México from Nayarit to southern Veracruz (Figs. 1, 2). The volcanic rocks generally are linked to subduction of oceanic lithosphere of the Cocos and Rivera plates (Nixon, 1982; Nixon and others, 1987), but in several respects the TMVB is atypical of continental arcs. Compared to the Central American arc, for instance, the TMVB is broader, less continuous, and farther from the trench (Robin, 1982). Few intermediate-depth earthquakes that may be attributed to a subducted slab have been recorded beneath the TMVB, but an inclined band of earthquake foci that dips shallowly beneath southern México to a maximum depth of about 100 km probably marks the Wadati-Benioff zone (Bevis and Isacks, 1984; Burbach and others, 1984; LeFevre and McNally, 1985). The TMVB may also mark a nascent plate or crustal block boundary along which southern México is moving eastward (Shurbet and Cebull, 1984) or, more probably, westward (Johnson, 1987; Urrutia-Fucugauchi and Böhnel, 1988; DeMets and Stein, 1990) with respect to northern México. The TMVB is marked by very high heat flow (Ziagos and others, 1985) and very low shear velocities in the lower lithosphere (Gomberg and Masters, 1988). Recent studies of the neotectonics of the belt include those by Campos-Enríques and others (1990), Martínez-Reyes and Nieto-Samaniego (1990), Suter (1991), and Suter and others (1992), and papers in a special volume of Geofísica Internacional (e.g., Verma, 1987). Recent work on the age, geology, petrology, and geochemistry of specific volcanoes within the belt includes that of Dobson and Mahood (1985), Nixon (1989), Luhr and Carmichael (1990), Nelson (1990), Ferrari and others (1991), and Pasquaré and others (1991).

North American intraplate deformation

Although the Gulf of Mexico and eastern México generally are considered to be part of North America, active tectonism and volcanism imply relative motion between the two regions. North-trending fold hinges and east-vergent thrusts in Cretaceous and Cenozoic strata off the coast of Tamaulipas and northern Veracruz ("Mexican Ridges Foldbelt" of Buffler and others, 1979; "Cordillera Ordóñez" of de Cserna, 1981) probably record latest Cenozoic east-west shortening between eastern México and the Gulf of Mexico basin (Fig. 1). A younger, probably modern, phase of northeast-southwest shortening is indicated by the northeastward overthrusting of this offshore fold belt by the continental shelf (de Cserna, 1981, 1989). Late Cenozoic northeast-southwest shortening in southern México also is indicated by a southwest-verging fold and thrust belt on the northeast side of the Chiapas Massif and by active northwest-southeast-trending folds offshore northern Tabasco and Campeche (Fig. 1) (de Cserna, 1989).

Miocene to Quaternary, calc-alkalic to alkalic, siliceous to basaltic volcanic rocks that crop out along the Gulf of Mexico coast from Tamaulipas to San Andres Tuxtla (Fig. 2) have been attributed to intraplate rifting, but new geochemical studies imply eruption in a backarc setting during subduction of the Cocos plate (López-Infanzón and Nelson, 1990; Nelson and others, 1991).

Significance of regional geophysical studies

Few regional studies of geophysical characteristics of México have been published. On the basis of a short, unreversed refraction profile between northern Guanajuato and eastern Durango, Meyer and others (1961) inferred a crustal thickness of 37 to 47 km beneath the southeastern Sierra Madre Occidental. Fix (1975) inferred an average crustal thickness of about 30 km for central México based on arrival times of waves from earthquakes in Chiapas at receivers in the southwestern United States. Based on measurements of surface wave phase velocity and travel times at three newly installed long-period seismic stations in northern México, Gomberg and others (1988) inferred an average crustal thickness of about 40 km, corroborating the results of Meyer and others (1963), and an average thickness of a high-velocity mantle lid of about 30 to 40 km. The indicated lithospheric thickness of 70 to 80 km is significantly thicker than in the Basin and Range province of the western United States, implying less thinning and thus less extension in northern México, but is significantly thinner than in cratonal areas of the United States, implying that northern México may not be underlain by thick Precambrian continental lithosphere. The crustal and lithospheric thickness estimates are consistent with gravity models of Bouguer gravity anomalies in northern México (Aiken and others, 1988; Schellhorn and others, 1991). Gomberg and others (1988) also noted the presence of a low-velocity zone for shear waves (S_N as low as 4.0 km/sec) above 250 km in central México.

Gravity data from the vicinity of the Baja California peninsula are interpreted to indicate a southward decrease in crustal thickness from about 28 km at the border to about 20 km near the tip, a lithospheric thickness of 50 to 55 km, and thinning of the lithosphere on both sides of the Gulf of California (Aiken and others, 1988; Couch and others, 1991). Positive low-pass filtered gravity anomalies trend north-northwest on the Pacific continental shelf of Baja California between 30° and 23°N and in mainland Baja from north of the international border at least as far south as 26°N and perhaps to the tip of the peninsula at 23°N (Couch and others, 1991). In detail, the continental shelf anomaly appears to consist of at least three en echelon domains that may indicate dextral offset on northwest-striking strike-slip faults. Few of the linear magnetic anomalies on the Pacific continental shelf correspond to bathymetric and gravity lineaments (Ness and others, 1991). A well-defined paleotrench at the foot of the continental slope is indicated by a gravity low between 29° and 22°30′N, and a remnant of oceanic crust subducted at this trench probably extends eastward 100 to 150 km beneath the continental slope (Couch and others, 1991).

The compressional velocity structure of parts of Sonora and Chihuahua is interpreted to indicate crustal thickness of 36 km and total lithosphere thickness of 70 to 76 km (Gomberg and others, 1989), estimates consistent with gravity models of anomalies in Sonora (Aiken and others, 1988; Schellhorn and others, 1991). The crust beneath the Sierra Madre del Sur in southern México has been interpreted to be about 45 ± 4 km thick (Valdes and others, 1986).

Heat flow measurements have been reported from México by Smith (1974), Smith and others (1979), Ziagos and others (1985), and Prol-Ledesma and Juárez (1986). Although coverage is not complete, it seems clear that most of northern and central México, including Sonora, eastern Chihuahua, Sinaloa, Durango, Zacatecas, Guanajuato, and San Luis Potosí, is characterized by high heat flow values (38 to 190 mW/m², with most values 75 to 125 mW/m²). Lower values were obtained east of the Sierra Madre Oriental (most values 25 to 75 mW/m²), in the Sierra Madre del Sur (13 to 45 mW/m²), and in the Baja California peninsula (most values 35 to 50 mW/m²). The region of higher heat flow values roughly corresponds to the region affected by late Cenozoic extension in the Basin and Range province and along the eastern margin of the Gulf of California. The apparent southward increase in crustal and lithospheric thickness and southward decrease in heat flow from the Basin and Range province to southern México may indicate a southward decrease in the net thinning and extension of the lithosphere (Gomberg and others, 1989). Alternatively, these differences may reflect different starting thicknesses, crustal rheology, or both.

Other gravity, magnetic, and seismic data, particularly those interpreted in terms of problems of a more local, rather than a regional, scale, and all paleomagnetic data are presented in discussions of specific terranes.

TERRANE DESCRIPTIONS

We have identified and characterized 17 tectonostratigraphic terranes (hereafter called terranes) in México and northern Central America (Fig. 3). Our usage of the term terrane is from Howell and others (1985, p. 4): "a fault-bounded package of

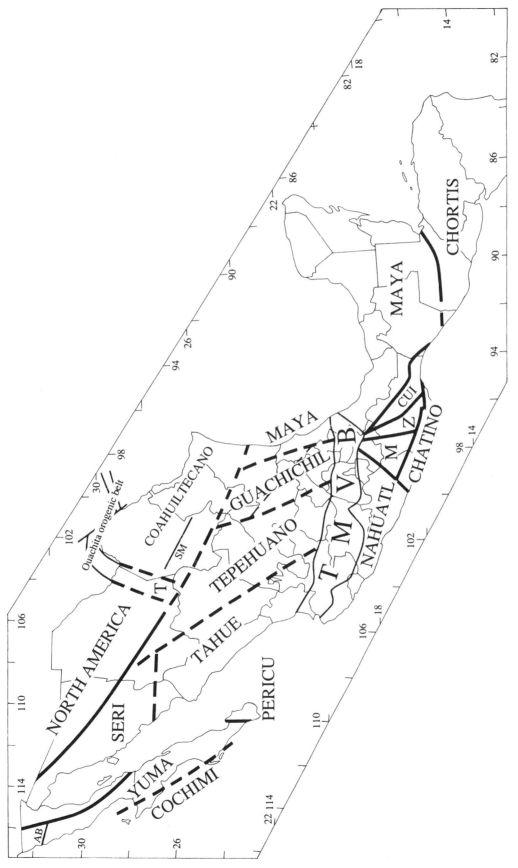

Figure 3. Terrane map of México and northern Central America. State boundaries as in Figure 2. Terrane boundaries (heavy lines) dashed where inferred. Terrane abbreviations: CUI, Cuicateco; M, Mixteco; T, Tarahumara; Z, Zapoteco. Other abbreviations: AB, Agua Blanca fault (Baja California); SM, San Marcos fault (Coahuila); TMVB, Trans-Mexican Volcanic Belt.

rocks of regional extent characterized by a geologic history which differs from that of neighboring terranes." Figure 3 is modified from a similar division that was developed at the Institute of Geology (Instituto de Geología) at the University of México (Universidad Nacional Autónoma de México) in México City and that was used in Continent-Ocean Transects H-1 and H-3 (Ortega-Gutiérrez and others, 1990; Mitre-Salazar and others, 1991). With the exception of North America, terrane names were selected on the basis of indigenous pre-Columbian cultures.

The distinctive characteristics of most Mexican terranes are to be found in their pre-Cretaceous geologic record. In Part 2, we develop the hypothesis that most of México, excepting its western and southern margins, behaved as a structurally intact, little-deformed mass throughout the Cretaceous and Cenozoic. Exposures of Jurassic and older rocks are sparse in some terranes due to widespread, overlapping Cretaceous sedimentary rocks and Cenozoic volcanic and sedimentary rocks. In these cases, we delimited terranes based on the scattered exposures of older rocks and on geophysical and isotopic data that help constrain the nature of the subsurface. However, we expect that future work will lead to the recognition that some of these terranes are composite, as we have formally proposed for other Mexican terranes.

Some earth scientists have expressed reservations about the term terrane, which they deem less precise than terms such as block, sliver, fragment, and nappe, and about the concept of terrane analysis, which they view as a merely descriptive routine that avoids interpretation of the genetic significance of lithotectonic assemblages (Sengor and Dewey, 1990). These complaints are unfounded, however, provided that the ultimate goal of terrane analysis is not the splitting of orogenic belts into myriad terranes but rather the evaluation of the genetic significance of assemblages within individual terranes and the genetic relations between and among terranes. When properly performed, terrane analysis can stimulate breakthroughs in understanding the evolution of complex regions, e.g., the North American Cordillera, and we believe that the Mexican region would be well served by such a strategy. To this end, this volume includes not only chiefly factual terrane descriptions (bulk of Part 1), but also a detailed interpretive model of the genetic relations among the terranes (Part 2).

Our terrane division is in some respects similar to an earlier division developed by Campa-Uranga and Coney (1983) and Coney and Campa-Uranga (1987). However, many aspects of their terrane division, such as terrane boundaries, descriptions of constituent rocks, and interpretations of constituent rocks and terrane-bounding faults, have been changed, clarified, or invalidated by the hundreds of scientific contributions that have been published in the last decade. Presentation of the data in Part 1 of this work would be rendered disruptively opaque if we were to express all of these advances in terms of revised terrane boundaries and definitions (e.g., "that part of the Guerrero terrane north of the Trans-Mexican Volcanic Belt and east of the Gulf of California" instead of our Tahué terrane). On this basis we have implemented the new terrane names used in the DNAG Transects. We do not expect that our division of Mexican terranes will

go unchallenged and unchanged in the future, and we realize that a new set of terrane names may cause consternation in some quarters, but we have concluded that the potential for confusion is minimized by avoiding a scheme that splices two sets of terrane names. In the few cases that the descriptions and boundaries of the terranes in our division are very similar to those of the Campa-Uranga and Coney division (e.g., Yuma-Santa Ana, Cochimi-Vizcaíno), we anticipate future reference to the older names, which have precedence. In the course of our terrane descriptions, we note the geographic and geologic relation of each terrane to earlier terrane nomenclature.

Below, we describe the terranes shown in Figure 3 in alphabetical order. Where possible, each terrane description includes the following elements: (1) a brief interpretive geologic or tectonic history of the major rock units, focusing on those features that distinguish a given terrane from other terranes; (2) a review and synthesis of published and unpublished data for all major rock units, including basement rocks, stratigraphic units from oldest to youngest, and plutonic rocks; (3) a review of available geophysical data; (4) a schematic tectonostratigraphic column that, in some cases, also shows geographic variation in the form of a structure section; and (5) a table of radiometric data including dates, system used, errors, and sources. In some instances we cite synthesis articles rather than unpublished theses and reports or particularly hard-to-find original sources. The interested reader is directed to the syntheses for reference to such works.

The natural resources of petroleum and economic mineral deposits are keystones of the Mexican economy. México ranks seventh in worldwide crude oil reserves and third in production (Beck and Thrush, 1991). México is the world's leading producer of silver and bismuth; is among the top five producers of barite, fluorospar, graphite, lead, lime, and molybdenum; and ranks among the top 10 producers of antimony, cadmium, copper, manganese, gypsum, mercury, mica, ammonia, salt, sulfur, zinc, and asbestos (Minerals Yearbook, 1990). Summarizing the wealth of available information on these topics is beyond the scope of this work, but recent summary volumes of possible interest include a recent DNAG volume entitled *Economic Geology of México* (Salas, 1991) and a special volume of the journal *Economic Geology* (1988, v. 83).

The DNAG time scale (Palmer, 1983) is used throughout this work. Where necessary, K-Ar ages have been recalculated using the revised constants of Steiger and Jager (1977). In order to simplify the reference section, we have taken the following liberties with the potentially confusing variety of ways in which the Latin American double surname is referenced: where known, both surnames are spelled out and hyphenated, even where the original citation abbreviated or omitted the maternal surname or omitted the hyphen. Symbols used in Figures 5 through 21 are those used in the DNAG Continent/Ocean Transects (Fig. 4).

Chatino terrane

The Chatino terrane consists mainly of orthogneiss and metasedimentary rocks derived from protoliths of unknown age that

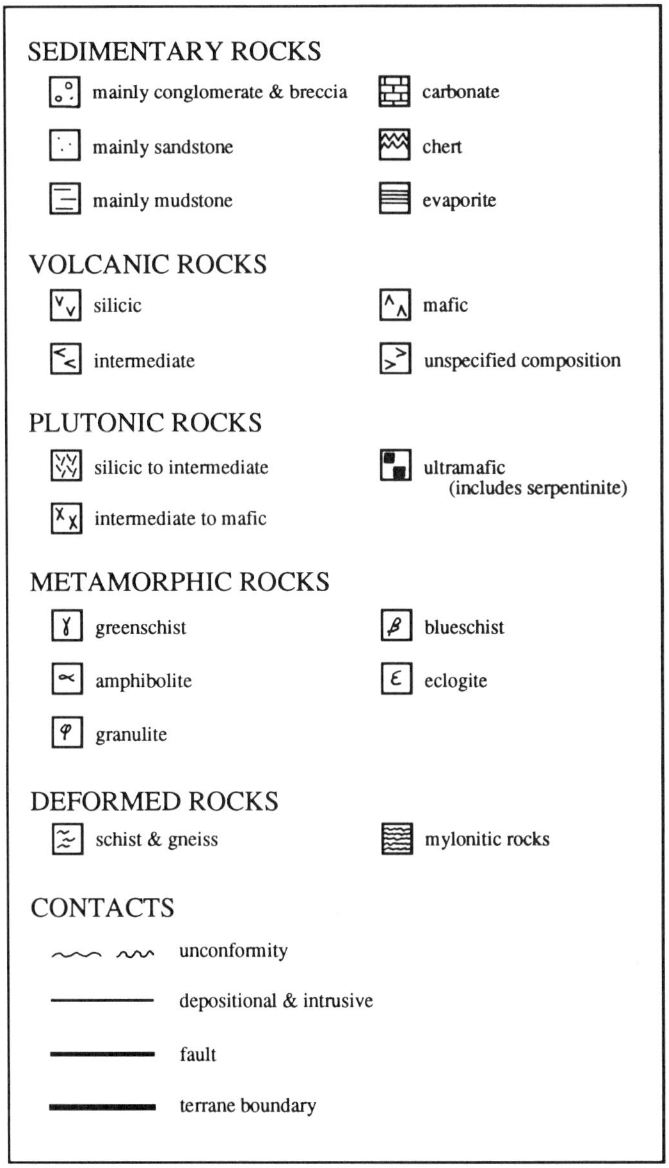

SEDIMENTARY ROCKS

- ⊙° mainly conglomerate & breccia
- ⊞ carbonate
- ⊡ mainly sandstone
- ⋈ chert
- ⊟ mainly mudstone
- ▤ evaporite

VOLCANIC ROCKS

- ᵛᵥ silicic
- ᴬ∧ mafic
- ᐸ∢ intermediate
- ⊳ unspecified composition

PLUTONIC ROCKS

- ᵛⱽᵥ silicic to intermediate
- ▪ ultramafic (includes serpentinite)
- ˣₓ intermediate to mafic

METAMORPHIC ROCKS

- γ greenschist
- β blueschist
- ∝ amphibolite
- ε eclogite
- φ granulite

DEFORMED ROCKS

- ≋ schist & gneiss
- ≋ mylonitic rocks

CONTACTS

- ∿∿ ∿∿ unconformity
- ——— depositional & intrusive
- ——— fault
- ━━━ terrane boundary

Figure 4. Legend of symbols for Figures 5 through 21. Geologic time abbreviations: pϵ, Precambrian; ϵ, Cambrian; O, Ordovician; S, Silurian; D, Devonian; M, Mississippian; ℙ, Pennsylvanian; P, Permian; Tr, Triassic; J, Jurassic; K, Cretaceous; T, Tertiary; Q, Quaternary; u, upper; m, middle; l, lower.

were repeatedly intruded and locally migmatized during the Mesozoic and Cenozoic (de Cserna, 1965; Klesse, 1968; Ortega-Gutiérrez, 1981a). At its northern margin, the terrane is faulted against the Mixteco and Zapoteco terranes; at its western margin, the contact with the Nahuatl terrane is obscured by Cenozoic plutons; and to the south, it abuts the late Cenozoic accretionary prism above the subducting Cocos plate. The Chatino terrane corresponds to the Xolapa terrane of Campa-Uranga and Coney (1983).

The oldest unit in the Chatino terrane is the Xolapa Com-

plex (Fig. 5), which includes amphibolite-facies migmatite, orthogneiss, amphibolite, pelitic schist, biotite schist, and marble (Ortega-Gutiérrez, 1981a; Alaniz-Alvarez and Ortega-Gutiérrez, 1988). Sedimentary protoliths are interpreted to have been interbedded graywackes, pelitic rocks, and carbonates. Some migmatite records the incomplete anatexis of pelitic sedimentary rocks and carbonates. Orthogneiss probably was derived from pre- or synkinematic tonalitic intrusives that yield Jurassic to mid-Cretaceous metamorphic ages (Table 1).

A Precambrian or Paleozoic age of the sedimentary protoliths generally is assumed or inferred (de Cserna, 1971; Carfantan, 1983), but few geochronologic data constrain that age. The protoliths clearly are older than the cross-cutting Jurassic-Cretaceous orthogneiss bodies. The protoliths probably were derived, at least in part, from a 1.6- to 1.3-Ga source, based on U-Pb dates from euhedral, probably igneous, zircons and Sm-Nd model ages (Table 1) (Robinson and others, 1989; Morán-Zenteno and others, 1990a, b, 1991; Robinson, 1991).

The strike of foliation and trend of lineations in onland exposures of the Xolapa Complex is roughly west-northwest (Ortega-Gutiérrez, 1981a; Robinson, 1991). According to preliminary refraction and gravimetric studies, the Xolapa Complex is 15 to 20 km thick along the coast (Nava and others, 1988). Drilling for DSDP Leg 66 at site 489, 30 km offshore at longitude 99°W, bottomed in biotite–hornblende–quartz schist, garnet–muscovite schist, and muscovite–chlorite quartzite that probably are part of the Xolapa Complex (Watkins and others, 1981). In southeasternmost Guerrero, the Xolapa Complex is overlain nonconformably by Late Cretaceous–Paleogene and late Miocene–Pleistocene marine clastic rocks (Durham and others, 1981).

The Xolapa Complex is intruded by widespread undeformed Tertiary granitoids (Fig. 5; Table 1), granitic pegmatites, and mafic dike swarms. The ages of individual granitic plutons decrease from about 45 Ma in the intrusive suite near Acapulco (Acapulco batholith) to about 12 Ma near Puerto Angel, 400 km to the southeast (Damon and Coney, 1983; Guerrero-García, 1989; D. Morán-Zenteno, 1992). The intrusive suite near Acapulco appears to be significantly younger than the 60- to 55-Ma intrusive rocks in the adjacent part of the Mixteco terrane that commonly are included in the Acapulco intrusive suite (Table 8). Diorite recovered from site 493 of DSDP Leg 66, about 20 km offshore at longitude 99°W (Bellon and others, 1981), probably is correlative with the undeformed granitoids exposed on land.

Chortis terrane

Although the Chortis terrane (Chortis block of Dengo, 1975) does not include rocks of México, we discuss it because of its probable interaction with terranes of México during Mesozoic and Cenozoic time. Following Dengo (1985) and Donnelly and others (1990a), we consider the Chortis terrane to include Guatemala south of the Motagua fault, Honduras, Nicaragua north of about 12°N, El Salvador, and the submarine Nicaragua Rise (Fig. 2).

CHATINO

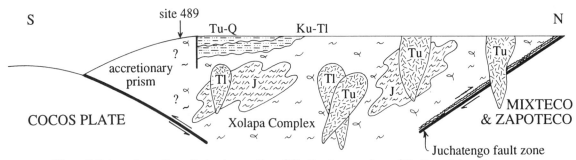

Figure 5. Schematic north-south structure section of Chatino terrane. Ages of Tertiary plutons decrease from west to east.

TABLE 1. CHATINO TERRANE RADIOMETRIC DATA

Sample	System	Mineral*	Date (Ma)			Reference†	Comments
Xolapa Complex							
Paragneiss	U-Pb	zr	1,525	±	170	1	Discordia intercepts
			78	±	35		
Metasedimentary rocks	Rb-Sr		308	±	5	2	Isochron age
Paragneiss	Rb-Sr	b	240	±	50	3	
Gneissic granitoid	Rb-Sr	wr	185	±	84	4	87Sr/86Sr; 0.7056
Gneissic granitoid	U-Pb	zr	160	±	3	4	Concordant age
Gneissic tonalite near Acapulco	Rb-Sr	wr	144	±	7	2	Isochron age; 87Sr/86Sr: 0.7030
	Rb-Sr	b	30–28			5	
Gneissic tonalite near Acapulco	Rb-Sr	wr	138	±	12	2	Isochron age; 87Sr/86Sr: 0.7048
	Rb-Sr	b	25	±	1	5	
Gneissic tonalite near Acapulco	Rb-Sr	wr	128	±	7	2	Isochron age; 87Sr/86Sr: 0.7049
	Rb-Sr	b	32–31			5	
Undeformed diorite near Acapulco	Rb-Sr	wr	136	±	11	2	Isochron age; 87Sr/86Sr: 0.7038
Biotite gneiss	K-Ar	b	44	±	7	6	
Biotite schist	K-Ar	b	38	±	2	6	
Paragneiss	Rb-Sr	b	32			3	
Undeformed intrusive rocks							
Acapulco granitoids	Rb-Sr	kf	80			7	Maximum age
Acapulco granitoids	Rb-Sr	b	48	±	1	8	
Acapulco granitoids	Rb-Sr	wr	43	±	7	8	Apparent isochron
Felsic-mafic plutons, Acapulco Batholith	Rb-Sr	b	43–26			5	
Acapulco granitoids	K-Ar		43	±	1	9	
Acapulco paragneiss	K-Ar		43			8	
Puerto Angel tonalite	U-Pb	zr	40			10	Concordant age
Diorite, DSDP Leg 66	K-Ar	wr	36	±	2	11	
Diorite, DSDP Leg 66	K-Ar	wr	35	±	2	11	
Diorite, DSDP Leg 66	K-Ar	wr	34	±	2	11	
Acapulco paragneiss	Rb-Sr	b	32	±	1	8	
Granitoid	K-Ar		30	±	1	9	
Granodiorite	Rb-Sr	b	24	±	1	8	
Puerto Angel gneiss	Rb-Sr		24	±	1	8	
Diorite	Rb-Sr	b	18.5	±	5	8	
Puerto Angel gneiss	Rb-Sr		15	±	1	8	
Puerto Angel gneiss	Rb-Sr		11	±	1	8	

*Mineral abbreviations: b = biotite; kf = potassium feldspar; wr = whole rock; zr = zircon.
†1 = Robinson and others, 1990; 2 = Morán-Zenteno and others, 1990b; 3 = Halpern and others, 1974; 4 = Guerrero-García and others, 1978; Morán-Zenteno, 1992; 6= de Cserna and others, 1962; 7 = Fries and Rincón-Orta, 1965; 8 = Guerrero-García, 1975; 9 = Böhnel and others, 1989; 10 = Robinson and others, 1989; 11 = Bellon and others, 1981.

The Chortis terrane consists of deformed Paleozoic-Precambrian(?) metamorphic basement unconformably overlain by a variety of Mesozoic and Cenozoic sedimentary and volcanic rocks. Basement once may have been part of Pangean continental crust (South America), or it may represent an exotic crustal fragment or fragments that accreted to western Pangea in the late Paleozoic or early Mesozoic. Mesozoic and Cenozoic cover strata indicate Triassic(?) to Jurassic clastic sedimentation in nonmarine, coastal, and shelf environments, deposition of Cretaceous carbonates in shallow epeirogenic seas, mid-Cretaceous uplift, Late Cretaceous–Paleogene orogenesis, and Cenozoic volcanism. Volcanic rocks and volcaniclastic detritus are present throughout the Mesozoic and Cenozoic section, implying proximity to a volcanic arc. The Chortis terrane probably was part of continental México during the Mesozoic, based on the similarity of its Mesozoic history to that of the Maya and Mixteco terranes, but its location at any particular time and the nature of its interaction with any particular part of México are not well understood. The current location of the Chortis terrane probably is due to hundreds of kilometers of Cenozoic eastward displacement with respect to southern México.

Basement rocks. The northern margin of the Chortis block consists of metamorphic and plutonic basement rocks that crop out between the Motagua fault zone and Cayman Trough to the north and the Jocotán-Chamelecón and Aguán faults in Guatemala and northern Honduras to the south. Basement rocks of the Chortis terrane include older, higher grade metamorphic rocks generally termed the Las Ovejas Complex, and younger, lower grade metamorphic rocks generally called the San Diego Phyllite (Fig. 6) (Horne and others, 1976a; Horne and others, in Donnelly and others, 1990a). The Las Ovejas Complex includes amphibolite-facies quartzofeldspathic gneiss, two-mica schist, subordinate amphibolite and marble, and foliated cross-cutting plutonic rocks that are pervasively mylonitized and isoclinally folded. Protoliths of the Las Ovejas Complex may be as old as Precambrian

or early Paleozoic, and metamorphism may have accompanied late Paleozoic plutonism (Table 2).

The San Diego Phyllite consists of greenschist-facies phyllite, schist, and slate with thin interbeds of quartzite that are overlain unconformably by Early Cretaceous limestone and cut by plutons of uncertain but probable Cretaceous and Tertiary age. The single generation of penetrative structures in the San Diego Phyllite is similar in style and geometry to the youngest fabric in the Las Ovejas Complex; thus, the phyllite may have been deposited unconformably on Las Ovejas basement prior to deformation.

South of the Jocotán-Chamelecón and Aguán faults, basement rocks in the Chortis terrane include scattered outcrops of greenschist- and lower amphibolite–facies phyllite, mica schist, graphitic schist, quartzite, metaconglomerate, marble, and metabasite. This unit includes isolated outcrops that locally are known as the Petén Formation, the Cacaguapa Schist, the Palacaguina Formation, and unnamed units (Fig. 6); these rocks are similar to, but neither clearly correlative with nor differentiable from, the San Diego Phyllite north of the Jocotán-Chamelecón and Aguán faults (Carpenter, 1954; Fakundiny, 1970; Mills and Hugh, 1974; Simonson, 1977; Horne and others, in Donnelly and others, 1990a). In east-central Honduras, deformed adamellite plutons that crop out near the schists yielded early Mesozoic Rb-Sr ages, suggesting that deformation of the schists and plutons may have been Early Triassic or older (Table 2). In most areas, exposures of phyllite or schist are overlain unconformably by unmetamorphosed Jurassic strata and intruded by undeformed plutons at least as old as 140 ± 15 Ma (Table 2). These isolated outcrops have been correlated among themselves, with the San Diego Phyllite north of the Jocotán-Chamelecón fault, and with the Santa Rosa Group of the Maya terrane (Burkart and others, 1973), but such correlations are conjectural at best (Dengo, 1985; Horne and others, in Donnelly and others, 1990a).

Basement rocks reported in wells drilled along the coast of

CHORTIS

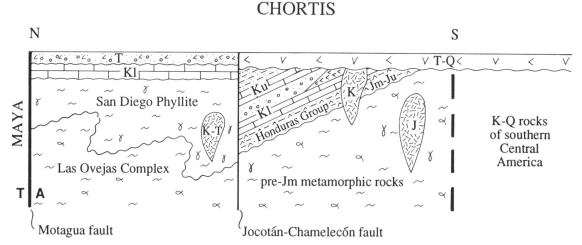

Figure 6. Schematic north-south structure section of Chortis terrane. Unknown nature and orientation of contact with pre-Quaternary rocks of southern Central America (the latter not covered in this volume).

TABLE 2. CHORTIS TERRANE RADIOMETRIC DATA

Sample	System	Mineral*	Date (Ma)	References[†]	Comments
Basement rocks					
Metaigneous rocks in Las Ovejas Complex	Rb-Sr		720 ± 260	1	3-pt isochron; samples may not be cogenetic
Metaigneous rocks, intrude Las Ovejas	Rb-Sr		305 ± 12	1	4-pt isochron
Adamellite gneiss	Rb-Sr	wr	230–203	2	Minimum ages assuming $^{87}Sr/^{86}Sr_i$
			150–125	2	= 0.704; uncertain relation to basement rocks
Intrusive rocks					
Granodiorite	Rb-Sr		150 ± 13	1	4-pt isochron; intruces Las Ovejas
Dipilto batholith	Rb-Sr	wr	140 ± 15	2	4-pt isochron; intrusion age; $^{87}Sr/^{86}Sr_i$; 0.7031
San Ignacio adamellite, central Honduras	K-Ar	h	123 ± 2	3	Source: F. McDowell
	K-Ar	b	117 ± 2	3	Source: F. McDowell
	K-Ar	b	114 ± 2	3	Minimum age
Tonalite pluton, northern Honduras	K-Ar	h	95 ± 2	4	
		b	76 ± 2	4	
Tonalite plutons, northern Honduras	K-Ar	b	83 ± 2	4	
		b	76 ± 2	4	
Tonalite pluton, northern Honduras		wm	59 ± 1	4	
	K-Ar	h	74 ± 2	4	
		b	58 ± 1	4	
Minas de Oro granodiorite, central Honduras	K-Ar	b	62 ± 1	3	
	K-Ar	b	59 ± 2	3	Source: F. McDowell
	K-Ar	h	55 ± 2	3	Source: F. McDowell
Adamellite	Rb-Sr	wr	60	1	Intrudes 720 Ma complex; assumed $^{87}Sr/^{86}Sr_i$ = 0.703
Dacite stock	K-Ar	b	60 ± 1	3	Cogenetic w/volc rocks?
Granodiorite	K-Ar	b	37 ± 1	4	Radiogenic Ar loss?
Igneous rocks in Motagua fault zone					
Diorite	K-Ar	h	104 ± 6	5	Xenolith in granite
Granite	K-Ar	b	95 ± 3	5	
Chiquimula batholith, granite	Rb-Sr	b, wr	95 ± 1	6	Isochron age
	K-Ar	b	84 ± 2	6	
Chiquimula batholith, mafic-intermediate	Rb-Sr		50 ± 5	6	Isochron age; $^{87}Sr/^{86}Sr_i$; 0.706
Granitoids	$^{40}Ar/^{39}Ar$		35	7	Source: J. Sutter; 3 samples
Ignimbrite	K-Ar	p	17 ± 1	5	

*Mineral abbreviations: b = biotite; h = hornblende; p = plagioclase; wr = whole rock.
[†]1 = Horne and others, 1976a; 2 = G. Horne and Clark, unpublished; 3 = Gose, 1985; 4 = Horne and others, 1976c; 5 = Ritchie and McDowell, 1979; 6 = Clemons and Long, 1971; 7 = Donnelly and others, 1990a.

Nicaragua and on the Nicaragua Rise include "metamorphic rocks" that may be correlative with the San Diego Phyllite, andesite of possible Cretaceous to Paleogene age, and Eocene granodiorite (Table 2) (Arden, 1975). The eastward extent of pre-Mesozoic basement rocks beneath the Nicaragua Rise is unknown.

Mesozoic and Cenozoic rocks. Metamorphic basement is overlain nonconformably by Mesozoic sedimentary and less common volcanic rocks south of the Jocotán-Chamelecón fault (Fig. 6) (Horne and others, in Donnelly and others, 1990a). Shallow marine clastic rocks of uncertain but possible early Mesozoic age crop out locally but have uncertain regional extent and correlation. The oldest widespread Mesozoic unit is the Honduras Group, which consists of siliciclastic marine strata and sparse interbedded volcanic rocks ranging in age from at least as old as Bajocian (early Middle Jurassic) to Early Cretaceous (Delevoryas and Srivastava, 1981; Ritchie and Finch, 1985; Gordon, 1989; Donnelly and others, 1990a). Previous correlations of part of the Honduras Group with the Todos Santos Formation of the southern Maya terrane should be abandoned (Horne and others, in Donnelly and others, 1990a). Successively younger units include Barremian to Late Albian limestone and shaly limestone, sparsely distributed andesitic volcanic and volcaniclastic rocks in central Honduras of inferred mid-Cretaceous age, coarse-grained mid-Cretaceous(?) red beds, Cenomanian limestone, and fine-grained Late Cretaceous to Paleogene(?) red beds. These Mesozoic strata were faulted, folded, and eroded prior to the deposition of unconformably overlying mid-Tertiary volcanic rocks (Carpenter, 1954; Simonson, 1977; Finch, 1981; Horne and others, in Donnelly and others, 1990a).

Basement rocks and cover strata as young as early Late Cretaceous are intruded by widespread, undeformed, silicic to mafic, calc-alkalic plutons. Plutonism appears to have begun at least as early as the Late Jurassic and to have continued until earliest Tertiary time (Table 2).

Paleogene red beds, tuff, and minor andesite (Subinal Formation) that crop out in southeastern Guatemala and northern El Salvador are similar to coeval rocks in the Maya terrane, but possible correlations are complicated by rapid lateral facies and thickness changes (Deaton and Burkart, 1984a; Donnelly, 1989; Donnelly and others, 1990a). The Cenozoic stratigraphy of the Chortis block is dominated by volcanic rocks (Reynolds, 1980; Weyl, 1980; Donnelly and others, 1990a, b). Andesitic lavas, tuff, and breccia of the Matagalpa and Morazán Formations are probably Oligocene and Eocene(?) in age. Early Miocene to middle Miocene siliceous ignimbrites as thick as 2 km are locally overlain by late Miocene to Pliocene basalts and andesites. Other Cenozoic rocks include local late Miocene to Pleistocene red beds and Quaternary stratovolcanoes near the Pacific margin.

Structural and geophysical data. The modern volcanic arc lies within the Nicaraguan Depression and Median Trough, a pronounced margin-parallel graben cut by transverse strike-slip faults (Carr, 1976; Weyl, 1980). In the northwestern part of the terrane, volcanic vents and normal faults that are aligned approximately north-south, i.e., at an angle of about 50° to the trench and arc (Carr, 1976), indicate either east-west extension due to internal deformation of the Caribbean plate or the existence of a separate microplate wedged among the Caribbean, North America, and Cocos plates (Burkart and Self, 1985; Guzmán-Speziale and others, 1989).

Seismic refraction and gravity studies indicate that continental crust of the Chortis terrane is 35 to 40 km thick beneath northwestern Central America, thinning to about 20 km beneath the Honduras Rise (Couch and Woodcock, 1981; Kim and others, 1982; Case and others, 1990). Seismic reflection data from offshore southwestern Guatemala (Ladd and others, 1982) and a +100-mgal gravity anomaly in the same region (Couch and others, 1985) indicate the presence of oceanic crustal rocks that may be roughly correlative with the Nicoya Complex of Costa Rica (Aubouin and others, 1982) or with the El Tambor Group in the Maya-Chortis boundary zone (Donnelly and others, 1990b).

Paleomagnetic studies of Jurassic and Cretaceous red beds and limestone and Early Cretaceous and Paleogene magmatic rocks indicate that the Chortis terrane apparently rotated more than 100° clockwise during the Jurassic and Early Cretaceous, and then rotated more than 100° counterclockwise during the Late Cretaceous and Paleogene (Gose, 1985).

Coahuiltecano terrane

The Coahuiltecano terrane contains Paleozoic low-grade metamorphic rocks and Paleozoic arc-derived flysch and related arc volcanic rocks. These units may correspond to the forearc and arc, respectively, of the Gondwana supercontinent that collided with southern North America in the late Paleozoic during the Ouachita orogeny. Alternatively, the arc-related assemblage may be an arc fragment that is exotic with respect to Gondwana. The Paleozoic rocks were intruded by Triassic calc-alkalic plutons and overlapped by Late Jurassic and Cretaceous platformal rocks that cover most of the terrane. The Coahuiltecano terrane is equivalent to the Coahuila terrane of Coney and Campa-Uranga (1987). It is also correlative, at least to the extent that it is a remnant of Gondwanan continental crust, with the submarine Sabine terrane in the northern Gulf of Mexico (U.S. Geodynamics Committee, 1989) and the composite Maya terrane as defined in this work.

Pre-Jurassic rocks. Paleozoic or Paleozoic(?) metamorphic rocks are known from wells in the northern and central parts of the terrane. Granitic gneiss in the PEMEX #1 La Perla well in easternmost Coahuila yielded a mid-Paleozoic Rb-Sr whole-rock date (Table 3), and fine-grained, strongly deformed, low-grade metamorphic rocks of unknown but inferred Paleozoic age have been reported from five PEMEX wells in eastern Coahuila and northern Nuevo León (Flawn and others, 1961). Cretaceous conglomerate in southern Coahuila contains clasts of mid-Paleozoic schist (Table 3). Similar rocks may underlie most of the northern and central Coahuiltecano terrane and, with the Sabine terrane to the east, are inferred to be part of the Gondwanan forearc that was stranded during Mesozoic rifting and drifting. Basement of this forearc may be Gondwanan continental crust (Fig. 7), but such a relation is purely conjectural.

Paleozoic volcaniclastic flysch and Triassic granitoids crop out south of the San Marcos fault in the southern part of the terrane (Fig. 3). Late Pennsylvanian to Permian strata of the Las Delicias basin (Fig. 7) consist chiefly of mass-gravity marine deposits that contain sand- to boulder-sized fragments of volcanic rocks, siliciclastic rocks, and limestone (King and others, 1944; Wardlaw and others, 1979; McKee and others, 1988, 1990). Volcanic and carbonate bank and reef detritus indicate that the Las Delicias basin was adjacent to an active calc-alkalic magmatic arc fringed by carbonate banks from mid(?)-Pennsylvanian to Late Permian time (Jones and others, 1986; McKee and others, 1988, 1990). The arc probably was constructed on continental crust (Fig. 7), based on the occurrence of clasts of Devonian schist and Triassic gneiss and granite in Cretaceous conglomerate (Table 3). Volcanic rocks originally mapped as intrusive rocks (King and others, 1944) may be the north edge of the arc and may form the depositional basement of the Las Delicias basin (Fig. 7) (McKee and others, 1988). Late Paleozoic strata of the Las Delicias basin are intruded by Triassic granodiorite (Fig. 7) (Table 3). The Las Delicias basin underwent east-west to southeast-northwest shortening prior to the deposition of overlying Late Jurassic strata (King and others, 1944; McKee and others, 1988).

Triassic granodiorite also crops out north of the San Marcos fault near Potrero de la Mula (Table 3). PEMEX wells in east-central Coahuila and northern Nuevo León bottomed in igneous

TABLE 3. COAHUILTECANO TERRANE RADIOMETRIC DATA

Sample	System	Mineral*	Date (Ma)	References[†]	Comments
Paleozoic-Triassic rocks					
Schist cobbles in Cretaceous conglomerate	Rb-Sr	wr	370 ± 5	1	Model ages, assumed $^{87}Sr/^{86}Sr_i$:
			387 ± 4	1	0.705
Granitic gneiss, PEMEX No. 1 La Perla well	Rb-Sr	wr	358 ± 70	2	Assumed $^{87}Sr/^{86}Sr_i$: 0.706
	Rb-Sr	wr	277 ± 60	2	Assumed $^{87}Sr/^{86}Sr_i$: 0.713
Granodiorite in Acatia-Delicias region	K-Ar	b	266 ± 21	3	Source: M. Múgica
	K-Ar	h	256 ± 20	3	Source: M. Múgica
Cataclastic granodiorite Delicias basin	Rb-Sr	wm	240 ± 2	1	Model age; crops out along San Marcos fault
Plutonic basement in well near Parras	K-Ar?		236	4	
Gneiss cobble in Cretaceous conglomerate	Rb-Sr	wr, b, wm	230 ± 3	1	3-pt isochron $^{87}Sr/^{86}Sr_i$: 0.7082
Granite clasts in Jur(?)-K(?) conglomerate	Rb-Sr	wr	225 ± 4	1	$^{87}Sr/^{86}Sr_i$: 0.7063
Granodiorite, Potrero de la Mula	Rb-Sr	wr	213 ± 14	5	9-pt isochron
Granodiorite, Potrero de la Mula	K-Ar	h	212 ± 4	2	
Granodiorite, Delicias basin	K-Ar	b	210 ± 4	2	Intrudes upper Pz section
	K-Ar	b	206 ± 4	2	
Cenozoic magmatic rocks					
Andesite, northern Nuevo Leon	K-Ar	h	44 ± 1	6, 7	
Hornblende diorite, eastern Coahuila	K-Ar	wr	44 ± 1	6, 7	
Andesite, eastern Coahuila	K-Ar	wr	40 ± 1	6, 7	
	K-Ar	wr	36 ± 1	6, 7	
Diorite, eastern Coahuila	K-Ar	wr	40 ± 1	6, 7	
	K-Ar	b	38 ± 1	6, 7	
Granodiorite, eastern Coahuila	K-Ar	b	38 ± 1	6, 7	
Syenite, Coahuila	K-Ar	b	35 ± 1	8	
Alkalic intrusives, San Carlos, Tamaulipas	K-Ar	h, wm	30 ± 1	9	3 samples

*Mineral abbreviations: b = biotite; h = hornblende; wm = white mica; wr = whole rock.
[†]1 = McKee and others, 1990; 2 = Denison and others, 1969; 3 = López-Infanzón, 1986; 4 = Wilson, 1990; 5 = Jones and other, 1984; 6 = C. R. Sewell and R. E. Denison, unpublished data; 7 = Also see Sewell, 1968; 8 = C. Henry, unpublished data; 9 = Bloomfield and Cepeda, 1973.

rocks similar to those near Potrero de la Mula (Flawn and others, 1961; Wilson, 1990), implying that much of the terrane is underlain by Permian(?) and Triassic granitoids.

Jurassic-Tertiary sedimentary rocks. Pre-Cretaceous clastic rocks near Monterrey were mapped as Triassic and correlated with the Upper Triassic–Lower Jurassic Huizachal Formation in the Guachichil terrane by Lopéz-Ramos (1985, p. 274–275), but are mapped as Jurassic and Cretaceous rocks on the 1:1,000,000 geologic map of México (INEGI, 1980). Rhyolitic tuff and breccia that have yielded an unpublished Early Jurassic age crop out in the Sierra de Mojada, Chihuahua, in the southeastern corner of the Coahuiltecano terrane (McKee and others, 1990).

Basement rocks in the Coahuiltecano are overlain nonconformably by Oxfordian limestone and shale (Zuloaga Group of Götte and Michalzik, 1991) that also overlap the Guachichil and Tepehuano terranes. Emergent topographic highs, possibly produced by transtension during the Jurassic opening of the Gulf of Mexico, included the Burro-Picachos platform in northern Coahuila; the Tamaulipas platform in Nuevo León and Tamaulipas, which was submerged by the beginning of Cretaceous time; and the Coahuila peninsula (uplifted Las Delicias basin) in south-

central Coahuila, which was isolated as Coahuila Island in Barremian time and submerged by Albian time (Smith, 1981; Salvador, 1987; Mitre-Salazar and others, 1991; pp. 105, 113). Late Jurassic strata are overlain by Cretaceous platform carbonates, interplatform basinal limestone, fine-grained clastic rocks, and evaporites; local coal seams of Maastrichtian age; a local tektite-and shocked quartz-bearing, 3-m-thick clastic unit at the K-T boundary that may have been deposited by a tsunami caused by an impact in the Yucatán region (Smit and others, 1992); and Paleogene marl and nonmarine clastic rocks. The late Mesozoic paleogeographic evolution of this region is discussed by Enos (1983), Young (1983), López-Ramos (1985, p. 198–233), and de Cserna (1989).

A thick sequence of Albian to Maastrichtian carbonates, shales, and coarser clastic rocks crops out in the Parras basin (Imlay, 1936; Murray and others, 1962), which is elongate east to west along the south side of the Coahuila platform (Part 2). Cenomanian-Maastrichtian flysch in the Parras basin was derived, at least in part, from a calc-alkalic volcanic arc to the west (Tardy and Maury, 1973). The trace of the Mojave-Sonora Megashear is inferred to cut diagonally beneath the Cretaceous sedimentary rocks of the Parras basin from northwest to south-

COAHUILTECANO

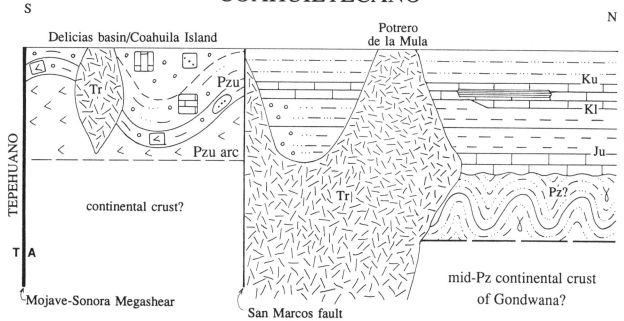

Figure 7. Schematic north-south structure section of Coahuiltecano terrane. Steepness of nonconformity between Triassic pluton and younger Mesozoic sedimentary rocks due to large vertical exaggeration. Nature and age of crystalline basement unknown. Effects of Laramide deformation have been omitted.

east (p. 78). Displacement on the San Marcos fault (Fig. 7), a possible splay or strand of the megashear, produced syntectonic Jurassic and Neocomian conglomerate that contains clasts of Paleozoic schist and Paleozoic-Triassic granite and granitic gneiss (Table 3) (McKee and others, 1990).

Cretaceous and older rocks in the Coahuiltecano terrane are intruded by numerous silicic to intermediate plutons, laccoliths, dikes, and sills of inferred Tertiary age (Kellum and others, 1932; INEGI, 1980). Although these plutons commonly are interpreted to be coeval with Laramide deformation (e.g., Kellum and others, 1932; López-Ramos, 1985), many and perhaps most cut Laramide structures (C. Henry, unpublished data), and similar rocks in the adjacent Tepehuano terrane are post-Laramide.

Thick Paleocene to Miocene nonmarine and marine clastic rocks that have been mapped and drilled in the Burgos basin in northern Tamaulipas and the La Popa basin in Nuevo León (McBride and others, 1974; López-Ramos, 1985, p. 248–256; Vega-Vera and others, 1989) are correlative with strata in southern Texas (e.g., Wilcox and Frio: Frio Formations). Cenozoic volcanic rocks include siliceous mid-Tertiary rocks in northern Coahuila near the Rio Grande, Oligocene alkalic volcanic rocks in Tamaulipas, and sparsely distributed Quaternary basalts (Table 3) (INEGI, 1980; Hubberten and Nick, 1986; Nick, 1988).

Structural and geophysical data. The southern boundary of the Coahuiltecano terrane is obscured by thrust sheets of Jurassic to Late Cretaceous strata that underwent as much as 45 km of northward to northeastward transport during latest Cretaceous to middle Eocene Laramide orogenesis (de Cserna, 1956; Tardy,

1975; Padilla y Sánchez, 1985, 1986; Quintero-Legorreta and Aranda-García, 1985; Gotte, 1988, 1990). The east-west structural grain and marked curvature of the fold and thrust belt in this region probably was caused by divergence of thrust sheets around the margins of Coahuila Island, which apparently served as a rigid backstop during thrusting. South-dipping thrusts sole into a basal décollement at depth.

North- to northwest-striking high-angle normal faults in northwestern Coahuila formed during mid- to late Tertiary extension.

Paleomagnetic data from Triassic and Jurassic strata have been interpreted to indicate about 130° of counterclockwise rotation of parts of the southern Coahuiltecano terrane and adjacent Guachichil terrane in Early and Middle Jurassic time (Gose and others, 1982). Early Cretaceous strata in southernmost Coahuila may have been rotated about 35° counterclockwise during Laramide orogenesis (Kleist and others, 1984) or 10° to 15° counterclockwise in post-Eocene time (Nowicki and others, 1990).

Cochimí terrane

Oceanic rocks of Mesozoic ophiolite, island arc, mélange, and blueschist terranes probably underlie the continental shelf of North America from the southern California Transverse Ranges to the tip of Baja California. North of about 30°N, the distribution of these rocks has been disrupted by late Cenozoic transtension in the California Continental Borderland province (Crouch, 1979; Normark and others, 1987; Legg and others, 1991; Sedlock and Hamilton, 1991). In this Special Paper we define the Cochimí terrane as a composite terrane that includes outcrops and subcrops of oceanic rocks south of 30°N, approximately the same as the Vizcaíno terrane of Campa-Uranga and Coney (1983) and Coney and Campa-Uranga (1987).

Exposures of the Cochimí terrane are discontinuous but spectacular. On Isla Cedros, Islas San Benito, and the Vizcaíno Peninsula (Fig. 2), rocks of the Cochimí terrane are assigned to the Choyal, Vizcaíno Sur, Vizcaíno Norte, and Western Baja subterranes (called terranes by earlier workers); on Isla Magdalena and Isla Santa Margarita (Fig. 2), they are provisionally assigned to arc, ophiolite, and subduction complex terranes (Sedlock, 1993). In this section we describe the subterranes of the Cochimí terrane and then summarize geophysical data that indicate disruption and translation in a Late Cretaceous and Paleogene forearc.

Cedros-Benitos-Vizcaíno region. In the Cedros-Benitos-Vizcaíno region, the Cochimí terrane consists of three structural units (Fig. 8): an upper plate consisting of arc and ophiolite rocks of the Choyal, Vizcaíno Norte, and Vizcaíno Sur subterranes; a lower plate consisting of regionally metamorphosed blueschists of the Western Baja subterrane; and an unnamed intervening serpentinite-matrix mélange (Sedlock, 1988b).

The Choyal subterrane includes a well-preserved Middle Jurassic arc/ophiolite complex consisting of mafic to silicic volcanic rocks intruded by granitoids, ophiolite that probably formed during intraarc extension, and concordantly overlying volcanic rocks and clastic rocks derived solely from volcanic sources (Fig. 8; Table 4) (Kilmer, 1984; Kimbrough, 1984, 1985; Busby-Spera, 1988). The arc/ophiolite complex is overlain by Middle and Late Jurassic clastic rocks derived mainly from terrigenous sources including Paleozoic (probably Pennsylvanian) limestone, metasedimentary rocks including quartzite, and quartzose sandstone. The abrupt change in provenance has been interpreted to indicate that the Choyal subterrane was juxtaposed with or accreted to a continental mass, presumably North America, in early Late Jurassic time (Boles and Landis, 1984).

The Vizcaíno Norte subterrane includes Late Triassic ophiolite, conformably overlying tuffaceous Late Triassic sedimentary rocks that contain radiolarians and the megafossil *Monotis,* and Late Jurassic–Upper Cretaceous coarse volcanogenic rocks containing granitoid clasts that have Middle Proterozoic and Late Jurassic U-Pb discordia intercepts (Fig. 8; Table 4) (Hickey, 1984; Moore, 1985; Kimbrough and others, 1987). The Vizcaíno Sur subterrane includes Late Triassic ophiolite; conformably overlying Late Triassic chert, limestone, breccia, and sandstone that contain radiolarians and the megafossils *Monotis* and *Halobia*; pre–Late Jurassic volcanic and volcaniclastic rocks; and Middle Jurassic to Early Cretaceous tonalite (Fig. 8, Table 4) (Pessagno and others, 1979; Moore, 1985). The Vizcaíno Norte and Sur subterranes are fragments of island arcs that probably formed on Late Triassic oceanic crust and were accreted to North America by latest Jurassic or earliest Cretaceous time, based on a provenance change from purely volcanogenic to

partly or mainly siliciclastic (Moore, 1985). In Early Cretaceous(?) time, the Vizcaíno Norte and Sur subterranes were juxtaposed with one another along a fault zone that contains the Sierra Placeres mélange (Fig. 8, Table 4) (Moore, 1985).

All three arc/ophiolite subterranes are overlapped by the Valle Formation, which consists of Albian-Campanian siliciclastic turbidites that probably accumulated in a forearc basin setting (Kilmer, 1979; Patterson, 1984; Boles, 1986). Much of the Cretaceous section was deposited syntectonically, based on the presence of large basement blocks within olistostromes, asymmetric subsidence of the basin, and slide blocks as long as 100 m (Busby-Spera and others, 1988; Smith and Busby-Spera, 1992). Miocene and Pliocene shallow marine strata unconformably overlie arc/ophiolite rocks and Cretaceous siliciclastic turbidites on Isla Cedros and the Vizcaíno Peninsula (Fig. 8) (Kilmer, 1979; Smith, 1984). Latest Miocene fossil marine vertebrates on Cedros form one of the most diverse assemblages in the North Pacific realm (Barnes, 1992).

All upper plate units are cut by shallowly to moderately dipping normal faults and vein systems of Late Cretaceous and possibly Cenozoic age, and contractional structures are very rare (Sedlock, 1988c). Preliminary studies of fault plane striations imply changing stress directions during the Late Cretaceous and Cenozoic (R. Sedlock and D. Larue, unpublished data), but more work is needed to ascertain this trend. Upper plate normal faults and vein systems sole into major faults that separate the upper and lower plates.

Lower plate blueschists of the Western Baja subterrane (Fig. 8) are exposed on Isla Cedros and Islas San Benito, where they structurally underlie Jurassic, Cretaceous, and Cenozoic upper plate rocks (Sedlock, 1988b). Protoliths include ocean-floor basalt, siliciclastic metasedimentary rocks, ribbon radiolarian chert ranging in age from Late Triassic to mid-Cretaceous, and rare limestone (Sedlock and Isozaki, 1990). Contractional structures such as cleavage, folds, and thrust faults developed during peak blueschist metamorphic conditions of 5 to 8[+] kbar and 170° to 300°C in a subduction zone beneath western North America in the late Early Cretaceous (about 115 to 105 Ma). Contractional structures are cut by younger normal faults and carbonate-quartz vein systems that probably developed during slow uplift in the Late Cretaceous and Tertiary (Table 4) (Sedlock, 1988a, c, 1992, 1993; Baldwin and Harrison, 1989).

Serpentinite-matrix mélange occupies major fault zones between the upper and lower plates (Fig. 8). The mélange consists of diverse tectonic blocks up to 1 km long in a strongly foliated chrysotile-lizardite matrix. Petrologic and radiometric studies indicate that blocks were derived from several source terranes, including Middle Jurassic high-pressure amphibolite with a local mid-Cretaceous blueschist overprint, eclogite of uncertain age with a mid-Cretaceous blueschist overprint, coarse-grained mid-Cretaceous blueschists distinct from the lower plate Western Baja terrane, greenschists of latest Jurassic or earliest Cretaceous age, orthogneiss of uncertain age, and serpentinized ultramafic rocks of uncertain age (Moore, 1986; Sedlock, 1988c; Table 4). The

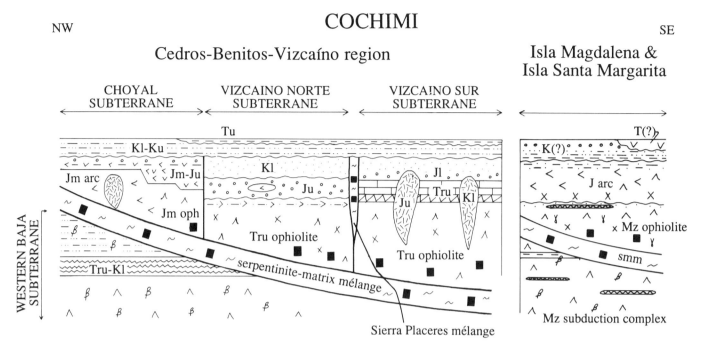

Figure 8. Schematic northwest-southeast structure section of Cochimí composite terrane. Relations between Cedros-Benitos-Vizcaíno region and Isla Magdalena–Isla Santa Margarita region unknown. In each region, serpentinite-matrix mélange separates footwall blueschist-facies subduction complex from hanging-wall ophiolitic, arc, and forearc basin rocks. Normal faults in upper plate and all deformation in lower plate not depicted.

fault zones between upper and lower plates were interpreted as thrust faults or strike-slip faults by earlier workers, but are reinterpreted as normal faults for several reasons (Sedlock, 1988c, 1992, 1993). (1) The plate-bounding faults dip 0° to 55°. (2) Upper plate normal faults and vein systems merge with plate-bounding faults, indicating a synkinematic origin. (3) Strains in the upper plate and the youngest strains in the lower plate are extensional. (4) Geobarometric estimates of wall rocks indicate "pressure gaps" of 1 to 6 kbar across these faults, with lower pressure rocks always in the upper plate; this relation requires tectonic thinning of 3 to 20 km of the crust and net normal displacements of 5 to 40 km. Mid-Cretaceous to Paleogene uplift of the lower plate blueschists probably was accommodated by major normal displacement on these shallowly dipping faults during synsubduction extension in the forearc (Sedlock, 1987, 1988, 1992, 1993).

Magdalena-Margarita region. Isla Santa Margarita and southern Isla Magdalena, islands west of mainland Baja between 25° and 24°N, consist chiefly of metamorphosed and deformed Mesozoic(?) rocks of oceanic origin, with subordinate Tertiary volcanic rocks and dikes. The islands are separated from unmetamorphosed Tertiary sedimentary rocks of the Yuma terrane on mainland Baja by a buried northwest-striking fault. Reconnaissance geologic studies were undertaken on southern Isla Magdalena by Blake and others (1984) and on Isla Santa Margarita by Forman and others (1971), Rangin and Carrillo (1978), and

Rangin (1978). The identity and nature of geologic map units and contacts have been modified on both islands, particularly on Isla Santa Margarita, on the basis of detailed mapping and structural, petrologic, and geochemical work (Sedlock, 1993).

As in the Cedros-Benitos-Vizcaíno region, Mesozoic rocks in the Magdalena-Margarita region are divided into three structural units: an upper plate consisting of ophiolite, arc, and forearc basin rocks and locally garnetiferous amphibolite; a lower plate subduction complex; and serpentinite-matrix mélange that crops out along intervening, shallowly dipping faults (Fig. 8) (Sedlock, 1993). The subduction complex and serpentinite-matrix mélange crop out only on Isla Santa Margarita. Ophiolitic, arc, and forearc basin rocks and amphibolite crop out on both islands, and possible consanguinity of each geographic pair (e.g., Magdalena arc rocks, Santa Margarita arc rocks) is indicated by mesoscopic igneous, sedimentary, and tectonic structures, metamorphic assemblages, and inferred P-T conditions. Ongoing biostratigraphic, geochemical, and geochronologic studies will test the correlation of units between islands. Rocks of both islands are grouped together in the discussion below.

Upper plate ophiolitic rocks include metamorphosed ultramafic rocks, gabbro, diabase, mafic volcanic rocks, chert, and clastic sedimentary rocks that crop out on southern Isla Magdalena and central Isla Santa Margarita. All rocks underwent ductile contraction as indicated by penetrative foliation and local lineation, open to isoclinal folding, and thrust faulting. Primary miner-

R. L. Sedlock and Others

TABLE 4. COCHIMÍ TERRANE RADIOMETRIC DATA

Sample	System	Mineral*	Date (Ma)	References†	Comments
Choyal terrane					
Plagiogranite	U-Pb	zr	173	1	Concordant age; minimum age of ophiolite
Granitoids	U-Pb	zr	166–160	1	Several samples
Plag-hornblende tuff	U-Pb	zr	166	1	Concordant age
Granitoids	^{40}Ar/^{39}Ar	kf	162, 161	2	2 samples; isochron ages
Hornblende tuff	^{40}Ar/^{39}Ar	p	~160	2	
Andesitic tuff	K-Ar	h	159 ± 5	3	
Plutonic clast in mid-K conglomerate	^{40}Ar/^{39}Ar	kf	158 ± 1	2	Isochron age
Quartz diorite (float)	K-Ar		148 ± 6	4	
Vizcaíno Norte terrane					
Plagiogranite	U-Pb	zr	~220	1	
Albitite in ophiolite	U-Pb	kf, sph	220 ± 2	5	Concordant age
Granitoid clasts in Upper Jurassic strata	U-Pb	zr	1,340 ± 3;	6	Discordia intercepts
			150 ± 3	6	
	K-Ar	b	155 ± 5	6	
Andesite dikes	K-Ar	p	128 ± 2	7	Intrude Upper Jurassic strata
			116 ± 5	7	
Andesite dike	K-Ar	wr	125	8	Intrudes Upper Jurassic strata
Vizcaíno Sur terrane					
Tonalite, granodiorite	U-Pb	zr	154, 151,	9,1	Several samples; minimum ages
			140, 127	9,1	of emplacement
Tonalite	K-Ar	b	154 ± 3	7	
Tonalite	K-Ar	h	143 ± 3	10	
Tuff in mid-Cretaceous flysch	K-Ar	b	103 ± 2	3	
Blocks in Sierra Placeres mélange					
Hornblende tonalite	^{207}Pb/^{206}Pb	zr	161	1	Lower intercept
	U-Pb	zr	149	1	
Amphibolite	K-Ar	h	140 ± 21	11	Source: D. Krummenacher
Hornblende porphyry	K-Ar	h	126 ± 4	11	Source: D. Krummenacher
Western Baja terrane					
Hornblende quartz diorite	^{40}Ar/^{39}Ar	h	153 ± 11	12	Protolith age
Conglomerate clast	^{40}Ar/^{39}Ar	wm	100 ± 14	12	Age of blueschist metam
Metasandstone, San Benitos	^{40}Ar/^{39}Ar	wm	113 ± 1	12	Isochron age
Metasandstone, Cedros	^{40}Ar/^{39}Ar	wm	109 ± 1	11	Plateau age
Metasandstone, Cedros	^{40}Ar/^{39}Ar	kf	~75	2	Cooling age
Plutonic clast in conglomerate	^{40}Ar/^{39}Ar	kf	20	12	Cooling age
Blocks in serpentinite-matrix mélange, Cedros-Vizcaíno region					
Garnet amphibolite, Cedros	^{40}Ar/^{39}Ar	h	167 ± 6	13	Plateau age;
	FTA	ap	55	14	Cooling age
Garnet amphibolite, Cedros	^{40}Ar/^{39}Ar	h	166 ± 2	13	Isochron age
Epidote amphibolite	^{40}Ar/^{39}Ar	h	170 ± 1	14	Isochron age
Epidote amphibolites	^{40}Ar/^{39}Ar	h	175–150;	14	2 samples; complex release
			~165	14	spectra
Epidote-amphibolite facies	^{40}Ar/^{39}Ar	wm	172–150	14	Convex-up pattern;
metasedimentary rocks	FTA	ap	99	14	cooling age
	^{40}Ar/^{39}Ar	wm	~170	14	Plateau age;
	FTA	ap	125	14	cooling age
	^{40}Ar/^{39}Ar	wm	168–152	14	Cooling gradient;
	FTA	ap	164	14	concordant cooling age
Amphibolites, Cedros	Rb-Sr	p, wm	151 ± 18	2	2 samples; 4-pt isochron; ^{87}Sr/^{86}Sr$_i$: 0.7048
Amphibolites, San Benitos	K-Ar	h, wm	148 ± 5	15	2 samples

TABLE 4. COCHIMI TERRANE RADIOMETRIC DATA (continued)

Sample	System	Mineral*	Date (Ma)	References[†]	Comments
Blocks in serpentinite-matrix mélange, Cedros-Vizcaíno region (continued)					
Epidote amphibolite-facies blocks	$^{40}Ar/^{39}Ar$	amph	170–115	14	Apparent ages; M_1 ~170; M_2 ~115
overprinted by blueschist assemblage,	$^{40}Ar/^{39}Ar$	amph	161–96	14	Apparent ages; M_1 ~161; M_2 ~96
Cedros and San Benitos	$^{40}Ar/^{39}Ar$	amph	161 ±12; 64 ± 5	14	Isochron ages from two degassing phases?
	$^{40}Ar/^{39}Ar$	wm	173–119	14	Apparent ages; M_1 ~173; M_2 ~119
	$^{40}Ar/^{39}Ar$	wm	171–142	14	Apparent ages; M_1 ~171; M_2 ~142
	$^{40}Ar/^{39}Ar$	wm	169–89	14	Apparent ages; M_1 ~169; M_2 ~89
	$^{40}Ar/^{39}Ar$	wm	168–59	14	Apparent ages; M_1 ~168; M_2 ~59
	$^{40}Ar/^{39}Ar$	wm	167–76	14	Apparent ages; M_1 ~76; M_2 ~167
Blueschist, Vizcaíno	K-Ar	na	173 ± 69	11	Source: D. Krummenacher
Greenschist, Vizcaíno	$^{40}Ar/^{39}Ar$	h	~140	2	Apparent age
Eclogite, blueschist overprint, Cedros	$^{40}Ar/^{39}Ar$	wm	115–105	12	Slow cooling gradient; cooling age
	FTA	ap	32 ± 4	12	
Blueschists, Cedros	$^{40}Ar/^{39}Ar$	wm	115–95	12	Four samples; Ar loss profiles: slow cooling
	FTA	ap	22 ± 3	12	Cooling age
Blueschists, Cedros	K-Ar	wm	110 ± 2	15	2 samples
Blueschist, San Benitos	K-Ar	na	104 ± 2	15	
Blueschist, Cedros	$^{40}Ar/^{39}Ar$	na	103 ± 4	14	Isochron age; degassing of wm intergrowths
Blueschists, Cedros	$^{40}Ar/^{39}Ar$	na	95 ± 1; 94 ± 1	14 12	Isochron age; degassing of wm intergrowths
Blueschist, Cedros	Rb-Sr	wr, wm, p, na	99 ± 13	12	Isochron age; $^{87}Sr/^{86}Sr_i$: 0.7050
Blueschist, Cedros	K-Ar	na	94 ± 4	15	
Magdalena and Santa Margarita Islands					
Amphibolite, southern Magdalena	K-Ar	h	138 ± 3	16	Base of ophiolite?
Amphibolite, southern Magdalena	K-Ar	h	133 ± 6	16	Block in mélange
Garnet-amphibolite, central Santa Margarita	K-Ar	h	134 ± 6	16	Block in mélange
Lamprophyre dike, central Santa Margarita	K-Ar	b	29 ± 1	16	Intrudes ophiolite

*Mineral abbreviations: amph = amphibole; b = biotite; h = hornblende; kf = potassium feldspar; na = sodic amphibole; p = plagioclase; sph = sphene; wm = white mica; wr = whole rock; zr = zircon.

[†]1 = Kimbrough, 1982; 2 = Baldwin, 1988; 3 = Gastil and others, 1978; 4 = Suppe and Armstrong, 1972; 5 = Barnes and Mattinson, 1981; 6 = Kimbrough and others, 1987; 7 = Minch and others, 1976; 8 = Robinson, 1975; 9 = Barnes, 1982; 10 = Troughton, 1974; 11 = Moore, 1985; 12 = Baldwin and Harrison, 1989; 13 = Baldwin and others, 1990; 14 = Baldwin and Harrison, 1992; 15 = Suppe and Armstrong, 1972; 16 = Forman and others, 1971.

alogy and textures were obliterated by greenschist-facies metamorphism, including 98 to 100% serpentinization of ultramafic phases, that outlasted contractional deformation. Contractional structures in all rock types are cut by later normal faults and vein systems, which in turn are cut by Oligocene lamprophyre dikes (Table 4). The ages of protoliths, contractional deformation, and metamorphism are not known.

Upper plate arc rocks that crop out on southern Isla Magdalena and southern Isla Santa Margarita include layered and massive gabbro; a mafic to intermediate dike and sill complex; mafic to intermediate pillow lavas, massive flows, and breccia; and tuffaceous volcaniclastic sedimentary rocks. Arc rocks lack a penetrative fabric and were statically metamorphosed at low to moderate temperature and low pressure. The sedimentary rocks

contain Jurassic fauna (Blake and others, 1984), but ages of the magmatic rocks are unknown. All rock types are cut by normal faults and vein systems that are inferred to be older than the Oligocene dikes that cut similar structures in the ophiolite. The ages of igneous protoliths and metamorphism are not known.

Upper plate forearc basin rocks consist of small, isolated, fault-bounded sequences of unmetamorphosed deep-water conglomerate and rhythmically interbedded terrigeneous turbidites that crop out on southern Isla Magdalena and in several places on Isla Santa Margarita. These rocks strongly resemble the mid-Cretaceous Valle Formation in the Cedros-Benitos-Vizcaíno region and thus are assigned a provisional Cretaceous age and interpreted as submarine fan deposits in a forearc basin. Contractional structures are absent, but all outcrops are cut by normal faults with displacements of at least 5 m; extension is inferred to be older than the Oligocene dikes in the ophiolite. Locally, the Cretaceous(?) strata depositionally overlie upper plate ophiolitic and arc rocks, but most contacts are moderately dipping normal faults that juxtapose hanging-wall Cretaceous(?) strata with footwall ophiolitic or arc rocks.

Upper plate amphibolite crops out in coherent 1- to 5-km sheets at Cabo San Lázaro on northern Isla Magdalena and near Puerto Alcatraz on northern Isla Santa Margarita. These sheets are juxtaposed with upper plate ophiolitic rocks along shallowly dipping faults interpreted as thrusts. The amphibolite bodies are strongly lineated and locally strongly foliated, with the metamorphic assemblage hornblende + plagioclase + sphene + apatite ± garnet, with retrogressive epidote. Hornblende from amphibolites in both areas yielded earliest Cretaceous K-Ar cooling ages (Table 4).

Lower plate subduction complex rocks crop out on northern and southern Isla Santa Margarita, with an outcrop area much less than proposed by Rangin and Carrillo (1978) and Rangin (1978). The subduction complex consists of weakly to moderately foliated and metamorphosed pillow and massive metabasite, radiolarian chert, and interbedded red and green argillite, metabasite(tuff?), and rare limestone. The "graywackes" reported by Rangin and Carrillo (1978) and Rangin (1978) are greenschist-facies metasedimentary rocks of the upper plate ophiolite and gabbros, dikes, volcanic rocks, and volcaniclastic rocks of the upper plate arc (Sedlock, 1993). High-pressure metamorphic minerals include lawsonite, aragonite, and sodic amphibole in veins, and incipient sodic clinopyroxene in the groundmass of metabasites. Deformation features include weakly to moderately developed foliation subparallel to bedding, thrust faults, rare folds, and normal faults that cut all other structures. The ages of protoliths, metamorphism, and deformation are unknown. The subduction complex was interpreted as the upper level of the adjacent ophiolite by Rangin and Carrillo (1978) and Rangin (1978), but the presence of high-pressure phases such as lawsonite, aragonite, and sodic amphibole indicates that these rocks probably were metamorphosed at depths of 10 to 20 km in a subduction zone.

The subduction complex is separated from upper plate rocks by major fault zones, up to 25 m thick, that contain serpentinite-matrix mélange. Mélange also crops out along some faults within the upper plate. The mélange consists of ultramafic, amphibolite, mica schist, and greenschist metasedimentary blocks in a strongly foliated lizardite-chrysotile matrix. Hornblende from an amphibolite block in central Isla Magdalena yielded an earliest Cretaceous K-Ar cooling age (Table 4). Ongoing petrologic, geochemical, and geochronologic studies will address the source terranes of these blocks. As in the Cedros-Benitos-Vizcaíno region, these major faults are interpreted as normal faults along which the footwall blueschists were exhumed during extension of the hanging wall ophiolitic, arc, and forearc basin rocks.

Unmetamorphosed, undeformed calc-alkalic rhyodacites form isolated, steep-sided peaks on the southwestern coast of Isla

Santa Margarita (Rangin and Carrillo, 1978; Rangin, 1978). Interpretation of these rocks as large blocks in serpentinite-matrix mélange is invalid because they are not cut by serpentine or carbonate veins or bounded by metasomatic rinds, as are other blocks in the mélange. Instead, the volcanic rocks probably were erupted directly onto exposed mélange, as indicated by the alteration of serpentinite at the contact. The volcanic rocks are provisionally interpreted to be Miocene in light of their similarity to parts of the Comondú Formation in Baja California Sur (see Yuma terrane).

Geophysical data. South of about 30°N, low-pass filtered gravity data reveal a north-northwest–striking positive gravity anomaly, attributed to "gabbro intrusions" and "Franciscan-like rocks," that coincides almost exactly with the mapped extent of the Cochimí terrane (Couch and others, 1991). Shorter wavelength gravity data show three en echelon positive anomalies on the continental shelf and slope between 26° and 23°N that may record several hundred kilometers of dextral separation (Couch and others, 1991). Faults inferred between these en echelon anomalies strike more westerly than and may be truncated by splays of the Tosco-Abreojos fault zone to the west, which accommodated much of the tangential component of Pacific–North America relative motion in the Miocene (p. 118). We suggest that dextral displacement on the inferred faults and resulting northward translation of outboard slices of Cochimí, Franciscan, and related rocks accommodated the tangential component of Late Cretaceous to Paleogene right-oblique convergence, transform motion in the Miocene Pacific–North America plate boundary, or both.

Paleomagnetic data from Triassic and Cretaceous sedimentary rocks of the upper plate, in conjunction with data from other terranes of the Baja California peninsula, are interpreted to indicate at least 10° and perhaps 20° of northward transport of Baja California between about 90 and 40 Ma. Relative paleolatitudes, which correspond to the northward latitudinal displacement since the time of magnetization (see Lund and others, 1991b), include 18° ± 11.3° for Triassic chert, sandstone, limestone, and pillow basalt in the Vizcaíno Sur terrane (Hagstrum and others, 1985); 19.5° ± 6.3° for the mid-Cretaceous Valle Formation on Isla Cedros (D. Smith and C. Busby-Spera, 1991, unpublished manuscript); and 15.6° ± 7.1° for the Valle Formation on the Vizcaíno Peninsula (Patterson, 1984; Hagstrum and others, 1985; D. Smith and C. Busby-Spera, unpublished manuscript). These studies also calculated net clockwise rotations of about 55° for the Triassic strata and 15° to 40° for the Cretaceous rocks.

Paleomagnetic data from bedded chert of the Western Baja subterrane, which is in fault contact with the bedded rocks of the upper plate, imply about 25° of northward transport and 55° clockwise rotation since the mid-Cretaceous accretion of the terrane to North America (Hagstrum and Sedlock, 1990, 1992). These and other paleomagnetic data from Baja are discussed more fully on pages 80–81.

Tectonic history. The Cochimí terrane in the Cedros-Benitos-Vizcaíno region was constructed during Late Jurassic to

Cenozoic time. The Choyal, Vizcaíno Norte, and Vizcaíno Sur subterranes are fossil arc and ophiolite complexes that were deactivated and attached to the continent and to one another during Late Jurassic and Early Cretaceous time, and overlapped by turbidites in the mid-Cretaceous. The major fault zones between upper and lower plates are interpreted as normal faults at which lower plate blueschists have been uplifted, atop which upper plate rocks have been brittely extended, and within which serpentinite-matrix mélange has been formed or emplaced (Sedlock, 1988b, c). Uplift and normal faulting probably began in the mid-Cretaceous, as indicated by facies analysis of the mid-Cretaceous overlap sequence in the upper plate (Busby-Spera and Boles, 1986; Smith, 1987; Busby-Spera and others, 1988), and continued in Late Cretaceous and Tertiary time (Table 4) (Baldwin and Harrison, 1989). Paleomagnetic data indicate that the upper and lower plates underwent significant, but different, amounts of northward translation in mid-Cretaceous to Paleogene time. Available data cannot distinguish between two end-member scenarios: (1) Lower plate blueschists were translated up to 1,500 km northward in mid-Cretaceous time, juxtaposed with upper plate rocks of the Cochimí terrane (e.g., Choyal subterrane), and translated about 1,000 km northward with the upper plate rocks in the Late Cretaceous to Paleogene. (2) Upper plate rocks of the Cochimí terrane were translated 1,000 to 2,000 km northward in the Late Cretaceous to Paleogene; lower plate blueschists were translated up to 2,500 km northward between late Early Cretaceous and Paleogene time, perhaps in a position more outboard than that of the upper plate rocks; upper and lower plates were juxtaposed more or less in their current position during Late Cretaceous or Paleogene normal faulting.

The tectonic history of the Cochimí terrane in the Magdalena-Margarita region remains uncertain pending the outcome of ongoing geochronologic, biostratigraphic, and paleomagnetic studies. However, the gross similarity of its regional structure, deformation history, and metamorphic history to the Cedros-Benitos-Vizcaíno region strongly suggests a similar evolution.

Cuicateco terrane

The Cuicateco terrane is a west-dipping fault-bounded prism of strongly deformed Jurassic and Cretaceous oceanic and arc rocks that structurally overlies the Maya terrane and underlies the Zapoteco terrane. We provisionally infer that the volcanic and sedimentary protoliths of the Cuicateco terrane were deposited in a southward-opening Jurassic–Early Cretaceous basin of enigmatic origin, and that these protoliths were pervasively deformed and metamorphosed to greenschist facies during Late Cretaceous to Paleogene closure of the basin between the converging Zapoteco and Maya continental massifs.

Many aspects of the geology of the Cuicateco terrane are unresolved. Few radiometric data are available, and the distribution of and relations among major map units still have not been satisfactorily determined. We emphasize that our structure sec-

tions of the Cuicateco terrane (Fig. 9) are very simple, schematic representations of the distribution of the major rock units. The rock units of the Cuicateco terrane broadly correspond to those of the Juárez terrane of Campa-Uranga and Coney (1983) and Coney and Campa-Uranga (1987), but the placement of terrane boundaries differ. Specifically, rocks in the long, narrow, western arm of the Juárez terrane have been reassigned to the Zapoteco and Chatino terranes and the fault zone between them, as shown on Transect H-3 (Ortega-Gutiérrez and others, 1990).

Along Transect H-3, the Cuicateco terrane is divided into three shallowly dipping structural units (Ortega-Gutiérrez and others, 1990). In the structurally lowest unit, greenstone, lenses of gabbro and serpentinite, metatuff, and graywacke in the southern part of the terrane are interpreted as a disrupted ophiolite (Carfantan, 1983) and sedimentary cover that have been faulted onto the Maya terrane (Fig. 9, section A-B). Similar rocks have not been recognized in the northern part of the terrane.

The intermediate and most voluminous unit is an assemblage of strongly deformed but weakly metamorphosed flysch, tuff, black slate, and limestone that contains Berriasian-Valanginian microfossils and the Valanginian ammonite *Olcostephanus* (Carfantan, 1981; Ortega-Gutiérrez and González-Arreola, 1985). Calcareous conglomerate contains small pebbles of granulitic gneiss and phyllite that may have been derived from the Oaxaca Complex (Zapoteco terrane) and Acatlán Complex (Mixteco terrane). A single K-Ar date of 82.5 Ma obtained from a phyllite may indicate early Late Cretaceous metamorphism (Carfantan, 1983). Along Transect H-3 (Ortega-Gutiérrez and others, 1990) the intermediate unit locally is overthrust by sheared serpentinite associated with low-grade phyllite-quartzite and greenstone (Fig. 9, sections A-B, C-D). Protoliths of this unit accumulated in a Jurassic to earliest Cretaceous basin, called the Cuicateco basin in Part 2 of this volume. Basement of the basin does not crop out but is inferred to be oceanic, at least in the southern part of the terrane (Ortega-Gutiérrez and others, 1990).

The structurally highest unit of the terrane includes mylonitic mafic to silicic orthogneiss that crops out at the western boundary of the Cuicateco terrane within the Juárez suture, at which the Zapoteco terrane overthrust the Cuicateco terrane in the Late Cretaceous (Ortega-Gutiérrez and others, 1990). Unpublished K-Ar dates and new $^{40}Ar/^{39}Ar$ laser probe results (Table 5) have been interpreted to indicate Middle Jurassic intrusion and Early Cretaceous metamorphism and cataclasis (R. Múgica, personal communication, 1981; C. Pacheco, unpublished data, in Delgado-Argote, 1989) and earliest Cretaceous cooling past the hornblende blocking temperature (Delgado-Argote and others, 1992b). The plutonic protoliths may have formed in a short-lived continental arc (Delgado-Argote and others, 1992b). The Juárez suture also contains mylonitic rocks derived from anorthosite of the adjacent Zapoteco terrane (Delgado-Argote, 1989; Ortega-Gutiérrez and others, 1990).

In the southern Cuicateco terrane (Fig. 9, section E-F), Early Cretaceous flysch and tuff of the intermediate unit are juxtaposed at an inferred fault with the "Chontal arc" (Carfantan,

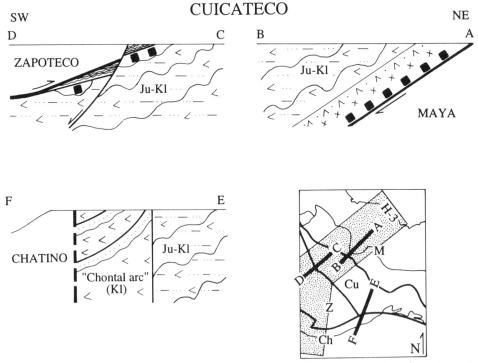

Figure 9. Schematic structure sections of Cuicateco terrane. Inset shows approximate locations of schematic section lines A-B, C-D, and E-F, transect H-3 (dot pattern), and terrane boundaries (heavy lines). Terrane abbreviations: Ch, Chatino; Cu, Cuicateco; M, Maya; Z, Zapoteco.

1981), which consists of andesite, volcaniclastic rocks, tuff, flysch, and black schists with intercalations of marble that contain Early Cretaceous fossils. We consider the Chontal arc a subterrane of the Cuicateco terrane, but future work may show that it is a separate terrane. The Chontal arc and unconformably overlying Campanian-Maastrichtian flysch are bounded on the southwest by an unexposed fault contact with the Chatino terrane. The intermediate Early Cretaceous flysch-tuff unit and the Early Cretaceous Chontal arc subterrane experienced two episodes of northeast-southwest shortening. The earlier episode probably was Turonian and was accompanied by synkinematic plutonism and weak regional metamorphism; the younger event, which also

affected Campanian-Maastrichtian flysch atop the Chontal arc, probably was of Laramide (latest Cretaceous–Paleogene) age (Carfantan, 1981, 1983).

In the northern part of the terrane, northwest of Transect H-3, strongly deformed metasedimentary and metavolcanic rocks and minor serpentinized ultramafic bodies near Tehuacán, Puebla (Fig. 2) also are considered part of the Cuicateco terrane, although their connection with rocks farther south is unclear (Carrasco, 1978; Carfantan, 1981; Delgado-Argote, 1988, 1989; Alzaga-Ruiz and Pano-Arciniega, 1989). The original stratigraphy cannot be reconstructed due to Laramide thrusting, folding, fabric development including local mylonitization, and meta-

TABLE 5. CUICATECO TERRANE RADIOMETRIC DATA

Sample	System	Mineral*	Date (Ma)	References[†]	Comments
Hornblende clinopyroxenite	$^{40}Ar/^{39}Ar$	h	134 ± 3	1	Plateau age
Hornblende diorite	$^{40}Ar/^{39}Ar$	h	132 ± 6	1	Plateau age
Phyllite	K-Ar		85	2	Source: O. Vila-Gómez; metamorphic age?

*Mineral abbreviation: h = hornblende.
[†] 1 = L. Delgado–Argote, unpublished data; 2 = Carfantan, 1981.

morphism to greenschist facies. Protoliths probably include, but may not be limited to, intermediate to silicic massive and pillowed lavas that locally contain gneissic xenoliths that probably were derived from the Oaxacan Complex of the Zapoteco terrane; intermediate to silicic tuff; volcaniclastic sandstone and conglomerate that locally contains clasts of the Oaxacan Complex; and fine-grained flysch, including black shales that may be correlative with Early Cretaceous strata in the intermediate unit of the Cuicateco terrane farther south (Delgado-Argote, 1988, 1989). The metatuffs contain tabular serpentinite bodies, derived from harzburgite or olivine pyroxenite, that probably rose diapirically from deeper crustal levels in the Cretaceous prior to pervasive deformation during Late Cretaceous–Paleogene (Laramide) shortening. Volcanic protoliths are most abundant along the western margin of the terrane, grading eastward into clastic rocks and perhaps into Early Cretaceous limestones of the Maya terrane that were overthrust by the Cuicateco terrane during Laramide orogenesis (Delgado-Argote, 1988, 1989); however, primary stratigraphic relations are not preserved. To the west, the metavolcanic rocks are adjacent to and are inferred to be cogenetic with the strongly deformed, locally mylonitic metagranitoids of the highest structural unit in the Cuicateco terrane.

Closure of the Cuicateco basin probably began by the end of the Early Cretaceous, based on the similarity of Albian strata in the Zapoteco, Cuicateco, and Maya terranes (Delgado-Argote, 1989). Maximum deformation and the peak of low-grade metamorphism probably occurred in the Turonian (Carfantan, 1983), but thrusting on the Juárez suture did not cease until Paleogene time, as indicated by undeformed Tertiary clastic and volcanic rocks (Ortega-Gutiérrez and others, 1990). Cenozoic rocks in the Cuicateco terrane include Oligocene to early Miocene(?) red beds intruded by mafic dikes and Miocene-Recent basalts. The Cuicateco-Zapoteco boundary was reactivated in mid-Tertiary time as the Oaxaca fault, a high-angle normal fault, along which accumulated Oligocene to early Miocene(?) red beds intruded by mafic dikes (Ortega-Gutiérrez, 1981; Delgado-Argote, 1989; Centeno-García and others, 1990). Miocene to Recent mafic to intermediate volcanic rocks are widely distributed across the northern Cuicateco terrane.

Guachichil terrane

Most of the Guachichil terrane is covered by Late Jurassic and Cretaceous carbonate and shallow marine clastic rocks similar to those elsewhere in eastern México. Pre-Jurassic rocks, exposed in two anticlinoria and encountered in several wells in southern Tamaulipas and northwestern Veracruz, include the following fault-bounded units: (1) Grenville gneiss that probably was rifted from southern North America in the latest Proterozoic or earliest Paleozoic; (2) early to middle Paleozoic miogeoclinal(?) sedimentary rocks; (3) Paleozoic metabasites and metasedimentary rocks that may have formed in a subduction complex (Granjeno Schist); and (4) Early Permian flysch. We provisionally distinguish northern and southern subterranes based on the

outcrop of Paleozoic sedimentary and metamorphic rocks. Both subterranes were attached to continental North America prior to Late Triassic to Middle Jurassic extension and volcanism that accompanied the rifting of Pangea. Late Jurassic to Cretaceous cover strata were folded and thrusted toward the north and east during latest Cretaceous to mid-Eocene Laramide orogenesis.

Precambrian gneiss. Middle Proterozoic (Grenville) gneiss crops out in both subterranes (Fig. 10, Table 6), and undated gneiss is known in several wells (López-Ramos, 1985, p. 395). In the northern Guachichil subterrane, near Ciudad Victoria, Tamaulipas (Fig. 2), the Novillo Gneiss consists of granulite-facies orthogneiss, paragneiss, amphibolite, and marble, and is intruded(?) by postkinematic plagiogranite (Table 6) (Carrillo-Bravo, 1961; Ortega-Gutiérrez, 1978c; Garrison and others, 1980; Castillo-Rodriguez, 1988; Cossio-Torres, 1988). In the southern Guachichil terrane, near Molango, Hidalgo (Fig. 2), the Huiznopala Gneiss consists of granulite-facies orthogneiss, paragneiss, and metaquartzite (Carrillo-Bravo, 1965; Fries and Rincón-Orta, 1965). Sm-Nd dates for both gneiss units are about 900 Ma, but peak metamorphism may have been attained about 1,000 Ma (Patchett and Ruiz, 1987).

Paleozoic sedimentary rocks. In the northern subterrane, Carrillo-Bravo (1961) described a roughly 1-km-thick sequence of Paleozoic marine sedimentary rocks consisting of conformable Cambrian (possibly latest Precambrian) to Early Silurian conglomerate, quartzite, and thin limestone; Silurian shale, sandstone, and limestone; Devonian chert, novaculite, shale, sandstone, and limestone; Early Mississippian sandstone and shale; and unconformably overlying Late Pennsylvanian limestone, sandstone, and shale. The Paleozoic sequence was internally deformed and faulted against other pre-Mesozoic units prior to the Late Triassic (Fig. 10). The sequence was correlated with the Ouachita orogenic belt in the Marathon region of west Texas (Flawn and others, 1961), but the strength of this correlation is waning in light of recent work in this region. The Cambrian(?) La Presa quartzite near the base of the section has been reinterpreted as a part of the Novillo Gneiss (C. Ramírez-Ramírez, unpublished data). The "Devonian" siliceous sedimentary rocks have been remapped as rhyolitic to rhyodacitic volcanic rocks of uncertain age (Gursky and Ramírez-Ramírez, 1986). Some parts of the section contain transported shallow water fauna of North American affinity, and much of the section may be olistostromal (Stewart, 1988; C. Ramírez-Ramírez, unpublished data). From these observations, we conclude that the Paleozoic strata in the northern Guachichil terrane were deposited in a basinal environment near the southern margin of North America, that they may not form a conformable, continuous vertical sequence, and that they do not correlate uniquely or perhaps even strongly with the Ouachita orogenic belt.

Paleozoic metamorphic rocks. In the northern subterrane near Ciudad Victoria, strongly deformed, interbedded schist, metabasite, and rare metachert that have been metamorphosed to greenschist facies and that structurally overlie serpentinite are known collectively as the Granjeno Schist (Fig. 10) (Carrillo-

GUACHICHIL

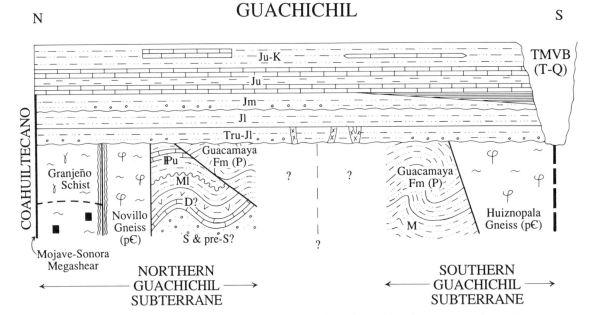

Figure 10. Composite tectonostratigraphic section and schematic north-south structure section of Guachichil composite terrane. Northern and southern Guachichil terranes defined by differences in pre-Permian rocks. Effects of Laramide deformation have been omitted.

TABLE 6. GUACHICHIL TERRANE RADIOMETRIC DATA

Sample	System	Mineral*	Date (Ma)	References†	Comments
Novillo Gneiss and plagiogranite					
Micaceous marble	K-Ar	ph	928 ± 18	1	
Granitic gneiss	K-Ar	h	919 ± 18	1	
Granitic gneiss	K-Ar	h	880 ± 17	1	
Gneiss	Rb-Sr	wr	1,140 ± 80	2	6-pt isochron; $^{87}Sr/^{86}Sr_i$: 0.7061
Gneiss	Rb-Sr	wr	860 ± 77	2	5-pt isochron; $^{87}Sr/^{86}Sr_i$: 0.7070
Gneiss	K-Ar	b	744 ± 25	3	
Plagiogranite	Rb-Sr	wr	774 ±256	2	5-pt isochron; $^{87}Sr/^{86}Sr_i$: 0.7037
Plagiogranite	Rb-Sr	wr	570 ±181	2	4-pt isochron; $^{87}Sr/^{86}Sr_i$: 0.7040
Granjeno Schist					
Schist	Rb-Sr	wr	452 ± 45	4	Corrections to de Cserna and
Schist	Rb-Sr	wr	373 ± 37	4	others, 1977
Pelitic schist	Rb-Sr	wr + wm	330 ± 35	2	Composite isochron
Pelitic schist	Rb-Sr	wr + wm	327 ± 31	2	Data of Denison and others, 1971
Pelitic schist	Rb-Sr	wr + wm	320 ± 12	2	
Schist	K-Ar	m	318 ± 10	3	
Pegmatite	K-Ar	m	313 ± 10	5	
Graphitic schist	K-Ar	m	311 ± 6	1	
Graphitic schist	K-Ar	m	300 ± 6	1	
Graphitic schist	K-Ar	m	299 ± 6	1	
Pelitic schist	Rb-Sr	wr	286 ± 66	2	
Graphitic schist	K-Ar	m	278 ± 5	1	
Schist	K-Ar	m	270 ± 8	6	
Schist	K-Ar	m	257 ± 8	6	

*Mineral abbreviations: b = biotite; h = hornblende; m = mica; ph = phlogopite; wm = white mica; wr = whole rock.
†1 = Denison and others, 1971; 2 = Garrison and others, 1980; 3 = Fries and others, 1962; 4 = de Cserna and Ortega-Gutiérrez, 1978; 5 = Fries and Rincón-Orta, 1965; 6 = de Cserna and others, 1977.

Bravo, 1961; de Cserna and others, 1977; Ortega-Gutiérrez, 1978c; Garrison and others, 1980; Castillo-Rodriguez, 1988). Schist that probably is correlative with the Granjeno Schist crops out beneath Triassic strata near Aramberri, Nuevo León (Denison and others, 1971; López-Ramos, 1985, p. 275) and has been encountered in several wells in southern Tamaulipas and northwestern Veracruz (López-Ramos, 1985, p. 395). Protoliths of the schist include pelites, volcaniclastic rocks, metabasites, and chert of unknown age; rare carbonates and coarse-grained siliciclastic rocks imply deep water accumulation at a distance from terrigenous sources (Castillo-Rodriguez, 1988).

These protoliths were subjected to at least three phases of deformation and metamorphism (Castillo-Rodriguez, 1988). During D_1/M_1, the rocks were penetratively foliated under greenschist(?)-facies conditions. D_2/M_2 caused north-northwest–trending isoclinal folds of D_1 and nearly complete recrystallization to a biotite-rich greenschist-facies assemblage. D_3/M_3 resulted in north- to northwest-trending folds, widespread crenulation of D_2 structures, and retrograde metamorphism (chlorite zone). K-Ar and Rb-Sr dates, chiefly from micas, imply a late Paleozoic age for M_2 metamorphism (Table 6). However, de Cserna and others (1977) and de Cserna and Ortega-Gutiérrez (1978) correlated schist clasts in the Silurian strata described above with the Granjeno Schist, requiring an early Paleozoic metamorphic age for the schist.

The contact of the Granjeno Schist with the Novillo Gneiss is a subvertical shear zone containing mylonite and ultramylonite produced from granulitic gneiss (Ortega-Gutiérrez, 1978c; Mitre-Salazar and others, 1991). Mineral dates in the Novillo Gneiss were not reset by the late Paleozoic metamorphic event that affected the Granjeno Schist, suggesting postmetamorphic juxtaposition of the two units.

Permian rocks. In both subterranes, the Lower Permian Guacamaya Formation consists of strongly folded but unmetamorphosed flysch, mudstone, and conglomerate containing a rich Wolfcampian-Leonardian faunal assemblage (Carrillo-Bravo, 1961, 1965; Pérez-Ramos, 1978). Deformation and fault juxtaposition with other pre-Mesozoic units are older than Late Triassic (Fig. 10). The formation generally coarsens upward, and the proportion of volcanic and skeletal carbonate rock fragments increases upward (Gursky and Michalzik, 1989). Single brachiopod and trilobite fossils of reported Mississippian age were collected from the base of the Guacamaya Formation in the southern subterrane (Carrillo-Bravo, 1965).

Mesozoic and Cenozoic strata. Pre-Mesozoic rocks of the Guachichil terrane are overlain unconformably by Triassic to Jurassic sedimentary and minor volcanic rocks (Fig. 10) that crop out locally and have been penetrated by many wells (Mixon and others, 1959; Carrillo-Bravo, 1961, 1965; Imlay, 1965; 1980; López-Ramos, 1972, 1983, 1985; Schmidt-Effing, 1980; Scott, 1984; Salvador, 1987). The oldest of these strata are unmetamorphosed, weakly deformed red beds of the Upper Triassic–Lower Jurassic Huizachal Formation and marine clastic rocks of the Sinemurian-Pliensbachian (Lower Jurassic) Huayacocotla For-

mation; in the northern part of the terrane, the two formations are combined in the La Boca Formation. The red beds were derived from a sedimentary-metamorphic source similar to the rocks described above, were deposited in fluvial and alluvial environments in an arid climate, and display southward to westward paleocurrent indicators (Michalzik, 1991). In the northern part of the terrane, the red beds contain rhyolite interbeds and are intruded by silicic to mafic dikes and sills that do not affect younger Jurassic strata in the region (Michalzik, 1991).

The Triassic–Early Jurassic red beds were locally folded and eroded prior to deposition of unconformably overlying Bajocian-Callovian (Middle Jurassic) red beds and evaporites that are named the Cahuasas Formation in the southern part of the terrane and the La Joya Formation of the Huizachal Group to the north. The La Joya Formation consists of a thick, coarse basal lag deposit overlain by a fining-upward sequence that records a change to quiet marine conditions by the early Late Jurassic (Michalzik, 1991). Primary volcanic and volcaniclastic rocks are absent.

Late Triassic to Middle Jurassic strata apparently were deposited in roughly north-south–trending extensional grabens (Schmidt-Effing, 1980; Salvador, 1987) that probably formed during the incipient breakup of Pangea (pp. 96–98). Early Jurassic fauna imply a marine connection with the Tethyan realm to the east or west, perhaps via a seaway between North America and South America (Scott, 1984; Taylor and others, 1984).

Callovian (late Middle Jurassic) calcarenites and marine shales of the Tepexic Formation in east-central México indicate marine connection between the Pacific basin and the Gulf of Mexico region (Imlay, 1980; Salvador, 1987). Wells have penetrated Middle Jurassic rhyolitic lavas and tuff near Tezuitlán, Puebla; they have also penetrated volcanic and plutonic rocks that yielded reported Early to Late Jurassic K-Ar ages in central Puebla and northern San Luis Potosí (López-Ramos, 1972; López-Infanzón, 1986).

The Guachichil, Tepehuano, and Coahuiltecano terranes are overlapped by widespread Oxfordian carbonate and shale of the Zuloaga Formation (or Zuloaga Group of Götte and Michalzik, 1991) and by a thick sequence of Late Jurassic–Late Cretaceous carbonate and fine-grained clastic rocks deposited in platformal and basinal environments (Salvador, 1987; Mitre-Salazar and others, 1991). Tuff is interbedded with Tithonian limestone and clastic rocks in the southern part of the Guachichil terrane (Longoria, 1984), but volcanogenic strata are absent from the Cretaceous section. These strata were folded and thrusted during Laramide orogenesis (see section below) and, particularly in the northern part of the terrane, intruded by scattered silicic to intermediate plutons, stocks, and dikes of probable Paleogene, post-Laramide age (INEGI, 1980). Cenozoic nonmarine clastic rocks occupy basins between antiformal crests.

Structural and geophysical data. Late Jurassic and Cretaceous strata throughout the Guachichil terrane were affected by east-vergent folding and thrusting of probable Laramide origin, producing north-south structural trends such as the anticlinoria that expose Precambrian and Paleozoic rocks (López-

Ramos, p. 324–343). At the eastern margin of the terrane, i.e., the frontal zone of the Sierra Madre Oriental fold and thrust belt, deformation probably is older than middle Eocene, and east-vergent thrust faults do not break to the surface but are inferred beneath the folded strata (Mossman and Viniegra-Osorio, 1976; Hose, 1982). In the southern part of the terrane, along the Querétaro-Hidalgo border and in San Luis Potosí, detailed structural studies indicate thin-skinned deformation, displacement along low-angle thrusts, and at least 40 km of eastward tectonic transport during late Maastrichtian to Paleocene time (Suter, 1984, 1987; Carrillo-Martínez, 1990). Thrust faults and deformed strata are intruded by granitoids that have yielded Paleogene K-Ar dates (Table 6).

Paleomagnetic data from Late Triassic(?) to Early Cretaceous rocks are interpreted to indicate that parts of the northern Guachichil terrane and the adjoining Coahuiltecano terrane underwent about 130° of counterclockwise rotation during Early and Middle Jurassic time and none thereafter (Gose and others, 1982). A small circle fit to the Late Triassic poles indicates that net counterclockwise rotation was less than about 100° (Urrutia-Fucugauchi and others, 1987).

Maya terrane

We informally divide the Maya terrane into three geographic provinces: the *northern province,* which includes southern Tamaulipas, Veracruz as far southeast as the Isthmus of Tehuantepec, and thin transitional crust along the western margin of the Gulf of Mexico; the *Yucatán platform,* which includes the Mexican states of Tabasco, Campeche, Quintana Roo, and Yucatán, northern Belize, northern Guatemala, and thinned transitional crust in the adjacent Gulf of Mexico and Yucatán basins; and the *southern province,* which includes central Guatemala, Chiapas, and northeastern Oaxaca. Basement rocks include disjunct outcrops and subcrops of Paleozoic and Precambrian(?) metamorphic rocks that are widely interpreted as Gondwanan continental crust stranded during the rifting of Pangea. This basement and overlying Pennsylvanian-Permian flysch were strongly deformed in the Permian, perhaps during Ouachitan orogenesis. A Permo-Triassic continental magmatic arc that formed in the northern and southern provinces of the Maya terrane probably was produced by eastward subduction of oceanic lithosphere of the Pacific basin. During late Middle to Late Jurassic opening of the Gulf of Mexico, the Yucatán platform and the southern province were displaced to the south-southeast with respect to the northern province along an enigmatic north-northwest–striking fault. Rocks on both sides of the fault apparently underwent differential rotation concomitant with this displacement. Post-Jurassic rotation and displacement of the Maya terrane is below the detection limit of paleomagnetic studies. Late Jurassic to Cenozoic strata were deposited on carbonate platforms and in shelf basins around the margins of the Gulf of Mexico. A Jurassic-Cretaceous ophiolite was accreted to the southern margin of the Maya terrane in the Maastrichtian.

The Maya terrane as defined in this volume broadly corresponds to the Maya terrane of Coney and Campa-Uranga (1987); it includes the Yucatán terrane (United States Geodynamics Committee, 1989) and the Maya block (Dengo, 1975). Ophiolitic and associated rocks that were accreted to the southern margin of the Maya terrane in the latest Cretaceous are considered a subterrane of the Maya terrane (El Tambor subterrane; see below). The Coahuiltecano terrane in northeastern México may be correlative or in fact continuous with the Maya terrane across a postulated eastward extension of the Mojave-Sonora Megashear.

Pre-Mesozoic basement rocks, general statement. Pre-Mesozoic basement rocks crop out only in the southern province of the Maya terrane (central Guatemala, Maya Mountains of Belize, northeastern Oaxaca, Chiapas). The nature, history, and age of basement rocks penetrated by wells in the Yucatán platform (Yucatán peninsula, northern Guatemala, Belize) and the northern province (Veracruz, southern Tamaulipas) are too poorly known to evaluate possible correlations among basement outcrops and subcrops throughout the terrane. Much more work is needed to determine whether the Maya is a composite terrane consisting of two or more pre-Mesozoic basement terranes overlapped by Mesozoic and Cenozoic strata.

Pre-Mesozoic basement rocks, southern province. Basement rocks crop out only in the southern part of the Maya terrane. Constituent units include, roughly from east to west, the Chuacús Group, the Santa Rosa Group and Chochal Formation, the Chiapas Massif, and unnamed metamorphic rocks in northeastern Oaxaca.

Chuacús Group (or Series) crops out only along the northern side of the Motagua fault in central Guatemala (Fig. 11); metasedimentary rocks in wells in the Yucatán peninsula may be correlative (see below). In central Guatemala, the Chuacús Group consists of quartz-mica schist, marble, mylonitized metagranitoids, and minor greenstone and quartzite derived from clastic sedimentary rocks, carbonates, granitoids, and minor volcanic rocks of uncertain but inferred Paleozoic or Proterozoic age (McBirney, 1963; Kesler and others, 1970; Anderson and others, 1973; Roper, 1978; Burkart and others, 1987; Donnelly and others, 1990a). A deformed Carboniferous granitoid in the Chuacús Group contains zircons that are probably Precambrian (Table 7). Garnet-biotite and staurolite-sillimanite assemblages indicate peak metamorphism at conditions of high greenschist to low garnet-amphibolite facies (Anderson and others, 1973; Clemons and others, 1974). The Chuacús Group was overprinted by retrograde greenschist-facies metamorphism and pervasively deformed during Maastrichtian north-south shortening (Sutter, 1979; Donnelly and others, 1990a).

An unmetamorphosed but strongly deformed sequence of late Paleozoic sedimentary and minor volcanic rocks crops out in the Maya Mountains in Belize and along the Polochic fault between eastern Guatemala and Chiapas, and was recognized in wells in northern Guatemala; metasedimentary rocks in wells in the northern Yucatán peninsula also may be correlative (see below) (Bateson and Hall, 1971, 1977; Viniegra-Osorio, 1971;

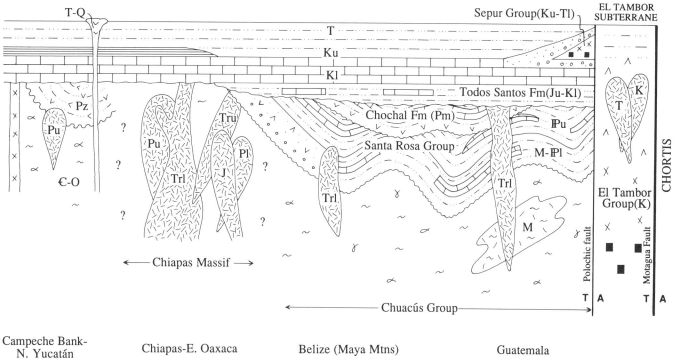

Figure 11. Composite tectonostratigraphic section of Maya terrane. Arrangement of geologic elements by geographic region (noted at bottom of figure) reflects projection into a roughly north-south structure section.

Bateson, 1972; Anderson and others, 1973; López-Ramos, 1983; Donnelly and others, 1990a). Mapped contacts between the Paleozoic strata and the Chuacús Group are faults, but locally the former may be unconformable on the latter (Weyl, 1980); we have inferred such a relation in Figure 11. Following the stratigraphic nomenclature of Donnelly and others (1990a) and Maurrasse (1990), the Paleozoic section is divided into the Carboniferous to Lower Permian Santa Rosa Group and the unconformably overlying mid-Permian Chochal Formation (Fig. 11). The Santa Rosa Group includes an undated lower unit of marine conglomerate and sandstone (flysch) with minor interbeds of tuffaceous sandstone, limestone, greenstone, and siliceous volcanic rocks, and a younger unit of fossiliferous Pennsylvanian to Early Permian limestone and shale. A Late Mississippian maximum age of the lower unit in Belize may be indicated by a 336-Ma Rb-Sr date from an adjacent pluton that apparently was intruded and eroded prior to deposition. In Belize, the lower and upper units are separated by a sequence of silicic lavas and pyroclastic rocks that apparently were erupted in latest Pennsylvanian to earliest Permian time (Table 7). The total thickness of the Santa Rosa Group is about 3 km in Guatemala and as much as 6 km in Belize. In Guatemala, the Santa Rosa Group is overlain along a 30° angular unconformity by mid-Permian limestone, dolomite, and minor shale of the Chochal Formation (Fig. 11),

indicating a late Early Permian (~Leonardian-Guadalupian) phase of deformation, uplift, and erosion.

The Chuacús Group and the late Paleozoic strata are intruded by late Paleozoic granitoids in the Maya Mountains of Belize and by Triassic granitoids in Belize and northern Guatemala (Table 7). A well in northern Belize penetrated granite of presumed late Paleozoic to Triassic age (Viniegra-Osorio, 1971). A mid-Jurassic ^{40}Ar/^{39}Ar plateau date from biotite in the Matanzas granite (Table 7) indicates late passage through the 250°C blocking temperature (Donnelly and others, 1990a).

In southwestern Chiapas and eastern Oaxaca, a complex assemblage of metaplutonic, metasedimentary, and plutonic rocks is generally referred to as the Chiapas Massif (Fig. 11). The metamorphic rocks were derived from sedimentary and plutonic protoliths of inferred late Proterozoic to early Paleozoic age, and are overlain nonconformably by undeformed Carboniferous to Permian(?) strata (Hernández, 1973; Dengo, 1985). Based on similar lithology and inferred age, these two units are similar to and possibly correlative with the Chuacús Group and Santa Rosa Group/Chochal Formation, respectively (Fig. 11). The Chiapas Massif is intruded by Permian, Triassic, Jurassic, and Cenozoic granitoids (Fig. 11) (Table 7 and unpublished data of Damon and others, 1981), and by pegmatite and granitoids of reported latest Proterozoic (Pantoja-Alor and others, 1974; López-Infanzón,

R. L. Sedlock and Others

TABLE 7. MAYA TERRANE RADIOMETRIC DATA

Sample	System	Mineral*	Date (Ma)	References[†]	Comments
Chuacús Group and metamorphosed intrusion					
Rabinal granite	U-Pb	zr	1,075 ±25;	1, 2	Discordia intercepts
			345 ± 20	1, 2	
Amphibolite	$^{40}Ar/^{39}Ar$	h	238	3	Source: J. Sutter; reflects late thermal event
Santa Rosa Group					
Siliceous volcanic rocks	Rb-Sr		285	4	Tight 4-pt isochron, one point omitted
Paleozoic–lower Mesozoic drill and dredge samples					
Granodiorite, granite,	K-Ar?		320	5	
tonalite, quartz diorite,	K-Ar?		273 ± 5	5	
and volcanic equivalents	K-Ar	b	264 ± 21	6	
(locally mylonitized) in	K-Ar	b	260 ± 20	6	
southern Tamaulipas and	K-Ar	b	258 ± 21	6	
Poza Rica region,	K-Ar	b	257 ± 21	6	2 samples
northern Veracruz	K-Ar	b	250 ± 20	6	2 samples
	K-Ar	b	247 ± 21	6	
	K-Ar	b	243 ± 19	6	
	K-Ar	b	241 ± 20	6	
	K-Ar	b	233 ± 19	6	
	K-Ar	wm	223 ± 18	6	
	K-Ar		212 ± 5	5	
	K-Ar	b	208 ± 16	6	
	K-Ar	h	187 ± 11	6	
	K-Ar?		183	5	2 samples
	K-Ar	b	179 ± 14	6	Latite
	K-Ar	b	173 ± 14	6	
Granodiorite near Tezuitlán, Puebla	K-Ar	b	252 ± 20	7	
	K-Ar	b	246 ± 7	7	
Meta-andesite, Yucatan			330, 290	1	2 samples
"Hornfels schist" in southern Tamaulipas	K-Ar	b	276 ± 22	6	
and Poza Rica region	K-Ar	b	272 ± 22	6	
	K-Ar	b	263 ± 21	6	
	K-Ar	b	177 ± 14	6	
Muscovite schist, central Veracruz	K-Ar	wm	269 ± 22	7	
Schist, northern Veracruz	K-Ar	m, b	242 ± 6	9	Cooling are?
Schist, southern Tamaulipas	K-Ar?		210 ± 6	5	Cooling age?
Gneiss, northern Veracruz	K-Ar?		192 ± 3	5	Cooling age?
Leg 77 phylite	$^{40}Ar/^{39}Ar$	wr	500 ± 8	10	Plateau age
Leg 77 amphibolite	$^{40}Ar/^{39}Ar$	h	501 ± 9	10	Plateau age
Leg 77 amphibolite	$^{40}Ar/^{39}Ar$	h	496 ± 8	10	Plateau age
Leg 77 gneiss	$^{40}Ar/^{39}Ar$	b	350	10	Plateau age
Leg 77 diabase	$^{40}Ar/^{39}Ar$	wr	190 ± 3	10	Plateau age
Leg 77 diabase	$^{40}Ar/^{39}Ar$	wr	164 ± 4	10	2 samples; plateau ages
Chiapas Massif					
Orthogneiss	K-Ar	b	288 ± 6	11	
Pluton near 19°40'N	K-Ar	K-Ar	Late Perm	11	Source: H. Palacios
Granite	Rb-Sr	wr	256 ± 10	11	10-pt isochron; $^{87}Sr/^{86}Sr_i$: 0.70453
Biotite granite	K-Ar	b	246 ± 5	11	
Granite	K-Ar	b	239 ± 5	11	
Orthogneiss	K-Ar	h	232 ± 5	11	
Granite	K-Ar	b	219 ± 4	11	
Granitoids	K-Ar	b	191–170	11	5 samples
Granitoids	K-Ar	b	6-2	12	Multiple samples

TABLE 7. MAYA TERRANE RADIOMETRIC DATA (continued)

Sample	System	Mineral*	Date (Ma)	References[†]	Comments
Cenozoic magmatic rocks north of TMVB					
Alkalic volcanic rocks, southern Tamaulipas	K-Ar	wr	28, 24, 21, 7	13	4 samples
Olivine basalt, northern Veracruz	K-Ar	wr	20 ± 1	13	Alkalic
Diorite, central Veracruz	K-Ar	wr	17 ± 1	13	Calc-alkalic
Alkalic basalt, central Veracruz	K-Ar	wr	14, 3	13	2 samples; transitional
Andesite and basalt andesite, Puebla	K-Ar	wr	9–1	13	7 samples; calc-alkalic
Basalt and ignimbrite, Hidalgo and C Veracruz	K-Ar	wr	8–2	13	12 samples; alkalic and transitional
Dacite, central Veracruz	K-Ar	wr	7	13	Calc-alkalic
Intrusive rocks in Belize and Guatemala					
Mountain Pine Ridge granite, Maya Mountains	Rb-Sr	wr, kf	320 ± 10	4	9-pt isochron; uncertain relation with Santa Rosa Group; yielded "Triassic" K-Ar date
Hummingbird and Sapote granites, Maya Mountains	K-Ar	b	237–227	4	4 samples; intrudes Santa Rosa Group
Undeformed Matanzas granite; intrudes Chuacús Group	Rb-Sr	wr, kf, p	227	3	Source: P. Pushkar
	^{40}Ar/^{39}Ar	m	213–212	3	Source: J. Sutter; 2 samples
	^{40}Ar/^{39}Ar	b	161	3	Source: J. Sutter
Granitoids in Polochic Valley, Guatemala	A-Ar		85–58	14	Source: MMA, Japan; 5 samples
	K-Ar	kf	68 ± 3	14	
Other rocks in Guatemala					
Amphibolite and diabase, El Tambor Group	K-Ar		59 ± 4	15	11-pt isochron; metamorphic age?
Andesite boulder in Eocene Subinal Formation	K-Ar	p	42 ± 2	16	
Welded tuff clasts in Miocene Colotenango beds	K-Ar	b, g	12–7	16	3 samples

*Mineral abbreviations: b = biotite; g = glass; h = hornblende; kf= potassium feldspar; m = mica; p = plagioclase; wm = white mica; wr = whole rock; zr = zircon.

[†]1 = Gombert and others, 1968; 2 = Mc Birney and Bass, 1969; 3 = Donnelly and others, 1990a; 4 = Bateson and Hall, 1977; 5 = López Rámos, 1972; 6 = Jacobo-Albarrán, 1986; 7 = López-Infanzón, 1985; 8 = Marshall, 1984; 9 = Denison and others, 1969; 10 = Schlager and others, 1984; 11 = Damon and others, 1981; 12 = Damon and Montesinos, 1978; 13 = Cantagrel and Robin, 1979; 14 = Burkart and others, 1987; 15 = Bertrand and others, 1978; 16 = Deaton and Burkart, 1984b.

1986; Pacheco-G. and Barba, 1986) and Late Cretaceous (Burkart, 1990) age.

In the eastern Sierra Juárez of northeastern Oaxaca, an area crossed by Transect H-3, the Maya terrane includes an unnamed subterrane of poorly understood metamorphic rocks that crops out east of the fault boundary with the Cuicateco terrane (not shown in Fig. 11) (Ortega-Gutiérrez and others, 1990). The unit consists of polydeformed, greenschist-facies metasedimentary and metaigneous rocks including phyllite, schist, gabbro, and rare metagabbro and serpentinite. Biotite and muscovite schists have yielded late Paleozoic K-Ar dates (S. Charleston, personal communication, 1981). These metamorphic rocks are not present along the Maya/Cuicateco boundary to the southeast of Transect H-3; in this region, the Maya terrane consists of weakly metamorphosed Albian-Cenomanian clastic rocks and marble that were strongly deformed in Late Cretaceous time (Carfantan, 1981).

Pre-Mesozoic basement rocks, Yucatán platform. Wells in the northern Yucatán peninsula bottomed in Paleozoic or Paleozoic(?) metavolcanic rocks, quartzite, and schist (López-Ramos, 1983). Metamorphic rocks yielded radiometric dates of

420 to 410 Ma with an inferred metamorphic event at 330 Ma (Dengo, 1969), and 290 ± 30 Ma (Viniegra-Osorio, 1971). These ages are consistent with the suggestion that the Yucatán platform is underlain by the Chuacús Group, the Santa Rosa Group, or both units (Fig. 11).

At the Catoche Knolls in the Gulf of Mexico, about 300 km northeast of the Yucatán peninsula, DSDP Leg 77 encountered gneiss that yielded early Paleozoic metamorphic or cooling ages, as well as Jurassic diabase (Table 7) (Dallmeyer, 1982; Schlager and others, 1984). These rocks may be remnants of rifted, thinned basement of the Maya terrane.

Pre-Mesozoic basement rocks, northern province. The only information about pre-Mesozoic basement in the northern province (Veracruz, southern Tamaulipas, and adjacent offshore regions) comes from the numerous wells that have been drilled through overlying Cenozoic and Mesozoic strata (López-Ramos, 1972, 1985). Many wells throughout the northern province bottomed in granitic basement and metamorphic rocks that, where dated, yielded ages ranging from Carboniferous to Jurassic (Table 7). The abundance of Permo-Triassic granitoids generally is taken as evidence of intrusion of a batholith of this age in eastern México (also see Coahuiltecano terrane); younger Triassic and Jurassic dates are probably cooling ages. Permian to Jurassic dates from some metamorphic rocks probably indicate thermal metamorphism of country rocks due to intrusion of the batholith (Table 7). However, some K-Ar dates indicate Carboniferous to Early Permian plutonism and thermal or regional metamorphism that probably predate intrusion of the Permo-Triassic batholith. We suggest that these older dates reflect magmatism and metamorphism accompanying late Paleozoic Ouachitan orogenesis.

Volcanic and plutonic rocks that yielded unpublished Early to Late Jurassic K-Ar dates are reported from wells throughout the state of Veracruz (López-Infanzón, 1986), but it is uncertain whether these are crystallization ages or cooling ages due to Permo-Triassic magmatism.

Mesozoic and Cenozoic rocks, northern province. Although Triassic through Middle Jurassic strata do not crop out in the Maya terrane, a few wells in northern Veracruz have penetrated nonmarine and shallow marine clastic rocks that are inferred to be correlative with the Upper Triassic–Lower Jurassic Huizachal Formation in the adjacent Guachichil terrane (Imlay and others, 1948; López-Ramos, 1972; Wilson, 1990). Callovian (late Middle Jurassic) shallow marine strata are known from a few wells near Tampico (Veracruz-Tamaulipas border), but most of the northern province was emergent until the Late Jurassic. Late Jurassic and Cretaceous platform carbonates, basinal carbonates, evaporites, and minor clastic rocks indicate deposition on topographic highs (e.g., southern Tamaulipas arch, Tuxpán platform) and in intervening deeper water basins. The complex distribution of facies along this eastern margin of the Gulf of México basin is markedly different from the Atlantic-type passive margin sequences that accumulated along the other margins of the basin (Winker and Buffler, 1988). The Cretaceous strata are overlain by Paleocene to Eocene marl, shale, and sandstone (Barker and Blow, 1976). The cumulative thickness of Late Jurassic to Paleogene strata beneath the Veracruz-Tamaulipas coastal plain ranges up to 10 km (Mossman and Viniegra-Osorio, 1976).

A single well (San Antonio 101) near Cucharas, Veracruz, encountered Early Jurassic strata that apparently were thermally metamorphosed by underlying granite, implying an Early Cretaceous or younger intrusion age (López-Ramos, 1972).

Late Jurassic to middle Eocene strata in the northern province of the Maya terrane underwent east-vergent folding and thrusting of probable Laramide origin and were intruded by poorly dated silicic to intermediate granitoids prior to the deposition of unconformably overlying late Eocene to Miocene marl and fine-grained marine clastic rocks (Barker and Blow, 1976; Mossman and Viniegra-Osorio, 1976). The belt of Laramide shortening may terminate to the south near the Isthmus of Tehuantepec (Viniegra-Osorio, 1971, 1981).

Miocene to Quaternary calcalkalic to alkalic, siliceous to basaltic volcanic rocks crop out in the northern province from southern Tamaulipas to San Andrés Tuxtla in southeastern Vera-

cruz (Table 7). Geochemical data from a middle Miocene calc-alkalic basaltic suite and a more alkalic, nepheline-normative late Miocene–Recent basaltic suite imply derivation of both suites from subduction-related magmas (López-Infanzón and Nelson, 1990; Nelson and others, 1991). A large positive gravity anomaly beneath San Andrés Tuxtla may indicate that the area is underlain by ultramafic rocks (Woollard and Monges-Caldera, 1956) or a gabbroic complex (Ortega-Gutiérrez and others, 1990).

Mesozoic and Cenozoic rocks, Yucatán platform and western part of southern province. In this section we consider Mesozoic and Cenozoic rocks of the Yucatán platform together with rocks of the western part of the southern province (Chiapas and northeastern Oaxaca). The entire region was emergent until the Callovian, when widespread evaporites (chiefly halite), commonly known as the Isthmian Salt, were deposited in Chiapas and in what is now Bahía de Campeche in the Gulf of México (Fig. 1) (Martin, 1980). The original thickness and lateral extent of these evaporites is poorly understood, but it is widely believed that the Isthmian Salt was once contiguous with the Louann Salt in the U.S. Gulf Coast (e.g., Salvador, 1987). The evaporites probably were deposited in a low-lying, gently subsiding region that periodically was flooded by marine water from the west (Pacific Ocean) through a gap near Tampico, Veracruz (Salvador, 1987), and subsequently separated into two bodies by southward displacement of the Yucatán region during Late Jurassic opening of the Gulf of México.

Salt accumulation was followed abruptly by transgression, subsidence, and deposition of Late Jurassic shallow marine carbonates and clastic rocks along the eastern margin of the Yucatán platform, in the Tabasco-Chiapas region, and in the northern province of the Maya terrane (see above). South of this subsiding basin in Chiapas and western Guatemala, eroded Paleozoic granitoids and metamorphic rocks are nonconformably overlain by the Oxfordian (Upper Jurassic) to Lower Cretaceous Todos Santos and San Ricardo Formations, which consist of discontinuous basal andesite overlain by a thick sequence of locally derived clastic rocks deposited in alluvial, fluvial, and lacustrine environments, carbonates, and minor evaporites (López-Ramos, 1973, 1975, 1983; Castro-Mora and others, 1975; Blair, 1987, 1988; Michaud, 1988). A Late Jurassic K-Ar date from andesite (Table 7) indicates a minimum age for the base of the sequence, which may be as old as Middle Jurassic (Michaud, 1988). A gross southward coarsening of average grain size in the Todos Santos Formation may indicate that the southern margin of the Yucatán platform was tectonically active in Late Jurassic to earliest Cretaceous time (Donnelly and others, 1990a), perhaps along the rifting, subsiding continental margin northwest of an actively opening proto-Caribbean ocean basin (p. 4).

Early Cretaceous massive dolomite and evaporites indicate that the Yucatán region was a reef-bounded platform extending as far east as the Isthmus of Tehuantepec and at least as far south as northern Guatemala (Viniegra-Osorio, 1981). The southern edge of this Early Cretaceous platform apparently was destroyed during latest Cretaceous collision along the southern margin of the Maya terrane, as indicated by the deposition of synorogenic flysch (see below) (Anderson and others, 1973, 1985; Clemons and others, 1974; Donnelly and others, 1990a).

Thin-bedded limestones accumulated on the Yucatán platform in the Late Cretaceous, indicating protracted flushing by open marine waters. Limestone deposition was interrupted by volcanism in eastern Chiapas and Guatemala, where Late Cretaceous volcanic rocks are reported from a backarc basin sequence (Burkart, 1990), and near Mérida in northwestern Yucatán, where wells penetrated a 400-m-thick sequence of submarine Late Cretaceous andesite and glass (Paine and Meyerhoff, 1970; López-Ramos, 1973, 1983). The andesitic volcanic rocks near Mérida occur within a multi-ring pattern recognized in proprietary aeromagnetic and gravity data, the Chixculub structure, that may be a buried K-T impact structure about 180 km in diameter (Penfield and Camargo, 1981). Glass and shocked quartz identified in some core samples are interpreted to indicate an ejecta blanket (Hildebrand and others, 1991), but examination of other cores from the proposed ejecta blanket revealed no shocked quartz, suggesting that either an impact breccia is not present at Chixculub or that the crater is much smaller than generally thought (Sharpton and others, 1991). Much more geophysical and geologic work is needed to understand the true significance of the Chixculub structure.

Tertiary strata in the southern province include Eocene nonmarine conglomerate, sandstone, and mudstone; mid-Tertiary ignimbrites; scattered Miocene-Quaternary nonmarine clastic rocks with interbedded ignimbrite (Table 7) and basalt; and Quaternary-Tertiary calc-alkalic volcanic rocks (Donnelly and others, 1990a). The Yucatán platform and Chiapas region did not undergo latest Cretaceous-Paleogene Laramide shortening. Late Cenozoic and probably active tectonism in eastern Chiapas and Oaxaca and western Tabasco includes northwest-trending folding and thrust faulting and left-lateral displacement on strike-slip faults that strike northeast-southwest, east-west, and southeast-northwest (Viniegra-Osorio, 1971, 1981; de Cserna, 1989; Delgado-Argote and Carballido-Sánchez, 1990). Modern deformation in this region may be due to instability of the diffuse Caribbean–North America–Cocos triple junction (p. 96). Active northeast-southwest shortening offshore the northern coast of Tabasco and Campeche is indicated by northwest-trending folds (de Cserna, 1989). The Isthmus of Tehuantepec purportedly is transected by the Salina Cruz fault, a high-angle structure of presumed Cenozoic age that was recognized in marine seismic reflection work in the Gulf of Mexico (Viniegra-Osorio, 1971), bu the fault has not been documented by gravity or geologic studies on land (see discussion in Salvador, 1988; Delgado-Argote and Carballido-Sánchez, 1990).

Mesozoic and Cenozoic rocks, eastern part of southern province. Along the southern margin of the Maya terrane, within and a few tens of kilometers north and south of the Motagua Valley in central Guatemala, disrupted, strongly deformed ultramafic, mafic, and pelagic and volcaniclastic sedimentary rocks are provisionally assigned to the El Tambor subterrane of the

Maya terrane (Fig. 11). Known as the El Tambor Group (old El Tambor Formation), these rocks are widely interpreted as a dismembered Cretaceous ophiolite and forearc assemblage (McBirney, 1963; Williams and others, 1964; McBirney and Bass, 1969; Lawrence, 1975, 1976; Muller, 1979; Rosenfeld, 1981; Donnelly and others, 1990a). Tenuous correlations have been proposed with ophiolitic rocks to the east (Islas de la Bahía) and south (Sierra de Omoa) and mafic rocks off the west coast of Guatemala (DSDP Legs 67 and 84) (McBirney and Bass, 1969; Horne and others, 1976a; Bourgois and others, 1984; Donnelly and others, 1990a).

Constituent rock types of the El Tambor Group are peridotite, gabbro, plagiogranite, diabase dikes, pillow basalt, chert and limestone, clastic sedimentary rocks, and serpentinite-matrix mélange. Abundant ultramafic rocks include serpentinite-matrix mélange with blocks of jadeitite, omphacite-bearing metabasite, and amphibolite, and slabs of partly to completely serpentinized peridotite up to 80 km long (McBirney, 1963; McBirney and others, 1967; Lawrence, 1975, 1976; Donnelly and others, 1990a). Gabbro is strongly deformed and metamorphosed to amphibolite facies. Basalts are midocean-ridge basalt (MORB) –like normative tholeiites, contain interbedded Valanginian to Cenomanian pelagic rocks, and grade upward into Early Cretaceous chert and mudstone and volcaniclastic wackes that contain fragments of andesite, dacite, and Aptian-Albian chert (Lawrence, 1975, 1976; Muller, 1979; Rosenfeld, 1981).

North-directed thrusting of latest Cretaceous age caused tectonic intercalation of the El Tambor Group with the Chuacús Group and with synorogenic late Campanian-Paleogene flysch (see below) (Wilson, 1974; Johnson and Muller, 1986; Donnelly and others, 1990a). Early Paleogene K-Ar dates (Table 7) from ophiolitic rocks may be metamorphic cooling ages related to this deformation. Gravity modeling suggests that the north-vergent thrusts root into a north-dipping amphibolite body that may be a remnant of subducted oceanic lithosphere (T. Donnelly, unpublished data). The El Tambor Group is widely interpreted as a fault-bounded, strongly deformed, dismembered Cretaceous ophiolite and forearc assemblage that was obducted onto the southern Maya terrane during Maastrichtian collision of Chortis or an island arc of uncertain identity (p. 109). The El Tambor Group probably does not represent obducted Caribbean crust, based on its many dissimilarities to widespread Cretaceous basalts that crop out around the margins of and probably underlie the Caribbean plate (Donnelly and others, 1973; Donnelly, 1989).

Several other Mesozoic to Paleogene units crop out near the Motagua and Polochic fault zones and are tentatively included in the El Tambor subterrane. Coarse terrigenous clastic detritus, ophiolitic debris, and arc-derived cobbles and pebbles in the Upper Campanian-Paleogene Sepur Group were derived from southern sources and locally are overthrust by ophiolitic El Tambor rocks (Rosenfeld, 1981; Donnelly and others, 1990a). Granitoids within and between the Polochic and Motagua fault zones have yielded Late Cretaceous to mid-Tertiary radiometric dates (Table 7). Within the Motagua Valley, fault-bounded units of uncertain origin and significance include (1) unmetamorphosed Campanian-Maastrichtian limestone and pelagic marl with blocks of limestone, arc rocks(?), and ophiolitic rocks up to 2 km across; (2) highly deformed, marmorized limestone of probable Late Cretaceous age; and (3) nonmarine clastic rocks with thin limestone interbeds (Donnelly and others, 1990a). The Eocene Subinal Formation consists of limestone and clastic rocks containing an Eocene andesite boulder (Table 7), and is similar to and possibly correlative with coeval rocks in the Chortis terrane (Deaton and Burkart, 1984a; Donnelly, 1989; Donnelly and others, 1990a). Miocene fluvial clastic rocks and volcanic rocks referred to as the Colotenango beds include clasts of middle to late Miocene ignimbrite (Table 7) (Deaton and Burkart, 1984a, b; Donnelly, 1989; Donnelly and others, 1990a).

Geophysical data. Models of total tectonic subsidence, seismic refraction data, and gravity data indicate that continental crust of the Maya terrane is 35 to 40 km thick along the Veracruz coast, and that transitional crust under the Yucatán platform is about 30 km thick (Dillon and others, 1973; Sawyer and others, 1991). The boundaries between these regions and the oceanic crust in the Gulf of Mexico are narrow zones of thin (10 to 30 km thick) transitional crust (Sawyer and others, 1991).

A gravity model for a profile that crosses the Chortis-Maya boundary indicates that continental crust is about 38 km thick beneath both terranes but about 15 km thicker in a 30-km-wide zone straddling the Motagua fault (T. Donnelly, unpublished data). At shallower depths the observed gravity profile is best fit by north-dipping slabs of ophiolitic rocks.

Marine geophysical studies have recognized late Cenozoic east-west folding and thrusting of Cretaceous and Cenozoic strata in the 600-km-long, north-south–trending Mexican Ridges foldbelt off the coast of Tamaulipas and northern Veracruz (Buffler and others, 1979). This belt was named the Cordillera Ordóñez by de Cserna (1981).

Paleomagnetic studies of rocks in the southern Maya terrane indicate late Paleozoic and early Mesozoic southward displacement, mid-Jurassic counterclockwise rotation, and post-Oxfordian tectonic stability. Permo-Triassic granitoids in the Chiapas Massif are interpreted to have undergone about 1,200 ± 900 km of southward latitudinal displacement and ~75° counterclockwise rotation (Molina-Garza and others, 1992). Paleomagnetism of Early Permian(?) strata in the Chiapas Massif originally was interpreted to indicate considerable southward displacement with respect to North America (Gose and Sánchez-Barreda, 1981), but Molina-Garza and others (1992) have reinterpreted the magnetism as a Jurassic overprint indicative of little displacement.

Paleomagnetic studies of Middle Jurassic to earliest Cretaceous sedimentary rocks (Todos Santos and San Ricardo Formations) and intrusives have interpreted negligible latitudinal displacement and 63° ± 11° of counterclockwise rotation during the Middle Jurassic, and negligible rotation and latitudinal displacement with respect to stable North America since the Oxfordian (Guerrero and others, 1990; Molina-Garza and others, 1992).

Mixteco terrane

Basement rocks of the Mixteco terrane record early Paleozoic subduction, early Paleozoic obduction of an ophiolite onto a subduction complex, early to middle Paleozoic collision of the oceanic rocks of the Mixteco terrane with continental crust of the Zapoteco terrane, middle to late Paleozoic deformation and metamorphism, and deposition of late Paleozoic synorogenic and postorogenic marine strata. Mesozoic epicontinental strata include Jurassic marine and nonmarine clastic rocks and Cretaceous carbonates. Paleogene and early Neogene volcanic rocks indicate proximity to an arc. Sparse paleomagnetic data may indicate large Jurassic displacements, but such an interpretation has not yet been supported by other geologic or geophysical data.

Acatlán Complex. The oldest unit in the Mixteco terrane is the Acatlán Complex, which is divided into the structurally lowermost Petlalcingo Subgroup, the structurally overlying Acateco Subgroup, and the Upper Devonian(?) Tecomate Formation, which overlaps the thrust contact between the other two units (Fig. 12) (Ortega-Gutiérrez, 1978a, 1981a, b). The Petlalcingo Subgroup consists of schist, amphibolite, quartzite, and phyllite that probably were derived from marine sedimentary rocks and intercalated mafic igneous rocks. The lower, partly migmatitic part of the subgroup (Magdalena migmatite) is more calcic and presumably derived from strata that is more carbonate-rich than the middle and upper parts of the subgroup (Chazumba and Cosoltepec Formations, respectively), which are dominantly metagraywackes intercalated with metapelite and metagabbro. Protoliths of all three units probably were derived from a Grenville source such as the Oaxacan Complex of the Zapoteco terrane (Table 8) (Ruiz and others, 1990; Yañez and others, 1991).

The Acateco Subgroup consists of the basal Xayacatlán Formation and the mylonitic Esperanza granitoids (Fig. 12) (Ortega-Gutiérrez, 1978a, 1981a, b). The Xayacatlán Formation contains serpentinized peridotites, eclogitized and amphibolitized metabasites, pelitic schist, and quartzite, and is interpreted as a dismembered ophiolite. The Esperanza granitoids consist of polymetamorphic mylonitic gneisses derived from tonalitic to granitic protoliths. Cataclastic granitoids correlated with the Esperanza also intrude the Oaxacan Complex of the Zapoteco terrane (F. Ortega-Gutiérrez, unpublished data). The Esperanza granitoids are interpreted to be products of partial crustal anatexis caused by the Early-Middle Devonian collision of the Mixteco and Zapoteco terranes on the basis of U-Pb, Rb-Sr, and Sm-Nd data (Table 8), the locally intrusive contact between the granitoids and ophiolitic rocks of the Xayacatlán Formation, and apparent syntectonic intrusion of both the Acatlán Complex and the Oaxacan Complex (Robinson and others, 1989; Yañez and others, 1991; F. Ortega-Gutiérrez, unpublished data).

The Tecomate Formation consists of arkosic metaclastic rocks, calcareous metapelites, and limestone, and contains clasts of the Esperanza granitoids. The depositional age is probably Late Devonian based on sparse fossils of possible post-

MIXTECO

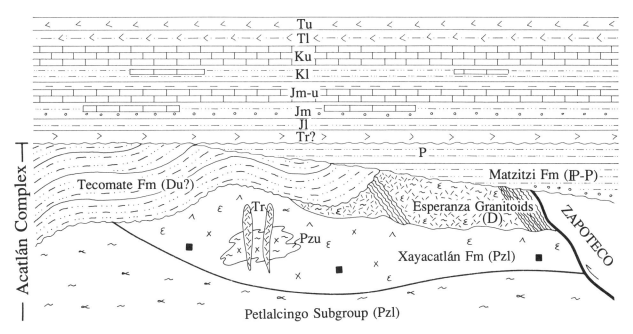

Figure 12. Schematic tectonostratigraphic section of Mixteco terrane. Unknown direction of motion on old thrust between lower plate Petlalcingo Subgroup and upper plate Xayacatlán Formation. Upper Devonian(?) Tecomate Formation overlaps this thrust. Pennsylvanian-Permian Matzitzi Formation is the oldest unit to physically overlap the Zapoteco-Mixteco contact.

Cambrian to pre-Mississippian age and the presence of clasts of the Esperanza granitoids (Ortega-Gutiérrez, 1978a, 1981a, b; Yañez and others, 1991). The Tecomate Formation depositionally overlaps the suture between the two structurally lower units of the Acatlán Complex, but was folded and metamorphosed prior to the deposition of Early Mississippian cover strata (Fig. 12). All three units of the Acatlán Complex have Nd crustal residence ages of about 1,700 to 1,400 Ma (Yañez and others, 1991).

The Acatlán Complex has undergone several phases of metamorphism and deformation (Ortega-Gutiérrez, 1974, 1979, 1981a, b; Yañez and others, 1991). The upper plate is inferred to have overthrust the lower plate in an early Paleozoic(?) subduction zone, causing high-pressure metamorphism (M_1: 8-12 kbar, 500° to 550°C) and isoclinal folding (D_1) of the upper plate in response to northwest-southeast shortening; minimum thrust displacement was 200 km. No record of this event has been identified radiometrically. The Early to Middle Devonian, probably syntectonic, intrusion of the Esperanza granitoids was concomitant with strong deformation of both plates of the Acatlán Complex that caused retrograde metamorphism (M_2) and isoclinal folding (D_2); the deformation records east-west shortening and westward vergence (Table 8). This event has been related to Acadian orogenesis in eastern North America (Ortega-Gutiérrez, 1981a, b; Ruiz and others, 1988b; Yañez and others, 1991). In

the early(?) Carboniferous, after latest Devonian deposition of the Tecomate Formation but prior to intrusion by the 287-Ma Totoltepec stock, the entire Acatlán Complex was folded, foliated, and domed during east-west shortening (D_3) and underwent coeval high-temperature metamorphism (M_3: 5-6 kbar, 700° to 750°C) and a later retrogression (M_4) to greenschist-facies assemblages. Incomplete resetting of K-Ar and Rb-Sr systems by the Carboniferous event may explain 350- to 320-Ma whole-rock and white mica dates obtained from the Acatlán Complex (Table 8). Carboniferous metamorphism and intrusion of the Totoltepec stock (Table 8) are attributed to collision of Gondwana and North America (Yañez and others, 1991).

Other Paleozoic rocks. In the northern part of the Mixteco terrane, the Acatlán Complex is overlain unconformably by Early Mississippian marine strata of the Patlanoaya Formation and presumably by unmetamorphosed continental sandstone, siltstone, and conglomerate of the Matzitzi Formation of Pennsylvanian and probable Permian age; however, the contact between the Acatlán Complex and the Matzitzi Formation is not clearly exposed (Silva-Pineda, 1970; Carrillo and Martínez, 1981; Villaseñor and others, 1987; Weber and others, 1987). The Matzitzi Formation is the oldest unmetamorphosed stratigraphic unit that physically overlaps the fault contact between the Mixteco and Zapoteco terranes. Permian marine sedimentary rocks that non-

R. L. Sedlock and Others

TABLE 8. MIXTECO TERRANE RADIOMETRIC DATA

Sample	System	Mineral*	Date (Ma)	References†	Comments
Lower plate Petlalcingo Subgroup					
Magdalena migmatite, granite	Sm-Nd	wr	1,870–1,320	1	Model ages
Magdalena migmatite, paragneiss	U-Pb	zr	1,187 ±52; 356 ±140	2	Discordia intercepts
Magdalena migmatite, amphibolite	Sm-Nd	wr	760-670	1	Model ages
Magdalena migmatite	Sm-Nd	wr, gt	204 ± 4	1	
Magdalena migmatite	Rb-Sr	wr, b	163 ± 2	1	
Cosoltepec Formation	Sm-Nd	wr	1,650–1,430	1	Model ages
Quartzite	U-Pb	zr	1,800, 360	3	Discordia; detrital zr
Chazumba Formation schist	Sm-Nd	wr	1,470–1,430	1	Model ages
Chazumba Formation schist	Sm-Nd	wr, gt	429 ± 50	1	
"Schist"	Rb-Sr	wr	386 ± 6	4	Isochron age
Chazumba Formation schist	Sm-Nd	wr, gt	349 ± 27	1	
Muscovite schist	K-Ar	wm	346 ± 28	5	
Muscovite schist	K-Ar	wm	328 ± 26	5	
Upper plate Acateco Subgroup					
Xayacatlán schist	Sm-Nd	wr	1,500–1,460	1	Model ages
Xayacatlán eclogite	Sm-Nd	wr	1,080–700	1	Model ages
Xayacatlán schist	Sm-Nd	wr, gt	416 ± 12	1	3-sample "isochron"
Xayacatlán eclogite	Sm-Nd	wr, gt	388 ± 44	1	
Xayacatlán Formation	Rb-Sr	wr	386 ± 6	1	5-pt isochron; source: R. Armstrong
Xayacatlán eclogite	Rb-Sr	wr, wm	332 ± 4	1	
Xayacatlán schist	Rb-Sr	wr, wm	318 ± 4	1	
Esperanza granitoids	Sm-Nd	wr	1,600–1,400	1	Model ages
Esperanza granitoids	U-Pb	zr	1,140 ±69; 425 ± 13	2	Discordia intercepts
Esperanza granitoids	U-Pb	zr	1,116 ±44; 371 ± 34	1	Discordia intercepts
Esperanza granitoids	Rb-Sr	kf	448 ±175	6	
Esperanza granitoids	Rb-Sr	wr	428 ± 24	7	Isochron age
Esperanza granitoids	Sm-Nd	wr	411 ±123	1	
Esperanza granitoids	U-Pb	zr	1,180; 360	3	Discordia intercepts
Esperanza granitoids	Rb-Sr	wr, wm	330 ± 5	1	
Other rocks					
Tecomate Formation	Sm-Nd	wr	1,710–1,410	1	Model ages
Totoltepec Stock:	Sm-Nd	wr	810-660	1	Model ages
Deformed trondjhemite	U-Pb	zr	287 ± 2	1	Concordant age
Totoltepc Stock	K-Ar	wm	278 ± 13	8	
Pegmatite	Rb-Sr	wm, p	283	7	
Cataclastic granite, southeastern	KAr	b	262 ± 21	8	
Mixteco terrane	K-Ar	b	241 ± 19	8	
(Meta?)granitoids, southeastern	K-Ar	wm	259 ± 21	8	
Mixteco terrane	K-Ar	h	251 ± 6	8	
San Miguel intrusives	SM-Nd	wr	660	1	Model ages
San Miguel intrusives	Rb-Sr		207, 173	9	
San Miguel intrusives	Rb-Sr	wr, wm	175 ± 3	1	
San Miguel intrusives	Sm-Nd	wr, gt	172 ± 1	1	
Andesitic breccia			67 ± 3	10	
Aplite pegmatite near Chatino boundary	Rb-Sr	wm	60 ± 1	11	Isochron age
Tierra Colorada pluton at Chatino boundary	Rb-Sr	wr	55 ± 1	12	8-pt isochron; $^{87}Sr/^{86}Sr_i$: 0.7039
Tuff interbedded with clastic rocks	K-Ar?		49 ± 8	13	Schlaepfer and Rincon-Orta, unpublished data
Andesite	K-Ar	wr	29 ± 1	13	
Silicic tuff	K-Ar	b	26 ± 1	13	

*Mineral abbreviations: b = biotite; gt = garnet; h = hornblende; kf = potassium feldspar; p = plagioclase; wm = white mica; wr = whole rock; zr = zircon.

†1 = Yañez and others, 1991; 2 = Robinson, 1991; 3 = Robinson and othes, 1989; 4 = de Cserna and others, 1980; 5 = López-Infanzón, 1986; 6 = Fries and Rincón-Orta, 1965; 7 = Halpern and others, 1974; 8 = Grajales and others, 1986; 9 = Ruiz-Castellanos, 1979; 10 = Chavez-Quirarte, 1982; 11 = Morán-Zenteno and others, 1990a; 12 = Morán-Zenteno and others, 1991; 13 = Ferrusquía-Villafranca, 1976.

conformably overlie the Acatlán Complex near Olinalá contain fauna that are provisionally correlated with Permian rocks in the Serí terrane near El Antimonio, Sonora (Corona-Esquivel, 1981; Enciso de la Vega, 1988).

At the southeastern edge of the Mixteco terrane, deformed and metamorphosed late Paleozoic marine clastic rocks, pelagic rocks(?), mafic volcanic rocks(?), and granitoids (mapped as Acatlán Complex on the H-3 Transect) are wedged between the Zapoteco terrane and the Chatino terrane (Grajales-Nishimura, 1988). These rocks may be a late Paleozoic forearc and arc assemblage that accumulated outboard of the Totoltepec stock and other Permian magmatic rocks in southeastern México, or a distinct terrane that was accreted to the Mixteco terrane prior to the accretion of the Chatino terrane. We provisionally call these oceanic rocks the Juchatengo subterrane of the Mixteco terrane.

Mesozoic and Cenozoic rocks. Granite in the Magdalena migmatite and silicic dikes of the San Miguel intrusive suite have yielded latest Triassic to Middle Jurassic ages (Table 8). The Acatlán Complex and overlying Paleozoic strata are overlain by Mesozoic epicontinental strata that are very similar to those in the adjacent Zapoteco terrane. Near Olinalá, Triassic(?) ignimbrite unconformably overlies Upper Permian strata and the Acatlán Complex (Fig. 12) (Corona-Esquivel, 1981). Jurassic strata include Toarcian(?) nonmarine sandstone, carbonaceous shale, and coal (Salvador, 1987); Aalenian-Bajocian quartz-cobble conglomerate that contains fragments of the Triassic(?) ignimbrite (Corona-Esquivel, 1981); and Bajocian-Callovian marine and nonmarine clastic rocks, coal, and carbonates whose fauna tie the Mixteco terrane to the Pacific margin near the Central Andes and to the western end of a seaway between North America and South America (Imlay, 1980; Westermann and others, 1984). Other Mesozoic strata include Callovian-Oxfordian carbonates and shale, Kimmeridgian-Tithonian marine clastic rocks, Neocomian carbonates and marine clastic rocks that contain reptilian fossils with affinities to Tethyan forms in Europe and South America, Neocomian-Aptian red beds, and Albian-Maastrichtian basinal carbonates that are partly coeval with the Morelos-Guerrero platform in the adjacent Nahuatl terrane (Ferrusquía-Villafranca, 1976; Salvador, 1987; Ferrusquía-Villafranca and Comas-Rodríguez, 1988). Campanian-Maastrichtian conglomerate and sandstone derived from the Juchatengo subterrane were deposited on that subterrane as well as on the Zapoteco terrane (Carfantan, 1986; F. Ortega-Gutiérrez, unpublished data).

Latest Cretaceous to Eocene Laramide orogenesis caused north-south–trending folds in Cretaceous and older rocks, serpentinization and plastic deformation of deep-level harzburgite, and the diapiric emplacement of the serpentinite into the upper part of the Acatlán Complex along reverse faults (Carballido-Sánchez and Delgado-Argote, 1989). Cretaceous and older rocks subsequently were overlain unconformably by Paleogene(?) conglomerate, sandstone, and shale that were derived from the Oaxacan Complex (Zapoteco terrane) and an unidentified subordinate volcanic source, and minor interbedded tuff (Table 8) (Ferrusquía-Villafranca, 1976). Along the southern margin of the terrane, the 60- to 55-Ma Tierra Colorada pluton (Table 8) intrudes not only Cretaceous limestones of the Nahuatl terrane, but also mylonites along the fault boundary with the Chatino terrane (p. 73). Other Cenozoic rocks include Oligocene(?) silicic ignimbrite, volcaniclastic rocks, andesitic lavas, and andesitic hypabyssal rocks (Table 8), and Miocene-Pliocene(?) lake deposits (Ferrusquía-Villafranca, 1976; Ortega-Gutiérrez and others, 1990).

Geophysical data. Paleomagnetic studies indicate that Albian and younger rocks have undergone little or no displacement of the Mixteco terrane with respect to stable North America (Morán-Zenteno and others, 1988). However, anomalous directions from Bathonian to Oxfordian rocks and from the Permian strata near Olinalá imply counterclockwise rotation and $15° \pm 8°$ of southward translation of the Mixteco terrane with respect to stable North America between Oxfordian to Albian time, about 160 to 110 Ma (Urrutia-Fucugauchi and others, 1987; Ortega-Guerrero and Urrutia-Fucugauchi, 1989). These results supersede an earlier estimate of 25° to 30° of southward translation (Morán-Zenteno and others, 1988).

Nahuatl terrane

The Nahuatl terrane consists of weakly to strongly deformed and metamorphosed sedimentary and magmatic rocks of Jurassic to Cretaceous age, as well as structurally lower and probably older strongly deformed metamorphic rocks in the eastern part of the terrane (Fig. 13). These rocks are intruded by numerous mid-Cretaceous and Tertiary plutons and overlain unconformably by Tertiary volcanic rocks. The Nahuatl terrane corresponds to the southern part of the Guerrero terrane of Campa-Uranga and Coney (1983) and Coney and Campa-Uranga (1987), and encompasses terrane and subterrane names proposed by other workers, as outlined below.

Many rock unit and terrane names have been published or proposed for rocks within the confines of the Nahuatl terrane, but data and observations are sparse and partly contradictory. In order to clarify matters for those who are familiar with these earlier works, the following paragraphs specify the components of the Nahuatl terrane in terms of the previously published nomenclature. The internal structure of the Nahuatl terrane is greatly simplified in Figure 13; many features and structures discussed below are not depicted because their regional geologic and tectonic relations are enigmatic.

Tierra Caliente Complex (TCC). The lower structural level of the Nahuatl terrane crops out in several regions in the eastern part of the terrane (Guerrero and México states). Prehnite-pumpellyite–, greenschist-, and lower amphibolite–facies metasedimentary and metavolcanic rocks that are known by many local names were informally named the Tierra Caliente Complex (TCC) by Ortega-Gutiérrez (1981a), who tentatively interpreted them as a magmatic arc/marginal basin assemblage. The TCC includes the Taxco Schist, Ayotusco Formation, and Taxco Viejo Greenstone in northern Guerrero, southern México state, and perhaps northeastern Michoacán; the Ixcuinatoyac and

38 R. L. Sedlock and Others

NAHUATL

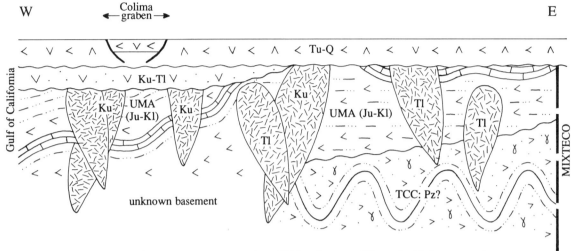

Figure 13. Schematic east-west structure section of Nahuatl terrane. Effects of mid-Cretaceous thrusting and Laramide deformation in Upper Mesozoic Assemblage (UMA) have been omitted. Deformation in Tierra Caliente Complex (TCC) is simplified; see text for discussion.

Chapolapa Formations in southern Guerrero; and metavolcanic and metasedimentary rocks in southwestern Guerrero. These rocks are not known in the western part of the terrane (Fig. 13).

The Taxco Schist, Ayotusco Formation, and Taxco Viejo Greenstone crop out over a large area near Arcelia and Taxco, northernmost Guerrero and Ixtapán de la Sal (Fig. 2) and Tejupilco, southern México state; isolated exposures of similar rocks in the late Cenozoic Trans-Mexican Volcanic Belt in northeastern Michoacán may be correlative. The Taxco Schist (Esquisto Taxco) consists of strongly foliated, greenschist-facies metapelite, quartzite and slate, and metavolcanic rocks of silicic to mafic composition (Fries, 1960; Díaz-García, 1980; de Cserna and Fries, 1981; de Cserna, 1982; Elías-Herrera, 1987, 1989). Near Tejupilco, about 50 km west of Ixtapán de la Sal, Taxco Schist includes phyllite and schist derived from mafic to silicic volcanic and pyroclastic rocks, carbonaceous pelite, flysch, and limestone (Elías-Herrera, 1987, 1989; Tolson, 1990). The metasedimentary and metavolcanic rocks have a structural thickness of about 2 km, and have been subjected to at least four phases of metamorphism (M_1 to M_4) and two episodes of penetrative deformation (D_1, D_2). M_1 and M_2 were coeval with D_1 and D_2, respectively, and attained greenschist-facies conditions (Elías-Herrera, 1987, 1989). M_3 was characterized by low-pressure, high-temperature lower amphibolite-facies metamorphism with geothermal gradients of 70° to 90°C/km, suggesting magmatic arc conditions (Elías-Herrera, 1989). M_4 was a very low grade retrogressive overprint coeval with kilometer-scale open folding. Structural basement does not crop out beneath the schist and phyllite but may be represented by nearby mylonitic, amphibolite-facies, adamellitic orthogneiss of continental or perhaps transitional crustal affinity (two-mica granites, pelitic xenoliths) (Elías-

Herrera, 1989). Dating of protoliths and metamorphic events has been frustrated by the complex thermal history; although dates as old as late Precambrian have been reported (Table 9), unpublished Rb-Sr studies imply that the orthogneiss protolith is late Paleozoic (R. L. Armstrong, in de Cserna, 1982) or Permo-Triassic (P. E. Damon, in Elías-Herrera, 1989). The structural and metamorphic history of the orthogneiss is similar to that of the schist and phyllite, and supports the interpretation that the sedimentary and volcanic protoliths were deposited in a continental arc environment (Elías-Herrera, 1989). However, contact relations are equivocal; a possible alternate interpretation is that the orthogneiss was faulted against the schist-phyllite unit after the two units had developed similar structural and metamorphic histories. According to the latter alternative, the schist-phyllite sequence may have been deposited within an island arc, rather than a continental arc.

The Taxco Schist is discordantly overlain along an unexposed contact by the Ayotusco Formation, which consists of strongly foliated, greenschist-facies carbonaceous slate, quartzite, and marble (Díaz-García, 1980). Protolith and metamorphic ages are unknown. The Ayotusco Formation probably has been mapped as Taxco Schist in many areas (de Cserna and Fries, 1981).

The Taxco Schist and Ayotusco Formation are discordantly overlain along unexposed contacts by the Taxco Viejo Greenstone (Roca Verde Taxco Viejo), which consists of weakly foliated, greenschist-facies lava, tuff, and lahar of andesitic to basaltic-andesitic composition (Fries, 1960; Díaz-García, 1980). Metamorphic minerals such as pumpellyite, phengite, piedmontite, and stilpnomelane(?) may indicate moderately high P/T conditions (Díaz-García, 1980; Ortega-Gutiérrez, 1981a). Direct fossil and radiometric age control are lacking, but based on its

TABLE 9. NAHUATL TERRANE RADIOMETRIC DATA

Sample	System	Mineral*	Date (Ma)	References[†]	Comments
Tierra Caliente Complex					
Metarhyolite	Pb-α	zr	1,020 ±100	1	
Gabbro, monzonite	Rb-Sr	wr	311 ± 30	2	Minimum age: $^{87}Sr/^{86}Sr_i$: 0.7037
Andesite, schist from Taxco Viejo Green-	K-Ar	wr	125 ± 5	3	Age of hydrothermal alteration?
stone, southern México state	K-Ar	wr	108 ± 5	3	
"Upper Mesozoic Assemblage"					
Low-grade fuchsite schist	Rb-Sr	wm	149 ± 64	4	
southwestern México state	K-Ar	wm	81 ± 5	4	
Quartz porphyry, southwestern Jalisco	K-Ar	p	114 ± 2	5	
Mafic dike, south-central Guerrero	$^{40}Ar/^{39}Ar$	h	112 ± 3	6	Plateau age
Welded tuff, southwestern Jalisco	K-Ar	kf, b	92 ± 2	5	
Welded tuff, southwestern Jalisco	K-Ar	kf, b	88 ± 2	5	
Welded tuff, southwestern Jalisco	K-Ar	kf, b	81 ± 4	5	
Welded tuff, southwestern Jalisco	K-Ar	kf, b	71 ± 1	5	
Silicic to intermediate volcanic rocks, W Guerrero	Rb-Sr	wr	69 ± 7	7	6-pt. isochron
Rhyolite, Nayarit	K-Ar	p	54 ± 2	5	
Intrusive rocks					
Mafic and ultramafic rocks,	$^{40}Ar/^{39}Ar$	h	114 ± 3	6	Plateau age
San Pedro Limon,	$^{40}Ar/^{39}Ar$	h	105 ± 1	6	Plateau age
southwestern México state;	$^{40}Ar/^{39}Ar$	h	102 ± 7	6	Plateau age
intrusive into Upper Mesozoic Assemblage	$^{40}Ar/^{39}Ar$	h	100 ± 3	6	Integrated age
Puerto Vallarta Batholith	Rb-Sr	wr	105	8	Emplacement age
Puerto Vallarta Batholith	K-Ar	h	91 ± 2	5	Cooling age
Puerto Vallarta Batholith	Rb-Sr	h	86 ± 2	8	Cooling age
Puerto Vallarta Batholith	Rb-Sr	b	83 ± 3	8	Cooling age
Puerto Vallarta Batholith	K-Ar	h, b	82–80	5	Cooling ages, 6 samples
Gabbro pegmatite, Nayarit	K-Ar	h	98 ± 3	5	
Granitoids, Nayarit and southwestern Jalisco	K-Ar	h, b, p	71–45	5	8 samples
Granite, southern Jalisco	Rb-Sr	wr	69 ± 1	8	
Quartz monzonite, western Guerrero	K-Ar	wm	63 ± 1	9	$^{87}Sr/^{86}Sr_i$: 0.7041
Granitoids, south-central Michoacán	K-Ar	b	57–44	10	5 samples
Tonalite, south-central Michoacán	Rb-Sr	wr	56 ± 5	11	
Granite, south-central Michoacán	K-Ar	b	55 ± 4	10	
			43 ± 5	10	
Quartz monzonite, central Jalisco	K-Ar	wr	54 ± 5	12	
Gabbro, southeastern Jalisco			53 ± 1	11	
Mineralized (Cu) breccia pipes, central Michoacán	K-Ar	h, p, wm, b	36–31	9	6 samples; $^{87}Sr/^{86}Sr_i$: 0.7039-0.7055 (four samples)
Granitoids near Petatlán, Guerrero	Rb-Sr	wr	33 ± 3	13	3-pt isochron
Diorite, Nayarit	K-Ar	h	27 ± 3	5	
Post-Laramide volcanic rocks					
Basalt, Balsas Group	K-Ar		42 ± 1	14	
Rhyolite near Taxco	K-Ar	wr, kf	36 ± 2	14	
Ignimbrite near Morelia	K-Ar	kf	33 ± 2	15	
Rhyolite near Morelia	K-Ar	kf	23 ± 1	15	
Ignimbrite near Morelia	K-Ar	kf	22, 21	15	
		p	18 ± 1	15	
Basalt near Morelia	K-Ar	wr	20 ± 1	15	
Rhyolite conglomerate	K-Ar	p	18 ± 1	16	
Andesite near Morelia	K-Ar	wr, p, b	18–8	15	6 samples
Gabbro	K-Ar	p	13 ± 2	16	
Basalt dike	K-Ar	p	13 ± 1	16	
Welded tuff	K-Ar	kf	11 ± 1	16	
Basalts	K-Ar	p	10–8	16	3 samples

*Mineral abbreviations: b = biotite; h = hornblende; kf = potassium feldspar; p = plagioclase; wm = white mica; wr = whole rock; zr = zircon.
[†]1 = de Cserna and others, 1974b; 2 = de Cserna and others, 1978; 3 = Urrutia-Fucugauchi and Linares, 1981; 4 = Fries and Rincón-Orta, 1965; 5 = Gastil and others, 1978; 6 = Delgada-Argote and others, 1992b; 7 = González and others, 1989; 8 = Köhler and others, 1992; 9 = Damon and others, 1983; 10 = Grajales-Nishimura and López-Infanzón, 1983; 11 = Pantoja-Alor, 1988; 12 = González and Martínez, 1989; 13 = González-Partida and others, 1989; 14 = de Cserna and Fries, 1981; 15 = Pasquaré and others, 1991; 16 = Gastil and others, 1979.

presumed youth with respect to the underlying Taxco Schist and Ayotusco Formation and the unconformable overlap by Tithonian and younger strata of the Upper Mesozoic Assemblage (see section below), the Taxco Viejo Greenstone has been assigned a provisional Late Triassic to Early Jurassic age (Díaz-García, 1980; de Cserna and Fries, 1981; de Cserna, 1982). The Taxco Viejo Greenstone differs from the underlying Taxco Schist and Ayotusco Formation by its weaker deformation and its marked hydrothermal alteration, implying that it may have evolved independently of the latter units until its juxtaposition with them (i.e., Taxco Viejo Greenstone as an allochthonous thrust sheet) prior to Tithonian overlap (de Cserna and Fries, 1981; de Cserna, 1982). Early Cretaceous dates from the Taxco Viejo Greenstone (Table 9) are inferred to correspond to the age of hydrothermal alteration (Campa-Uranga and others, 1974; Urrutia-Fucugauchi and Linares, 1981), but the presence of Taxco Viejo Greenstone clasts in overlying Tithonian-Neocomian strata (see below) implies an earlier age for the alteration.

We note here that an alternate interpretation of the stratigraphy of the Taxco-Tejupilco region has been proposed (Campa-Uranga and others, 1974; Campa-Uranga, 1978). In this interpretation, the protoliths of TCC as described above are not Paleozoic but rather are correlative or at least coeval with Late Jurassic to Early Cretaceous rocks that we assign to the Upper Mesozoic Assemblage (see below). These rocks were assigned to the Teloloapán-Ixtapán subterrane of the Guerrero terrane (Campa-Uranga and Coney, 1983; Coney and Campa-Uranga, 1987).

About 20 km southwest of Chilpancingo, southern Guerrero (Fig. 2), the TCC probably includes two greenschist-facies sequences of metavolcanic and metasedimentary rocks that may be separated by an unconformity (de Cserna, 1965; Klesse, 1968). The Ixcuinatoyac Formation consists of deformed quartzite, phyllite, metabasite, sedimentary sulfides, conglomerate, and serpentinite of unknown age that were deformed prior to intrusion of granitoids of unknown age. The unconformably overlying Chapolapa Formation consists of silicic to intermediate metavolcanic rocks and metaclastic rocks including conglomerate, sandstone, and phyllite of unknown age.

In the vicinity of Petatlán (Fig. 2), Zihuatanejo, and Puerto Escondido in southwestern Guerrero, near the western boundary of the Chatino terrane, the TCC consists of greenschist-facies metavolcanic and metasedimentary rocks that were derived from marine volcanic and volcaniclastic rocks, carbonaceous shale, and intercalated volcanic and sedimentary rocks and were deformed by northeast-vergent folds and thrusts prior to intrusion by gabbro and monzonite of possible late Paleozoic age (Table 9) (de Cserna and others, 1978; Delgado-Argote and others, 1986). These metamorphic rocks are overlain unconformably by younger rocks of the Nahuatl terrane (see below) along a contact locally marked by serpentinite.

In northeastern Michoacán in the Morelia-Zitácuaro area, the TCC may include a 1,500-m-thick sequence of rhythmically interbedded sandstone, siltstone, and shale affected by low-grade metamorphism (Pasquaré and others, 1991). Fossil plants indicate a Middle Jurassic age for part of the sequence (A. Islas and others, unpublished data, 1989; Pasquaré and others, 1991).

In summary, the name Tierra Caliente Complex refers to metabasites, metaandesites, and metasedimentary rocks in the structurally lower part of the eastern Nahuatl terrane. Protoliths of at least some of these rocks probably are of Paleozoic age. The nature of protoliths and metamorphism in the TCC may indicate the juxtaposition of arc (andesites; high-temperature metamorphism) and basinal (metabasites, flysch; low-grade metamorphism) assemblages, although other interpretations certainly are possible given the limited data base (Ortega-Gutiérrez, 1981). Many aspects of the complex structural and metamorphic history still are unresolved, but shallowly dipping to subhorizontal foliation, axial surfaces, and thrust faults probably record tectonic transport to the northeast (Ortega-Gutiérrez, 1981a; Robinson, 1990).

Upper Mesozoic Assemblage. The structurally higher unit of the Nahuatl terrane consists of generally strongly deformed but weakly metamorphosed Jurassic(?) to Cretaceous(?) sedimentary and volcanic strata, and overlying weakly deformed and metamorphosed late Early and Late Cretaceous carbonates and clastic strata. These rocks crop out throughout both the eastern and western parts of the Nahuatl terrane. First, we discuss outcrops of the Upper Mesozoic Assemblage in the eastern part of the terrane.

In western Morelos, northernmost Guerrero, and southern México state, the Upper Mesozoic Assemblage overlies the Taxco Schist and other rocks of the TCC along a contact that has been interpreted as a regionally extensive unconformity (Díaz-García, 1980) and, at least locally, as a fault (Tolson, 1990). The Tithonian (latest Jurassic) to Neocomian(?) Acuitlapán Formation consists of slightly to strongly deformed, low-grade metamorphic rocks derived from siliciclastic sedimentary rocks, limestone, and andesitic flows, tuff, and agglomerate (Fries, 1960; Campa-Uranga and others, 1974; de Cserna and Fries, 1981; de Cserna, 1982). Conglomerate contains hydrothermally altered clasts derived from the underlying Taxco Viejo Greenstone. In western Morelos and northern Guerrero, the Acuitlapán Formation is overlain with possible unconformity by Albian-Cenomanian platform carbonates and Turonian-Coniacian limestone grading to flysch (the Morelos-Guerrero platform and cover, including the Morelos Formation). In southern México state, the Acuitlapán Formation is overlain by Albian-Cenomanian basinal carbonates and Cenomanian-Coniacian andesitic to basaltic-andesitic lavas and pyroclastic rocks, associated mafic to ultramafic intrusives, flysch, and minor limestone (Fries, 1960; de Cserna and Fries, 1981; de Cserna, 1982; Delgado-Argote and others, 1992a). The post-Acuitlapán strata have been correlated with coeval platform and basinal carbonates throughout central and eastern México and with similar strata in the western Nahuatl terrane (see below).

Calcite c-axis fabrics in marbles of the Acuitlapán Formation indicate top-to-the-east tectonic transport; less deformed Albian-Turonian limestones lack crystallographic preferred orientation (Tolson, 1990). These observations imply a mid-Cretaceous episode of east-vergent thrusting. Both units underwent younger east-west to northeast-southwest shortening generally associated with the Laramide orogeny (Fries, 1960; Campa-Uranga, 1978; de Cserna and Fries, 1981; de Cserna, 1982). The age of Laramide deformation is bracketed by deformed Coniacian strata and postkinematic nonmarine volcanic and clastic rocks (Balsas Group) that are as young as late Eocene and as old as Maastrichtian (Table 9) (de Cserna, 1982). Some thrusts locally cut basement rocks of the TCC, but most are inferred to merge into a décollement near the base of the Upper Mesozoic Assemblage (de Cserna, 1982).

Near Petatlán, southern Guerrero (Fig. 2), the Upper Mesozoic Assemblage includes very weakly metamorphosed and deformed marine shale, sandstone, and limestone of possible Mesozoic age (de Cserna and others, 1978) and mid-Cretaceous volcanic and volcaniclastic rocks that may be cogenetic with adjacent plutonic rocks (Delgado-Argote and others, 1986). The volcanogenic rocks are weakly deformed, implying either that they were erupted after Laramide orogenesis, or that they are part of an arc terrane that collided with the southern Nahuatl terrane during Laramide orogenesis (Urrutia-Fucugauchi and Valencio, 1986). The volcanogenic rocks and associated mafic and ultramafic rocks were termed the Papanoa terrane by Coney and Campa-Uranga (1987), but recent remapping and geochronologic work by Delgado-Argote and others (1986, 1990, 1992b) indicate that the magmatic rocks are of Cretaceous age. It seems advisable to discontinue use of the term Papanoa terrane.

In the western part of the Nahuatl terrane (states of Michoacán, Colima, Jalisco, Nayarit), the Upper Mesozoic Assemblage includes Mesozoic volcanic and sedimentary rocks that have been assigned to a bewildering array of terranes and subterranes by different workers. (1) The Huetamo subterrane of the Guerrero terrane in Michoacán (Campa-Uranga and Coney, 1983; Coney and Campa-Uranga, 1987) consists of Late Jurassic marine volcaniclastic rocks overlain by Neocomian siliciclastic strata and tuff and Albian and Late Cretaceous siliciclastic strata and platform carbonates (Campa-Uranga, 1978). (2) Deformed but weakly metamorphosed andesitic flows, breccias and tuffs, and interbedded Albian limestone and siliciclastic strata in Michoacán and Colima have been called the Zihuatanejo subterrane of the Guerrero terrane (Campa-Uranga and Coney, 1983; Coney and Campa-Uranga, 1987) and the Colima terrane (Campa-Uranga, 1985a). (3) The Arteaga terrane in southeastern Michoacán (Coney and Campa-Uranga, 1987) reportedly consists of metamorphosed Middle to Late Triassic volcaniclastic rocks structurally juxtaposed with Paleozoic schists. Detrital zircons were derived from Grenville and early Paleozoic sources (K. Robinson, personal communication, 1991). (4) The Tumbiscatio terrane in southeastern Michoacán (Campa-Uranga, 1985a) geographically coincides with the Arteaga terrane but is said to consist of chert and metabasite that is at least partly Triassic. Gastil and others (1978) recognized (5) an unnamed sequence of interbedded graywacke and andesite of uncertain but possible Late Jurassic age in southern Nayarit and western Jalisco, and (6) an unnamed sequence of interbedded andesite, marble, and metasandstone of uncertain but possible Early Cretaceous age in southern Nayarit, Jalisco, and Colima. (7) In southern Colima, weakly deformed late Aptian–Albian deep-water to platformal limestone (Michaud and others, 1989) may be correlative with more deformed rocks in (1), (2), and (3) and with the Morelos Formation in the eastern part of the Nahuatl terrane. (8) In southeastern Jalisco, late Aptian andesitic conglomerate, limestone, and sandstone and early Albian rhyolitic to dacitic volcanogenic rocks, limestone, and sandstone (Pantoja-Alor, 1983; Pantoja-Alor and Estrada-Barraza, 1986) probably correspond to (6). (9) In Colima, Smith and others (1989) recognized Neocomian to Turonian(?) deposition, Turonian(?) uplift, and later strike-slip disruption of rocks that also probably correspond to (6). (10) In northeastern Michoacán in the Morelia-Zitácuaro region, Upper Mesozoic Assemblage rocks include Late Jurassic to Neocomian low-grade schist, pillow lavas, and metatuff overlain transitionally by unmetamorphosed Hauterivian to Albian volcaniclastic and terrigenous sedimentary rocks with intercalated bioclastic limestone (I. Israde and L. Martínez, unpublished data, 1986).

Based on the broad similarity of lithotype and, where known, depositional age, we provisionally assign the above rocks to a Late Jurassic–Early Cretaceous volcanic and sedimentary assemblage that we correlate with the Upper Mesozoic Assemblage in the eastern part of the Nahuatl terrane (Fig. 13). Weakly to moderately deformed Cretaceous limestone observed throughout the Nahuatl terrane probably was laterally continuous with the platform succession in eastern México. Triassic(?) rocks of the so-called Tumbiscatio terrane (4) may represent older, deeper parts of the terrane, a megablock within the terrane, or a distinct terrane of enigmatic origin. The definition of the Arteaga terrane (3) apparently is obsolete and we suggest that the name be abandoned.

Post-Laramide strata. The oldest post-Laramide rocks in the Nahuatl terrane are Maastrichtian to Eocene nonmarine coarse clastic rocks and intercalated rhyolite, andesite, and rarer basalt in the eastern part of the terrane known as the Balsas Group. Widespread Eocene to Oligocene rhyolitic to dacitic pyroclastic rocks (Table 9) may be correlative with Tertiary volcanic rocks of the Sierra Madre Occidental. Late Oligocene to late Miocene flows, marine volcaniclastic rocks, and less abundant intrusives of andesitic and subordinate mafic and silicic composition (Table 9) probably formed in an arc above the subducting Farallon (Cocos) plate (Edwards, 1955; Fries, 1960; Gastil and others, 1979; de Cserna and Fries, 1981; de Cserna, 1982). In Late Neogene to Quaternary time, voluminous calc-alkalic volcanic rocks were erupted in the Trans-Mexican Volcanic Belt on the northern margin of the terrane (Nixon, 1982; Nelson and Livieres, 1986; Nixon and others, 1987; Pasquaré and others, 1991). Starting about 4.5 to 4.0 Ma, alkalic and calc-alkalic volcanic rocks were erupted in the Colima and Zacoalco grabens in the western part of the terrane (Allan, 1986). Calc-alkalic rocks were derived from partial melting of heterogeneous mantle above the subducting Cocos and Rivera plates. Source magmas of alkalic rocks in the Colima graben included a large component of incompatible element-rich lherzolite dikes and metasomatic veins (Luhr and others, 1989), whereas alkalic and peralkalic magmas in the Zacoalco graben were derived from a source typical of an oceanic island (Verma and Nelson, 1989).

Intrusive rocks. Granitoids crop out in about one-third of the terrane, with a rough west to east decrease in crystallization age from mainly mid-Cretaceous in southern Nayarit, Jalisco, Colima, and western Michoacán to mainly Tertiary in eastern Michoacán and Guerrero (Fig. 13, Table 9) (Damon and Coney, 1983; Böhnel and others, 1989; Guerrero-García, 1989). Published geochronologic data are not available for many plutonic rocks in eastern Jalisco, Colima, Michoacán, and western Guerrero, but country rocks of these intrusions are mainly of Cretaceous and Paleogene age. Preliminary Nd isotopic data from plutons in the northwestern part of the terrane imply partial derivation from a Proterozoic(?) source and variable degrees of crustal contamination, whereas plutons in the eastern part of the terrane yield middle to late Paleozoic Nd model ages and show no evidence for crustal contamination (Schaaf and others, 1991).

Regional structural and geophysical data. The lateral continuity of late Early and Late Cretaceous clastic and carbonate strata indicate that the Nahuatl terrane was amalgamated and accreted to México prior to Late Cretaceous–Paleogene Laramide orogenesis that produced northeast-vergent folds and basement-involved thrusts (Johnson and others, 1990). Outliers of Jurassic-Cretaceous arc rocks within the Cretaceous carbonate platform of the Upper Mesozoic Assemblage are probably klippe stranded by northeast-vergent Laramide thrusting (Campa-Uranga, 1978). Early Miocene shortening is indicated by a large, open, asymmetric anticline in northeastern Michoacán that dips steeply west (Pasquaré and others, 1991).

Paleomagnetic data from volcanic rocks, sedimentary rocks, and granitoids imply negligible latitudinal displacement and minor local clockwise or counterclockwise rotation of the Nahuatl terrane since Early Cretaceous time (Böhnel and others, 1989). In the northwestern part of the terrane, post-Oligocene normal faults that strike northwest, east-west, and northeast are interpreted as products of north-south tension related to subduction of the Cocos plate to the south (de Cserna and Fries, 1981; de Cserna, 1982). In the same area, left-slip and rarer right-slip faults that strike north-south to north-northwest–south-southeast are inferred to be a result of Neogene transpression (Delgado-Argote and others, 1992a).

North America terrane

Proterozoic crystalline rocks of North American continental crust crop out locally in northern Sonora and Chihuahua and probably underlie the entire region. Paleozoic carbonate and siliciclastic rocks were deposited in shallow water on a south-facing shelf until the late Paleozoic, when foundering of the southern shelf edge resulted in the deposition of basinal flysch and pelagic rocks. Precambrian basement and late Paleozoic flysch were deformed and interleaved in the Permian(?), coeval with but perhaps unrelated to Ouachitan orogenesis. This region apparently was emergent during the Triassic and part of the Jurassic. The western part of the terrane hosted a Jurassic volcanic arc that continued northwest into southern Arizona and California, and also was the site of Early Cretaceous magmatism. The eastern part of the terrane accumulated Late Jurassic to Cretaceous carbonate, siliciclastic rocks, and evaporites in the Chihuahua Trough and fault-bounded Bisbee basin.

Precambrian basement rocks. In northern Sonora, North American continental basement consists of 1,700- to 1,660-Ma "eugeosynclinal" rocks that were deformed and metamorphosed to greenschist facies about 1,650 Ma (Fig. 14; Table 10); these rocks probably are correlative with the Pinal Schist in southern Arizona (Anderson and Silver, 1977b, 1981). These metamorphic rocks are intruded by undeformed, volumetrically abundant 1,460- to 1,410-Ma anorogenic granitoids that probably are part of a suite of similar rocks in the western and midcontinental United States (Anderson, 1983), and by much rarer 1,100-Ma anorogenic granites (Anderson and Silver, 1977a, b, 1981; Rodríguez-Castañeda, 1988).

In Chihuahua, known Precambrian basement includes granite gneiss encountered at depths of 4 to 5 km in two PEMEX wells; two small outcrops of fault-bounded metaplutonic rocks intruded by amphibolite dikes and pegmatites within Permian flysch in the Sierra del Cuervo, about 15 km north-northwest of Aldama; and small fault-bounded outcrops of undated amphibolite gneiss near Rancho El Carrizalillo, about 80 km east of Sierra del Cuervo (Thompson and others, 1978; Mauger and others, 1983; Quintero-Legorreta and Guerrero, 1985; Handschy and Dyer, 1987; Blount and others, 1988). Radiometric dates (Table 10) are similar to those of Grenville rocks in western and central Texas (Copeland and Bowring, 1988; Walker, 1992). Late Paleozoic and Triassic cooling ages probably indicate a regional low-grade thermal event in late Paleozoic time, perhaps associated with Ouachitan orogenesis (Denison and others, 1971; Mauger and others, 1983). Grenville rocks probably underlie much of northern and eastern Chihuahua (Fig. 14), as indicated by Sm-Nd and U-Pb studies of xenoliths (Table 10) (Ruiz and others, 1988b; Rudnick and Cameron, 1991). Thus, the Proterozoic southern margin of cratonal North America probably extended southwest from Texas into northern Chihuahua. The location and nature of the contact between the Grenville rocks of Chihuahua and the older crust of Sonora are not known (Fig. 14).

Paleozoic rocks. Scattered outcrops of Paleozoic strata in Chihuahua and northeastern Sonora are correlative with strata in adjacent Arizona and New Mexico, and record deposition along the southern margin of North America (Fig. 14). Similar rocks have been penetrated by numerous wells in northern Chihuahua (Thompson and others, 1978; López-Ramos, 1985, p. 156–159).

Ordovician to Permian carbonate, shale, and sandstone in north-central Chihuahua and northeastern Sonora were deposited in shallow-water shelf and platform environments in and on the

NORTH AMERICA

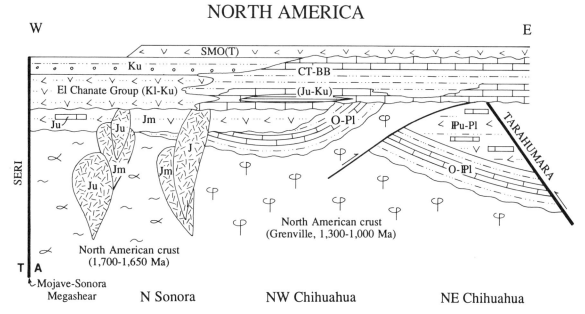

Figure 14. Composite tectonostratigraphic section and schematic east-west structure section of North American continental crust in northern México. Uncertain location and orientation of Grenville front, i.e., contact between Grenville and older basement rocks. SMO(T), Tertiary volcanic rocks of Sierra Madre Occidental. CT-BB, Upper Jurassic to Upper Cretaceous rocks of Chihuahua Trough and Bisbee basin. Effects of Cretaceous to Paleogene shortening have been omitted.

margins of the Pedregosa basin, which probably trended northwest from central Chihuahua into southern New Mexico and southeastern Arizona (Taliaferro, 1933; Imlay, 1939; Ramírez-M. and Acevedo-C., 1957; Bridges, 1964a; Díaz and Navarro-G., 1964; Navarro-G. and Tovar-R., 1974; King, 1975; Greenwood and others, 1977; Dyer, 1986; González-León, 1986; Armin, 1987; Pubellier and Rangin, 1987). During latest Pennsylvanian(?) and Early Permian time, the southern part of the Pedregosa basin in central Chihuahua subsided rapidly, as indicated by the deposition of thick basinal flysch and pelagic rocks; minor tuff and conglomerate composed exclusively of rhyolite clasts may indicate proximity to a cryptic Early Permian volcanic arc (de Cserna and others, 1968; Mellor and Breyer, 1981; Torres-Roldán and Wilson, 1986; Handschy and Dyer, 1987; Handschy and others, 1987). In the Sierra del Cuervo in central Chihuahua, late Paleozoic flysch was overridden in mid-Permian to Jurassic time by crystalline rocks of Grenville age along east-verging thrust faults (Fig. 14) (Handschy and Dyer, 1987; Handschy and others, 1987). A significant unconformity is inferred above the Precambrian and Paleozoic rocks because Triassic to Middle Jurassic strata are absent (Fig. 14). Latest Paleozoic K-Ar cooling ages reported from Precambrian granite and pegmatite in the Sierra del Cuervo region (Table 10) may reflect tectonism of this age.

Zircons collected from an orthogneiss xenolith in eastern Chihuahua indicate crystallization at 350 to 320 Ma followed by granulite metamorphism either shortly thereafter or at 190 Ma

(Rudnick and Cameron, 1991). Carboniferous and Permian zircons are absent from five xenoliths in this region, indicating that Ouachitan orogenesis had no effect on the lower crust in eastern Chihuahua, presumably because the thermal high was in the Ouachita interior zone to the east.

Mesozoic and Cenozoic rocks. In northern Sonora, crystalline Precambrian basement is overlain by Jurassic intermediate to silicic volcanic and hypabyssal rocks and sedimentary rocks (e.g., Fresnal Canyon sequence, Artesa sequence of Tosdal and others, 1990a; unnamed Jurassic volcanic rocks of Rodríguez-Castañeda, 1988) and intruded by Jurassic granitoids (Ko Vaya superunit of Tosdal and others, 1990a) (Table 10). In the Cucurpe-Tuape region in northeastern Sonora, Oxfordian (early Late Jurassic) volcaniclastic rocks, tuff, andesite, and shale grade upward into clastic rocks and limestone of uncertain but presumed Late Jurassic age (Rangin, 1978; Rodríguez-Castañeda, 1988). These Jurassic volcanic, plutonic, and volcaniclastic rocks are interpreted to have formed in a northwest-trending arc in southwestern North America that, in northwestern México, was the locus of Late Jurassic sinistral displacement on the Mojave-Sonora Megashear (Anderson and Silver, 1979; Tosdal and others, 1990a).

Tithonian to Albian basal conglomerate, limestone, fine-grained marine siliciclastic rocks, evaporites, and coal accumulated in the Chihuahua Trough and the fault-bounded Bisbee basin in Chihuahua and Sonora (Greenwood and others, 1977; Dickinson and others, 1986; Brown and Dyer, 1987; Araujo-

TABLE 10. RADIOMETRIC DATA FOR NORTH AMERICA IN NORTHERN MÉXICO

Sample	System	Mineral*	Date (Ma)	References[†]	Comments
Precambrian rocks					
Crystalline rocks, northeastern Sonora	U-Pb	zr	1,700–1,600	1	Unpublished data
Metamorphic rocks, northeastern Sonora	U-Pb	zr	1,680 ± 20	2	Metamorphosed about 1,650 Ma
Schist near Cucurpe, Sonora	U-Pb	zr	~1,675	3	Source: T. Anderson
Lower crustal xenoliths	Sm-Nd		1,600–1,300	4	Model ages
Granitoid near Cananea, Sonora	U-Pb	zr	1,440 ± 15	5	Upper intercept
Granitic plutons, Sonora	U-Pb	zr	1,440–1,410	2	
Orothogneiss xenolith, eastern Chihuahua	U-Pb	zr	1,370 ±180, 1,100 ±130	6	Crystallization age, Age of granulite metamorphism
Metaigneous rocks, Chihuahua	U-Pb	zr	1,328 ± 5	7	Discordant age
Granite gneiss, northwestern Chihuahua	Rb-Sr		1,327 ±242	8	PEMEX Chinos-1 well
Metaigneous rocks, Chihuahua	U-Pb	zr	1,280 ± 8	7	Discordant age
Granite, Sierra del Cuervo	K-Ar	kf	250 ± 21	9	Source: M. Múgica
Granite clasts, Aptian conglomerate, N Chihuahua	Rb-Sr	kf	1,270 ± 45	10	
Aibo granite, 20 km south	U-Pb	zr	1,110 ± 10	11	
of Caborca, Sonora	Rb-Sr	kf	710 ±100	12	
Amphibolite dikes, Chihuahua	K-Ar	h	1,027, 1,024	13	
	K-Ar	wm	349	13	
Pegmatite, Chihuahua	K-Ar	wm	940	13	
	K-Ar	kf	212	13	
Pegmatite, east-central Chihuahua	K-Ar	wm	267 ± 21	9	Source: M. Múgica
Granite gneiss, north-central Chihuahua	Rb-Sr		890 ± 32	8	PEMEX Moyotes-1 well
Metarhyolite clasts in Lower Cretaceous	Rb-Sr	wr	695 ± 10	10	7-pt isochron; assumed $^{87}Sr/^{86}Sr$: 0.706
Conglomerate, northern Chihuahua	Rb-Sr	wr	350 ± 15	14	Source: Shell
	Rb-Sr	wr, wm	287 ± 5	10	Isochron age
	Rb-Sr	wr, wm	287 ± 5	10	Isochron age
	K-Ar	wm	272 ± 5	10	
	K-Ar	wm	255 ± 5	10	
	K-Ar	wm	239 ± 9	10	
Upper Paleozoic and Mesozoic magmatic rocks					
Rhyolite clasts, Lower K congl, C Chihuahua	Rb-Sr	wr	246 ± 42	10	
Granitoid, northwestern Sonora	U-Pb	zr	225	11	
Volcanic rocks (Fresnal Canyon, Artesa sequences)	U-Pb	zr	>180–170	1	Unpublished data
			178–160	11	7 samples, N Sonora
Granitoids (Ko Vaya Superunit)	U-Pb	zr	175–150	1	Unpublished data
			177–149	11	6 samples, N Sonora
El Capitan, northwestern Sonora	U-Pb	zr	168 ± 5	15	
Rhyolite in Sonora and clasts in southern California, USA	U-Pb	zr	155 ± 3	16	Lower intercept, inferred crystallization age
Volcanic rocks, east-central Sonora	K-Ar		~140	17	
Rhyolite, Sonora	K-Ar		131, 127	18	Thermally altered
Rhyolite, Sonora	K-Ar		107	19	Interbedded with red beds
Granodiorite, central Chihuahua	K-Ar		89 ± 3	20	Source: J. Blount
Lamprophyre, central Chihuahua	K-Ar		89 ± 3	20	Source: J. Blount
Tuff and andesite, Sonora	K-Ar	b	86–83	21	4 samples
	K-Ar	h	81	21	
Granite gneiss, northern Sonora	U-Pb	zr	78 ± 3	22	
Granodiorite, northern Sonora	U-Pb	zr	74 ± 2	22	
Granites, northwestern Sonora	K-Ar	b	71, 68	23	Associated with Cu deposits
Quartz monzonite, north-central Sonora	K-Ar	b	69 ± 3	24	
Granitoids, north-central Sonora	U-Pb	zr	69 ± 1	5	
	U-Pb	zr	64 ± 3	5	
Cenozoic magmatic rocks					
Intermediate-silicic volcanics, EC Sonora	K-Ar	wr	75–52	17	
Mineralized pipe, northern Sonora	K-Ar	ph	60 ± 2	23	Associated with Cu deposits
Mineralized silicic-intermediate stocks,	K-Ar	wm	60 ± 2	25	
	K-Ar	wm	55 ± 1	25	
northeastern Sonora	K-Ar	wm	53 ± 1	25	
Granodiorite, north-central Sonora	K-Ar	wm	57 ± 1	26	Associated with W deposits

TABLE 10. RADIOMETRIC DATA FOR NORTH AMERICA IN NORTHERN MÉXICO (continued)

Sample	System	Mineral*	Date (Ma)	References[†]	Comments
Cenozoic magmatic rocks (continued)					
Granodiorites, north-central Sonora	$^{40}Ar/^{39}Ar$	h	57 ± 2	26	3 samples
Intrusive breccia, northern Sonora	K-Ar	b	57 ± 1	23	Associated with Cu deposits
Granitoids, northern Sonora	K-Ar	b	56 ± 2	23	3 samples; $^{87}Sr/^{86}Sr_i$: 0.7062-0.7070
Granitoids, northern Sonora	K-Ar	wm	56–50	23	5 samples; hydrothermal altered Cu deposits
Pegmatite, northern Sonora	K-Ar	b	55 ± 2	23	Associated with Cu deposits
Granite, northwestern Sonora	K-Ar	wm	53 ± 2	24	
Silicic plutonic and volcanic rocks, N Sonora	K-Ar		52–43	27	Source: P. Damon and others
Rhyolite stock, northern Sonora	K-Ar	kf	51 ± 1	23	Associated with Cu deposits; $^{87}Sr/^{86}Sr_i$: 0.7103
Granodiorite, north-central Sonora	Rb-Sr		50 ± 1	26	4-pt isochron; $^{87}Sr/^{86}Sr_i$: 0.7091
	$^{40}Ar/^{39}Ar$	h	48 ± 2	26	Associated with W deposits
	K-Ar	b	40 ± 1	26	
	$^{40}Ar/^{39}Ar$	b	37 ± 1	26	
Ignimbrites	K-Ar		50–40	28	Unpublished data
Rhyo(dacite), northern Chihuahua	K-Ar	h, b	48 ± 2	29	2 samples
Granodiorite, north-central Sonora	$^{40}Ar/^{39}Ar$	h	47 ± 1	26	Associated with W deposits
	$^{40}Ar/^{39}Ar$	b	37 ± 1	26	
Mineralized pipe, northern Sonora	K-Ar	wm	46 ± 1	23	Associated with Cu deposits
Silicic tuff and rhyolite, NC Chihuahua	K-Ar		>45–38	27	Sources: Alba and Chávez, Mauger, Capps
Silicic tuff, northern Chihuahua	K-Ar	kf + p	45 ± 1	30	
Andesite, northeastern Sonora	K-Ar		44 ± 1	31	Source: P. Damon
Rhyolite, central Chihuahua	K-Ar		44–37	32	Several samples
Pegmatite, north-central Sonora	K-Ar	wm	42 ± 1	26	Associated with W deposits
Silicic tuff and andesite, central Chihuahua	U-Pb		42–38	27	Source: F. McDowell
Granodiorite, northern Sonora	K-Ar	b	40 ± 1	23	Biotite associated with Cu mineralization
Volcanic rock, eastern Chihuahua	K-Ar	kf	40 ± 1	33	
Andesites from	K-Ar	p	40 ± 2	34	
lower volcanic sequence,			38 ± 7	34	
Chihuahua			32 ± 2	34	
Silicic tuff, northern Chihuahua	K-Ar	kf + p	39 ± 1	30	
Rhyolite, northern Chihuahua	K-Ar	kf + p	39–35	30	6 samples
Rhyolitic ignimbrite, Chihuahua	K-Ar	b	38 ± 2	34	
		p	38–35	34	3 samples
Pegmatite, north-central Sonora	K-Ar	b	38 ± 1	26	Associated with W deposits
Granite, Chihuahua	K-Ar	wm	37 ± 1	23	
Diabase, Chihuahua	K-Ar	wr	37 ± 1	35	2 samples
Skarn, southeastern Chihuahua	K-Ar	wm	37 ± 1	23	Associated with Cu deposits
Andesites from	U-Pb	zr	~37	34	F. McDowell, unpublished data
lower volcanic sequence	K-Ar		~35		
Skarns, north-central Sonora	K-Ar	b	36 ± 1	26	Associated with W deposits
	$^{40}Ar/^{39}Ar$	b	35 ± 2	26	2 samples
Skarn, central Sonora	K-Ar	wm	35 ± 1	26	Associated with W deposits
Quartz monzonite, Chihuahua	K-Ar	b	35 ± 1	35	
		kf, b, p	35–27		29 samples
Basaltic andesite, Chihuahua	K-kAr		33–25	36	F. McDowell, unpublished data
Basaltic andesite, E Sonora–W Chihuahua	K-Ar	p, wr	31–25	37	5 samples
Silicic stock, eastern Chihuahua	K-Ar	kf	31 ± 1	33	
Silicic tuff, eastern Chihuahua	K-Ar	kf	30 ± 1	33	2 samples
Volcanic rocks, eastern Chihuahua	K-Ar	kf	29 ± 1	33	4 samples
Basalt, northern Chihuahua	K-Ar	wr	29 ± 1	30	
Silicic tuff, northern Chihuahua	K-Ar	kf	29 ± 1	30	
Volcanic rocks	K-Ar	wr, wm, kf	29–26	35	4 samples
Silicic tuff, eastern Chihuahua	K-Ar	kf	28 ± 1	33	4 samples
Gabbro, eastern Sonora	K-Ar	p	27 ± 1	37	
Basaltic andesites, Chihuahua	K-Ar		26–24	36	Cameron and others, 1980; unpublished data

TABLE 10. RADIOMETRIC DATA FOR NORTH AMERICA IN NORTHERN MÉXICO (continued)

Sample	System	Mineral*	Date (Ma)	References[†]	Comments
Cenozoic metamorphic cooling ages, Caborca-Altar region, Sonora					
Metasedimentary schist	K-Ar	b	59 ± 3	12	Cooling ages for
	K-Ar	h	55 ± 3	38	Laramide metamorphism?
Metasedimentary schist	K-Ar	b, wm	17–15	38	4 samples; cooling ages of extensional metamorphism?
Synkinematic granite	K-Ar	wm	16	39	P. Damon, unpublished data

*Mineral abbreviations: b = biotite; h = hornblende; kf = potassium feldspar; p = plagioclase; ph = phlogopite; wm = white mica; wr = whole rock; zr = zircon.

[†]1 = Anderson and Silver, 1979; 2 = Anderson and Silver, 1977a; 3 = Rodriguez-Castañeda, 1988; 4 = Ruiz and others, 1988b; 5 = Anderson and Silver, 1977b; 6 = Rudnick and Cameron, 1991; 7 = Blount and others, 1988; 8 = Thompson and others, 1978; 9 = López-Infanzón, 1986; 10 = Denison and others, 1971; 11 = Anderson and others, 1979; 12 = Damon and others, 1962; 13 = Mauger and others, 1983; 14 = Bridges, 1971; 15 = Silver and others, 1969; 16 = Abbott and Smith, 1989; 17 = Pubellier and Rangin, 1987; 18 = Abbott and others, 1983; 19 = Jacques-Ayala and Potter, 1987; 20 = Handschy and Dyer, 1987; 21 = Grajales-Nishimura and others, 1990; 22 = Anderson and others, 1980; 23 = Damon and others, 1983; 24 = Fries and Rincón-Orta, 1965; 25 = Livingston, 1973; 26 = Mead and others, 1988; 27 = Aguirre and McDowell, 1991; 28 = McDowell and others, 1990; 29 = Marvin and others, 1988; 30 = Keller and others, 1982; 31 = Roldán-Quintana, 1982; 32 = Alba and Chávez, 1974; 33 = C. Henry, unpublished data; 34 = Wark and others, 1990; 35 = Shafiqullah and others, 1983; 36 = Cameron and others, 1989; 37 = Montigny and others, 1987; 38 = Hayama and others, 1984; 39 = Jacques-Ayala and others, 1990.

Mendieta and Estavillo-González, 1987; Pubellier and Rangin, 1987; Salvador, 1987; Rodríguez-Castañeda, 1988; Obregón-Andria and Arriaga-Arredondo, 1991). The asymmetric Chihuahua Trough consisted of a deep axial basin in eastern Chihuahua, a relatively narrow eastern margin in western Texas, and a broader western margin, in part built on the fossil Triassic-Jurassic magmatic arc, that sloped gently basinward from the active Late Jurassic–Cretaceous magmatic arc in Sonora (Coney, 1978; Dickinson and others, 1986; Araujo-Mendieta and Casar-González, 1987). Strata of the Lower Cretaceous Bisbee Group have been recognized as far west as Caborca, Sonora, where they interfinger with or are overlain by the Lower Cretaceous to lower Upper Cretaceous El Chanate Group and El Charro Complex, both of which include arc-derived rhyolite, andesite, tuff, and volcaniclastic rocks (Roldán-Quintana and González-León, 1979; Almazán-Vázquez and others, 1987; Jacques-Ayala and Potter, 1987; Jacques-Ayala, 1989; Jacques-Ayala and others, 1990; Scott and González-León, 1991).

Latest Cretaceous stratified rocks include nonmarine and shallow marine continental strata in northernmost Sonora and northwesternmost Chihuahua (Taliaferro, 1933; Hayes, 1970; Roldán-Quintana and González-León, 1979) and synorogenic(?) conglomerates containing clasts of andesite, rhyolite, quartz porphyry, granodiorite, chert, and quartzite clasts that are inferred to have been derived from thrust sheets uplifted during Laramide orogenesis (Jacques-Ayala and others, 1990). Magmatic rocks include Late Cretaceous to Eocene (Table 10) intermediate to silicic plutonic rocks and coeval volcanic rocks assigned to the "lower volcanic complex" of McDowell and Keizer (1977), and Oligocene rhyolitic ignimbrites and basaltic andesites of the "upper volcanic supergroup" of McDowell and Keizer (1977). Miocene-Pliocene conglomerate, mafic agglomerate and flows, and minor sandstone, shale, and limestone of the Baucarit Formation accumulated in small basins along active

normal faults (e.g., King, 1939; Roldán-Quintana, 1982; Rodríguez-Castañeda, 1988). Mid- to Late Cenozoic U-Pb ages determined from zircons in xenoliths in eastern Chihuahua probably indicate that the lower crust was subjected to granulite-facies metamorphism during the Oligocene or Pliocene-Pleistocene (Rudnick and Cameron, 1991).

Structural and geophysical data. Late Cretaceous to Paleogene folding and thrusting of North American rocks in northern México has been documented in northern and eastern Chihuahua (Bridges, 1964b; Lovejoy, 1980; Brown and Handschy, 1984; Corbitt, 1984; Brown and Dyer, 1987; Dyer and others, 1988) and in northern and central Sonora (Taliaferro, 1933; Roldán-Quintana, 1982; Haxel and others, 1984; Almazán-Vázquez, 1986; Rodríguez-Castañeda, 1988; Goodwin and Haxel, 1990; Jacques-Ayala and others, 1990; Nourse, 1990; Sosson and others, 1990). Cenozoic volcanic rocks between these two regions may obscure similar structural features.

In the Chihuahua tectonic belt, which roughly coincides with the Paleozoic Pedregosa basin and Mesozoic Chihuahua Trough, up to 80 km of eastward to northeastward transport of Cretaceous marine strata occurred in the latest Cretaceous to middle Eocene. This deformation episode is widely equated with the Laramide event in the United States. Thrusts probably sole into incompetent Late Jurassic evaporites, though basement was at least locally affected. Although vergence is predominantly northeastward, folds locally verge southwestward, especially along the margins of more rigid platformal blocks.

Contractional features in Sonora are divided into three episodes of regional extent: northeast-vergent thrusting and folding of Late Jurassic to possibly earliest Cretaceous age that may be synchronous with and caused by motion on the Mojave-Sonora Megashear (Rodríguez-Castañeda, 1990), east- to northeast-vergent isoclinal folding and thrusting of early Late Cretaceous (Cenomanian?) age that may be correlative with Sevier deformation, and late Late Cretaceous to Paleogene southwest-vergent folding and thrusting, as well as coeval plutonism, equated with Laramide deformation and magmatism (Roldán-Quintana, 1982; Pubellier and Rangin, 1987; Jacques-Ayala and others, 1990; Sosson and others, 1990). Many thrusts emplace Precambrian basement rocks over Mesozoic rocks (Rodríguez-Castañeda, 1988, 1990). The second and third episodes of shortening must be younger than deformed Early Cretaceous strata and older than mid- to late Tertiary extensional features. It is unclear whether these two deformation episodes are phases of a single protracted mid-Cretaceous to Paleogene (100 to 50 Ma?) compressional event. Near Altar, a major thrust of the older episode is inferred to be older than a "cross-cutting" intrusion that yielded K-Ar cooling ages of about 80 Ma (Table 10), but the contact is not exposed, and map relations permit the alternate interpretation that the intrusion is within the upper plate of, and thus older than, the thrust. According to this interpretation, all shortening occurred during late Late Cretaceous to Paleogene (Laramide) time.

Mid- to late Tertiary extensional deformation in northern México is temporally and kinematically similar to deformation in southern Arizona and California (e.g., Reynolds and others, 1988). Late Oligocene and early Miocene metamorphic core complexes in parts of north-central Sonora are characterized by top-to-the-southwest detachment faulting, displacement on shallowly dipping ductile shear zones, synextensional magmatism, and regional tilting of fault blocks (Davis and others, 1981; Goodwin and Haxel, 1990; Nourse, 1990). During middle Miocene to Quaternary time, high-angle normal faults associated with Basin and Range extension overprinted all older fabrics and formed north-northwest–trending ridges and valleys not only throughout northern Sonora and Chihuahua, but also in most of northern and central México (Roldán-Quintana and González-León, 1979).

A paleomagnetic study of Kimmeridgian red beds near Placer de Guadalupe concluded that the primary magnetization is masked by a strong Quaternary (Brunhes chron) overprint (Herrero-Bervera and others, 1990).

Pericú terrane

The Pericú terrane consists chiefly of a suite of prebatholithic metasedimentary and minor metaigneous rocks that was intruded by mafic to intermediate plutons in the late Early Cretaceous, strongly deformed with those plutonic rocks in the mid-Cretaceous, and intruded by undeformed granitoids in the Late Cretaceous. Late Cretaceous granitoids probably formed in the magmatic arc along the western margin of México, as did similar rocks in the Serí, Tahué, and Nahuatl terranes. The origin and early history of the prebatholithic rocks is poorly understood. The Pericú terrane probably was detached from western México and attached to the southern tip of Baja California prior to late Cenozoic opening of the Gulf of California.

To the east and south, thinned continental crust of the terrane is bounded by Miocene to Holocene oceanic crust (Fig. 15). On the west, the terrane boundary generally is called the La Paz fault, an enigmatic, partly buried feature that probably records at least two episodes of displacement (Aranda-Gómez and Pérez-Venzor, 1989). Recent work suggests that rocks of the Pericú terrane may underlie downdropped late Cenozoic strata west of the fault (Fig. 15), so the westward and northward extent of Pericú rocks is uncertain (A. Carrillo-Chávez, unpublished data).

The oldest rocks in the Pericú terrane are prebatholithic metasedimentary and minor metaigneous rocks that crop out chiefly in a narrow band between La Paz and Todos Santos (Fig. 2) along the western edge of the terrane (Fig. 15) (Ortega-Gutiérrez, 1982; Gastil, 1983; Aranda-Gómez and Pérez-Venzor, 1989; Murillo-Muñetón, 1991) Constituent rock types include schist, gneiss, phyllite, marble, amphibolite, slate, hornfels, migmatite, and skarn, and likely protoliths were shale, sandstone, marl, impure limestone, and mafic igneous rocks. Protolith ages and depositional environments are unknown. In the southern part of the belt, a typical assemblage in strongly foliated and lineated gneiss is quartz + plagioclase ± garnet ± hornblende ± potassium feldspar ± diopside ± epidote or zoisite, with accessory sphene, tourmaline, and zircon; a typical assemblage in strongly foliated phyllite is biotite + quartz + plagioclase ± andalusite ± sillimanite ± muscovite ± cordierite, with coexisting andalusite and sillimanite in several samples (Aranda-Gómez and Pérez-Venzor, 1989). Hornblende-sillimanite schist and diopside-hornblende-biotite quartzofeldspathic gneiss have been reported from the central part of the belt (Murillo-Muñetón, 1991). These assemblages indicate amphibolite-facies metamorphism of most of the metasedimentary suite, probably during intrusion of late Early Cretaceous mafic plutons and low-potassium granitoids (Table 11).

During the late Early Cretaceous and early Late Cretaceous, the metasedimentary suite and late Early Cretaceous plutonic rocks were penetratively deformed by roughly east-west compression. Evidence includes well-developed foliation, gneissic layering, mineral lineations, folds, strained early porphyroblasts of garnet and andalusite, boudinage, and widespread development of mylonite zones several meters to 2.5 km thick (Aranda-Gómez and Pérez-Venzor, 1989; Murillo-Muñetón, 1991). Mylonitic foliation dips east to southeast (Fig. 15), lineations in the foliation surface plunge shallowly south, and microstructures indicate left-oblique displacement (Aranda-Gómez and Pérez-Venzor, 1989). Strongly mylonitized garnetiferous gneiss near Todos Santos may be a pre–Late Cretaceous metamorphic core complex: mylonitic foliation dips very gently to the northwest, a strong mineral lineation plunges shallowly northwest, and pressure shadows of garnets indicate top-to-the-northwest transport. Enigmatic late Paleozoic and Triassic K-Ar dates obtained from the deformed plutonic rocks (Table 11) may indicate that the mid-Cretaceous deformation and metamorphism overprinted older fabrics and assemblages, but these dates are considered invalid by some workers (G. Gastil, personal communication, 1991).

Prebatholithic metasedimentary rocks of the Pericú terrane

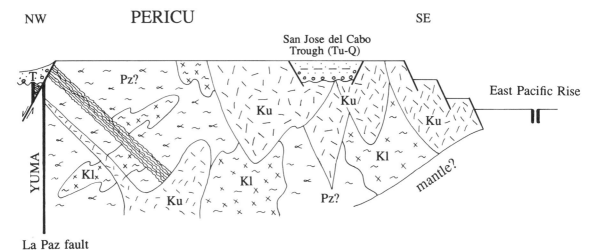

Figure 15. Schematic northwest-southeast structure section of Pericú terrane. Distribution of metasedimentary rocks (Pz?) at deep level is uncertain. Southeastern boundary of the terrane is transitional to oceanic lithosphere near East Pacific Rise. Feature labeled La Paz fault includes older subvertical(?) surface and younger, down-to-the-west normal fault.

are grossly similar to and may be displaced fragments of Paleozoic rocks in the Tahué terrane. On the other hand, they also resemble the Mesozoic prebatholithic rocks in the eastern subterrane of the Yuma terrane in Baja California and southern California, and may be continuous with buried basement rocks of the Yuma terrane west and north of the La Paz fault. Evaluation of possible correlations awaits U-Pb dating of zircon-rich metasedimentary and deformed plutonic rocks in the Pericú terrane.

The metasedimentary rocks and deformed plutonic rocks are cut by undeformed Late Cretaceous high-potassium granitoids that constitute the most abundant rock type in the terrane (Fig. 15; Table 11). Disseminated gold deposits occur in cataclastic tonalite and diorite in narrow fault zones (Carillo-Chávez, 1991). The granitoids are similar in age and lithology to granitoids in the Tahué, Yuma, and Serí terranes and probably formed in the magmatic arc along the western margin of México. K-Ar mineral ages from the metamorphic suite reflect thermal metamorphism due to Late Cretaceous intrusion (Table 11).

The metamorphic complex and granitoids are cut by mid-Tertiary(?) andesitic dikes and stocks and are nonconformably overlain by late Cenozoic nonmarine and marine clastic rocks. Miocene-Pliocene nonmarine to shallow marine strata that accumulated in the San José del Cabo trough (Fig. 15) indicate subsidence during middle to late Miocene time and shoaling during Pliocene and Quaternary time (McCloy, 1984).

The Pericú terrane is cut by numerous steeply dipping normal faults of uncertain but probable Cenozoic age (omitted from Fig. 15). The predominant strike directions in successively younger sets of faults are north to north-northwest, northeast, and northwest to west-northwest. The paucity of northwest to north-

TABLE 11. PERICÚ TERRANE RADIOMETRIC DATA

Sample	System	Mineral*	Date (Ma)	References[†]	Comments
Hornblende gneiss	K-Ar	h	225 ± 5	1	
Tonalite	K-Ar	h	335 ± 4	1	
Amphibolite	K-Ar	h	116 ± 6	2	Cooling age
Hornblende diorite	K-Ar	h	115 ± 2	3	
Undeformed granitoids	K-Ar		109–70	3	V. Frizzell and others
Undeformed granitoid	U-Pb	zr	93	3	unpublished data
Cataclastic granite	K-Ar	wm	100 ± 5	2	
		b	93 ± 5		
Undeformed granitoids	K-Ar	wm, b, h	91–64	2	6 samples
Mica schist	K-Ar	wm	87 ± 1	2	Cooling ages
		b	62 ± 3		
Augen gneiss	K-Ar	b	50 ± 4	2	Cooling age

*Mineral abbreviations: b = biotite; h = hornblende; wm = white mica; zr = zircon.
[†]1 = Altamirano-R., 1972; 2 = Murillo-Muñetón, 1991; 3 = Hausback, 1984.

northwest–striking normal faults typical of late Cenozoic Basin and Range faulting may be due to control by preexisting, chiefly north- to northeast-striking structures (Aranda-Gómez and Pérez-Venzor, 1988).

Serí terrane

The Serí terrane is distinguished by latest Proterozoic and Paleozoic shelfal and basinal rocks that were deposited on and outboard of Proterozoic North American continental crust. Basinal rocks, first deformed during Mississippian time, were deformed a second time during northward thrusting onto shelfal rocks in the Permo-Triassic. Paleomagnetic studies suggest that the western part of the Serí terrane may have been stripped from the eastern part in late Paleozoic or early Mesozoic time, only to return to, or nearly to, its departure point in the late Cretaceous and Paleogene. The Serí terrane hosted a continental magmatic arc during the Cretaceous and early Cenozoic. By 5, and perhaps as early as 12 Ma, the terrane was disrupted by extension and right-lateral faulting in the Gulf of California.

The Serí terrane is bounded by the Mojave-Sonora Megashear on the northeast, by a west-verging reverse fault at its boundary with the Yuma terrane to the west, and by an inferred fault contact with the Tahué terrane to the south (Figs. 3, 16). Parts of the Serí terrane have been included in the Caborca (Coney and Campa-Uranga, 1987), Ballenas (Gastil, 1985), and Cortés terranes (Champion and others, 1986; Howell and others, 1987).

Crystalline basement. Proterozoic crystalline basement rocks of the North American craton crop out in northwestern Sonora and are inferred to underlie the entire Serí terrane (Fig. 16; Table 12). Deformed quartz-rich schist and gneiss, quartzite, metarhyodacite, and amphibolite are cut by calc-alkalic plutons ranging in age from 1,750 to 1,710 Ma, and younger layered quartzofeldspathic and amphibolitic gneiss were deformed, metamorphosed, and cut by pegmatites about 1,685 to 1,645 Ma (Anderson and Silver, 1977a, 1981). These older rocks are intruded by volumetrically abundant anorogenic granitoids dated at about 1,450 Ma and the much rarer Aibo granites dated at about 1,110 Ma (Table 12) (Anderson and others, 1979; Anderson and Silver, 1981). Precambrian basement rocks of the Serí terrane probably were displaced southeastward on the Mojave-Sonora Megashear from a southwest-trending belt of similar rocks in the southwestern United States (Silver and Anderson, 1974, 1983; Anderson and Silver, 1979).

Crystalline Precambrian rocks are overlain nonconformably by latest Proterozoic and Paleozoic rocks that are assigned to two roughly defined units: (1) shallow-water shelfal ("miogeoclinal") strata that overlie Proterozoic basement nonconformably, and (2) deep-water basinal ("eugeoclinal") strata, south and west of the shelfal rocks, that were thrust above the shelfal rocks in the Permo-Triassic (Radelli and others, 1987; Stewart, 1988; Stewart and others, 1990). The shelfal strata bear a strong resemblance to coeval shelfal rocks in eastern and southern California and probably are part of a displaced fragment of south-

western North America. The basinal strata are grossly similar to Paleozoic basinal strata in southern Nevada, eastern California, and the Ouachita orogenic belt, but their stratigraphic and tectonic relation to these rocks is unknown (Poole and others, 1983; Stewart and others, 1984; 1990; Ketner, 1986, 1990; Poole and Madrid, 1986, 1988; Murchey, 1990). As more data become available, it may be useful to delineate these fault-bounded basinal strata as a distinct terrane.

Shelfal rocks. Latest Proterozoic to Middle Cambrian shelfal ("miogeoclinal") rocks near Caborca, Sonora include dolomite, quartzite, limestone, and rarer mafic volcanogenic rocks that are similar to rocks in the Death Valley region and in the San Bernardino Mountains of southern California (Stewart and others, 1984; McMenamin and others, 1992). Late Proterozoic strata, known as the Gamuza beds, lie nonconformably on the 1,110-Ma Aibo granite south of Caborca and are transitional upward to the Paleozoic section (Cooper and Arellano, 1946; Arellano, 1956; Anderson and Silver, 1981). East of Hermosillo, the Cambrian strata are overlain by Late Cambrian, Early Ordovician, Late Devonian, Mississippian, and Early Pennsylvanian strata, mainly carbonates (Fig. 16) (Stewart and others, 1984, 1990; Radelli and others, 1987). Cambrian to Permian shelfal strata also crop out in several other ranges in central Sonora (Ketner, 1986; Almazán-Vázquez, 1989; Stewart and others, 1990). Pennsylvanian-Permian depositional patterns indicate Early Permian foundering of the continental shelf (Stewart and others, 1990), as is recognized in the Pedregosa basin in New Mexico and Chihuahua (North America). Shelfal rocks that may be correlative with Paleozoic shelfal rocks in central and northwest Sonora crop out on the west coast of Sonora and perhaps on northern Isla Tiburón in the Gulf of California (Stewart and others, 1990), and in eastern Baja California (Miller and Dockum, 1983; Anderson, 1984; Gastil, 1985; Gastil and others, 1991).

Basinal rocks. Basinal ("eugeoclinal") strata are exposed in an east-west–trending belt in central Sonora and in isolated outcrops in western Sonora and Baja California. In central Sonora, these include Early and Middle Ordovician graptolitic shale, Late Ordovician and Silurian chert, shale, and dolostone, Late Devonian chert, clastic rocks, and bedded barite, Early Mississippian limestone turbidites, argillite, and chert, and unconformably overlying Late Mississippian clastic rocks and Pennsylvanian-Permian clastic rocks, limestone, chert, argillite, and bedded barite (Fig. 16) (Poole and others, 1983, 1990; (Ketner, 1986; Poole and Madrid, 1986, 1988; Radelli and others, 1987; Stewart and others, 1990). Detrital zircons collected from Late Devonian clastic rocks yielded a 1,675-Ma date, implying derivation from crystalline Precambrian basement (F. G. Poole, personal communication, 1990). Basinal rocks that may be correlative with Paleozoic basinal strata in central Sonora crop out in eastern Baja California and on islands in the Gulf of California (Gastil and Krummenacher, 1977a, b; Lothringer, 1984; Gastil, 1985; Griffith, 1987; Gastil and others, 1991). The contact between basinal rocks and shelfal rocks in eastern Baja California has not been recognized.

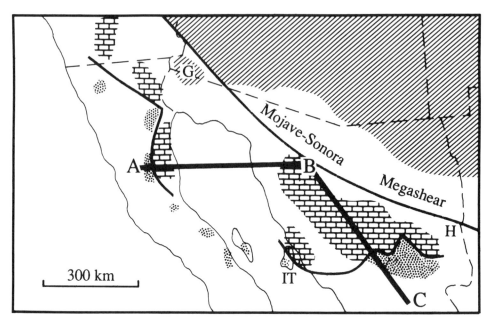

Figure 16 (on this and facing page). Schematic sections of Serí terrane. Location map shows approximate section lines; B is near Caborca, Sonora. Key to patterns on location map (modified after Stewart and others, 1990): ruled lines, cratonal upper Proterozoic–lower Paleozoic facies; brick, coeval shelfal (miogeoclinal) facies; dots, coeval basinal (eugeoclinal) facies. Abbreviations on location map: G, Gila Mountains–El Capitán area; H, Hermosillo; IT, Isla Tiburón.

Other pre–Middle Triassic rocks. Mid-Permian (Guadalupian) siltstone and limestone of the Monos Formation crop out near El Antimonio, about 40 km west of Caborca; the lower contact is not exposed (Brunner, 1979; González-León, 1979, 1980). These strata contain large fusulinids (*Parafusulina antimonioensis*) that are similar to fusulinids from accreted terranes in northern California and Washington state and unlike fusulinids in shelfal rocks of the Cordillera (Stewart and others, 1990). These strata probably were deposited outboard of the continental margin, but they are not correlative with basinal rocks elsewhere in the Serí terrane. In Baja California near 30°N, Early Permian to Early Triassic rocks record a transition from deep-water basinal rocks to shallow-water shelfal strata similar to shelfal strata in the southern Great Basin (Buch, 1984; Dellatre, 1984; Gastil, 1990).

Juxtaposition of basinal and shelfal strata. In early Late Mississippian time, Ordovician–Early Mississippian basinal rocks in central Sonora underwent northwest-southeast shortening. After deposition of Late Mississippian to Early Permian strata, basinal rocks of Ordovician to Early Permian age underwent north-northwest–south-southeast shortening in Late Permian to Middle Triassic time. During or immediately after the second phase of shortening, the basinal rocks were thrust over the shelf sequence (Poole and others, 1990; Stewart and others, 1990). The basinal allochthon is repeated by north-dipping thrust faults in at least one range (Bartolini and Stewart, 1990) and has been interpreted as a stack of thrust nappes (Radelli and others,

1987), but structural relations in most parts of the basinal and shelfal sequences are not yet understood. Most mapped contacts between the two sequences are high-angle faults of probable Cenozoic age, but the outcrop pattern indicates that the basinal sequence was transported over the shelfal sequence a minimum of 50 km northward, roughly normal to the inferred east-west trend of the latest Paleozoic continental margin in central Sonora (Fig. 16, location map) (Stewart, 1988; Poole and others, 1990; Stewart and others, 1990).

Westward from central Sonora, the boundary between the basinal and shelfal sequences probably continues across Isla Tiburón in the Gulf of California and into formerly adjacent eastern Baja California, from where it curves northward or northeastward back into Sonora between El Antimonio and Caborca (Fig. 16, location map) (Stewart, 1990; Stewart and others, 1990). North of Caborca, the boundary may be truncated and displaced northwestward by a left-lateral fault or faults such as the Mojave-Sonora Megashear, or it may have been originally continuous to the north-northwest, wrapping around a promontory of North America with little or no modification by later strike-slip displacement (Stewart and others, 1984, 1990; Stewart, 1990).

Mesozoic and Cenozoic rocks. Scattered outcrops of Early Jurassic and Triassic clastic rocks in northwestern and central Sonora once were considered part of a single formation (King, 1939), but the rocks subsequently have been divided into several discrete units. Near and southwest of El Antimonio,

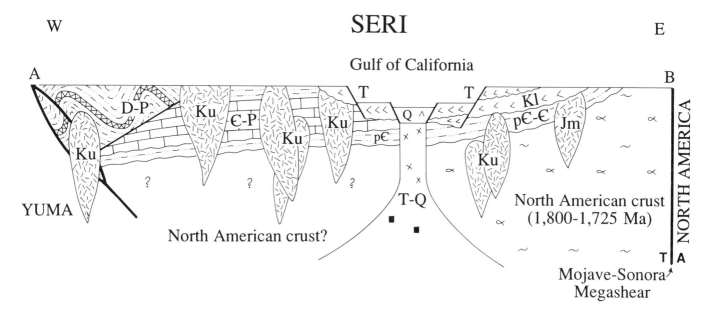

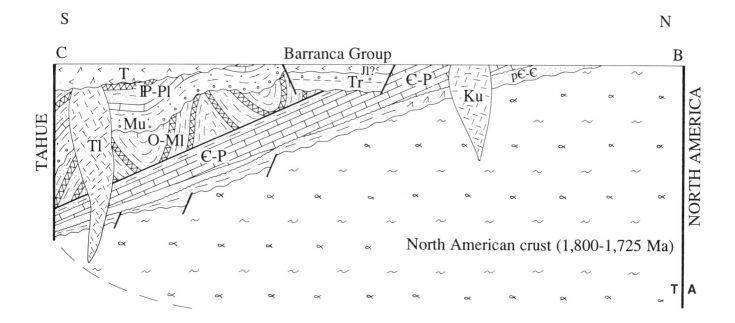

northwestern Sonora, the mid-Permian Monos Formation is overlain unconformably by marine clastic rocks and limestone of the Upper Triassic to Lower Jurassic El Antimonio Formation (White and Guiza, 1949; González-León, 1979, 1980). Carnian-Norian (Late Triassic) ammonites are broadly similar to fauna of the Hallstatt facies of Tethyan strata in Europe and Asia, and are dissimilar to Triassic fauna elsewhere in México, including accreted Triassic oceanic rocks on the Vizcaíno Peninsula (Cochimí terrane) (Tozer, 1982). Ichthyosaur remains from the Carnian-Norian strata belong to the same family (*Shastasaurus*) as fossils from Nevada, Oregon, Canada, Europe, Russia, China,

New Caledonia, and possibly New Zealand (Callaway and Massare, 1989). Strong stratigraphic and faunal similarities have been recognized between the El Antimonio Formation and the Upper Triassic Luning Formation in southwestern Nevada (Stanley and others, 1991). The Upper Triassic lower member of the El Antimonio Formation is overlain with slight unconformity by an upper member of coarse clastic rocks and shale of Hettangian-Sinemurian (Early Jurassic) and possibly late Early or even Middle Jurassic age (González-León, 1979, 1980). The Late Triassic strata and the underlying Permian Monos Formation probably were deposited outboard of the continental margin and have undergone significant displacement with respect to North America (Stewart and others, 1990).

Late Triassic marginal marine and nonmarine clastic rocks of the Barranca Group unconformably overlie Paleozoic strata in an east-west belt across central Sonora (Fig. 16) (King, 1939; Wilson and Rocha, 1949; Alencaster, 1961; González-León, 1979; Almazán-Vázaquez and others, 1987; Stewart and Roldán-Quintana, 1991). The Barranca Group includes fluvial and deltaic quartzose and feldspathic clastic rocks (red beds), Carnian (early Late Triassic) coal measures that contain Tethyan bivalves and ammonoids (Alencaster, 1961; Obregón-Andria and Arriaga-Arredondo, 1991), and Late Triassic(?) conglomerates that contain locally derived clasts of crystalline Precambrian, miogeoclinal, and eugeoclinal rocks (Cojan and Potter, 1991; Stewart and Roldán-Quintana, 1991). Volcanic rocks and detritus apparently are absent. Deformation is significantly less intense than in underlying Paleozoic strata. The Barranca Group crops out over a sizable region of east-central Sonora, but outcrops are chiefly confined to an east-west belt that is interpreted as an elongate rift basin that developed after the Late Permian–Middle Triassic juxtaposition of the eugeoclinal and miogeoclinal sequences (Stewart, 1988; Stewart and Roldán-Quintana, 1991).

In the Sierra López, 40 km northwest of Hermosillo, Sinemurian (Early Jurassic) clastic rocks interbedded with andesitic flows and tuffs unconformably overlie Paleozoic limestone (Avila-Angulo, 1990). In the Sierra de Santa Rosa, about 120 km north-northwest of Hermosillo, Early Jurassic(-Triassic?) clastic rocks, derived in part from a volcanic source, and minor limestone are conformably overlain by a thick sequence of andesitic flows and tuffs and volcaniclastic rocks that probably is Early to Middle Jurassic at its base and perhaps as young as Cretaceous at its top (Hardy, 1981). Jurassic volcanic rocks at both localities are compositionally and lithologically similar to, and may be correlative with, undated intermediate volcanic and volcaniclastic rocks that unconformably overlie the Barranca Group east of Hermosillo (Stewart and others, 1990). Early Jurassic clastic rocks and limestone that crop out about 120 km north of Hermosillo contain fauna similar to those in the Sierra de Santa Rosa (Flores, 1929). At Cerro Pozo Cerna, about 40 km west of Sierra de Santa Rosa, late Oxfordian–early Kimmeridgian corals and molluscs were identified in the upper part of a 2-km-thick sequence of clastic rocks (Beauvais and Stump, 1976).

Rare Middle Jurassic plutonic rocks have been reported from the Serí terrane (Anderson and Silver, 1979). These are interpreted as cogenetic and coeval with Middle to Late Jurassic granitoids in northern Sonora (North America) and thus part of the Jurassic arc of southwestern North America that probably hosted sinistral displacement on the Mojave-Sonora Megashear (Anderson and Silver, 1979; Tosdal and others, 1990a).

Diverse post-Jurassic magmatic rocks crop out in the Serí terrane (Fig. 16; Table 12). Early Cretaceous intermediate volcanic rocks are present in westernmost Sonora, and Late Cretaceous to Paleogene granitoids crop out in central Sonora and northeastern Baja California (e.g., Roldán-Quintana, 1991). Pre-

batholithic andesitic to rhyolitic lavas and tuffs in central Sonora probably were erupted in the Late Cretaceous to Paleogene (Roldán-Quintana, 1989). Oligocene silicic ignimbrites and basaltic andesites (e.g., Cochemé and Demant, 1991) are part of the "upper volcanic supergroup" of McDowell and Keizer (1977). Neogene calc-alkalic volcanic rocks along the west coast of Sonora, the east coast of northern Baja California, and on islands in the Gulf of California are remnants of a Neogene magmatic arc that also is preserved in Baja California Sur and Nayarit (Gastil and others, 1979). Neogene bimodal volcanic rocks in southern Sonora are inferred to be related to Basin and Range extension (Morales and others, 1990).

Nonmarine conglomerate and sandstone and mafic agglomerate and lavas of the Neogene Baucarit Formation accumulated in small basins along active normal faults throughout central Sonora (King, 1939; Hardy, 1981; Roldán-Quintana, 1982, 1989). K-Ar dating of volcanic rocks indicates that coarse clastic debris of the Baucarit Formation was deposited from late Oligocene to middle Miocene time (about 27 to 10 Ma), coincident with Basin and Range faulting in the region (McDowell and Roldán-Quintana, 1991; Bartolini and others, 1992) (Table 12). late Cenozoic marine clastic strata that record marine deposition in the subsiding Gulf of California rift basin were deposited as far north as Isla Tiburón by 13 Ma (Smith and others, 1985; Neuhaus and others, 1988).

Other structural and geophysical data. Much or all of the Serí terrane apparently underwent east-west shortening in Late Cretaceous time, though in many regions this deformation is masked by Tertiary volcanic rocks and extensional features. North-south–striking folds and thrust faults of probable Late Cretaceous age were mapped throughout central Sonora by King (1939). In several ranges south of Caborca, east-vergent thrusting of Jurassic strata and Precambrian basement occurred prior to the intrusion of dikes in the Late Cretaceous or Paleogene (Hardy, 1981). In northeastern Baja California, isoclinal overturned and recumbent folds indicate roughly north-south Late Cretaceous–Paleogene(?) shortening (Siem and Gastil, 1990).

Mid- to late Tertiary extensional deformation in the Serí terrane is temporally and kinematically similar to deformation in the southwestern United States (e.g., Reynolds and others, 1988) and in the North America terrane in México. Mid-Tertiary to active metamorphic core complexes have been recognized in the Sierra Mazatán, about 80 km east of Hermosillo, where ductilely deformed lower plate orthogneiss of Paleogene age (Table 12) is separated from upper plate Mississippian strata by mylonite (Davis and others, 1981), in the Sierra El Mayor at 32°N in northeastern Baja California, where shallowly dipping detachment faults that separate ductilely deformed footwall gneisses from brittlely extended Upper Miocene–Pleistocene rocks indicate west-northwest–east-southeast to northwest-southeast extension (Siem and Gastil, 1990, 1991), and at several points along the main Gulf of California escarpment between 31° and 29°N (Bryant and others, 1985; Gastil and Fenby, 1991). Tilting and high-angle normal faulting indicating east-northeast–west-south-

TABLE 12. SERÍ TERRANE RADIOMETRIC DATA

Sample	System	Mineral*	Date (Ma)	References†	Comments
Precambrian rocks					
Crystalline rocks, northwestern Sonora	U-Pb	zr	1,800–1,725	1	Unpublished data
Calc-alkalic plutons, northwestern Sonora	U-Pb	zr	1,750–1,710	2	
Layered gneiss, northwestern Sonora	U-Pb	zr	1,660 ± 15	2	Metamorphic ages
Anorogenic granitoids, southwestern Sonora	U-Pb	zr	~1,450	3	Unpublished data
Precambrian rocks in Caborca-Bamori area, northwestern Sonora					
Rhyodacite	U-Pb	zr	1,755 ± 20	3	Crystallization age
Granite gneiss	U-Pb	zr	1,745 ± 20	3	Intrusion age
Pegmatites	K-Ar	wm	1,703 ± 50	4	Accompanied regional intrusion
	K-Ar	wm	1,684 ± 50	4	and metamorphism about 1,675
	U-Pb	zr	1,680 ± 20	3	Ma
	U-Pb	zr	1,635 ± 20	3	
Schist intruded by pegmatite	K-Ar	wm	1,684 ± 50	4	
	Rb-Sr	wm	1,570	5	
Amphibolite	K-Ar	wm	1,654	5	
Aibó granite	U-Pb	zr	1,110 ± 10	6	
	Rb-Sr	kf	710 ±100	4	
Mesozoic and Cenozoic magmatic rocks					
El Capitan orthogneiss at San Luis, NW Sonora	U-Pb	zr	170 ± 3	7	
Andesite, western Sonora	K-Ar	p	143 ± 3	8	Source: G. Salas
Rhyolite, western Sonora	U-Pb	zr	142 ± 2	7	
Metarhyolite, western Sonora	U-Pb	zr	128 ± 2	7	
Granitoids, western Sonora	U-Pb	zr	100–82	7	Several plutons
Granitoids, northeastern Baja	U-Pb	zr	100–75	9	Many samples
Quartz monzonite, Caborca	K-Ar	b	87 ± 6	4	
Andesite, western Sonora	K-Ar	h	87 ± 2	8	Source: G. Salas
Granodiorite, Caborca	K-Ar	h	81 ± 2	10	Inferred to postdate thrusting
	K-Ar	b	79 ± 2	10	
Quartz monzonite, Caborca	K-Ar	b	72 ± 3	11	
Granite, central Sonora	K-Ar	b	70 ± 2	12	Source: G. Gastil
Granodiorite, west-central Sonora	K-Ar	b	65 ± 1	13	Associated with W deposits
Granodiorite, southeastern Sonora	K-Ar	h	63 ± 1	13	Associated with W deposits
	^{40}Ar/^{39}Ar	b	54 ± 1	13	
	K-Ar	b	53 ± 1	13	
Granite, Barita de Sonora	U-Pb	zr	62 ± 1	14	Source: R. E. Zartman
Mafic volcanic rocks, central Sonora	K-Ar	wr	62 ± 1	15	Lower volcanic complex
Granitoids, south-central Sonora	K-Ar	b, wm, h, wr	59–54	16	7 samples; associated with Cu deposits; ^{87}Sr/^{86}Sr$_i$: 0.7064-0.7079 (4 samples)
Granitoids, southeastern Sonora	Rb-Sr	wr	59 ± 5	13	5-pt. isochron; ^{87}Sr/^{86}Sr$_i$: 0.7058
Granitoids, central Sonora	U-Pb	zr	58 ± 3	17	
	U-Pb	zr	57 ± 3	17	
Granites, central Sonora	K-Ar	h	57, 52	12	Source: D. Mead
Quartz vein, northwestern Sonora	K-Ar	b	54 ± 1	13	Associated with W deposits
Granite, northwestern Sonora	K-Ar	wm	53 ± 2	11	
Granodiorite, east-central Sonora	K-Ar		50	18	Source: Damon and others
Granodiorite, west-central Sonora	^{40}Ar/^{39}Ar	b	48 ± 1	13	Associated with W deposits
	Rb-Sr		47 ± 2	13	4-pt. isochron; ^{87}Sr/^{86}Sr$_i$: 0.7068
	K-Ar	b	42 ± 1	13	
Pegmatite, central Sonora	K-Ar		42	18	Source: Damon and others
Andesite, central Sonora	K-Ar		33	19	Below Baucarit Formation
Silicic volcanic rocks, eastern Sonora	K-Ar	p, b	30–27	20	4 samples
Ignimbrite, central Sonora	K-Ar	p	27 ± 1	15	At base of Baucarit Formation
Mafic volcanic rocks, eastern Sonora	K-Ar	p	23 ± 1	20	
Volcanic rocks, Sonora and Baja	K-Ar	p, wr, b, h, kf	23–4	21	33 samples
Mafic volcanic rocks, central Sonora	K-Ar	wr	22 ± 1	22	At base of Baucarit Formation
Basalt, central Sonora	K-Ar	wr	14	15	Within Baucarit Formation
Calcic latite, central Sonora	K-Ar		16, 14	19	
Rhyolite, central Sonora	K-Ar	kf	14	15	Overlies Baucarit Formation
Ignimbrite, central Sonora	K-Ar		13, 12	19	2 samples
Latite, southern Sonora	K-Ar	p	~10	23	Part of bimodal suite

*Mineral abbreviations: b = biotite; h = hornblende; kf= potassium feldspar; p = plagioclase; wm = white mica; wr = whole rock; zr = zircon.
†1 = Anderson and Silver, 1979; 2 = Anderson and Silver, 1977a; 3 = Anderson and Silver, 1981; 4 = Damon and others, 1962; 5 = Livingston and Damon, 1968; 6 = Anderson and others, 1979; 7 = Anderson and others, 1969; 8 = Roldán-Quintana, 1986; 9 = Silver and others, 1979; 10 = DeJong and others, 1988; 11 = Fries and Rincón-Orta, 1965; 12 = Roldán-Quintana, 1989; 13 = Mead and others, 1988; 14 = Stewsart and others, 1990; 15 = F. McDowell, 1991; 16 = Damon and others, 1983; 17 = Anderson and others, 1980; 18 = Aguirre and McDoweell, 1991; 19 = Bartolini and others, 1992; 20 = Montigny and others, 1987; 21 = Gastil and others, 1979; 22 = Roldán-Quintana, 1979; 23 = Morales and others 1990.

west extension began by 17 Ma in parts of coastal Sonora (Gastil and Krummenacher, 1977), 12 Ma in coastal northeastern Baja California (Dokka and Merriam, 1982; Stock and Hodges, 1989, 1990; Stock, 1991), and 15 to 13 Ma on Isla Tiburón (Neuhaus and others, 1988). Late Cenozoic block faulting in central Sonora is geometrically similar to late Miocene to Recent Basin and Range extension in the southwestern United States (Roldán-Quintana and González-León, 1979).

The least principal stress and extension direction around the periphery of the gulf changed from east-northeast–west-south-west to roughly northwest-southeast about 6 Ma, roughly coeval with the initiation or acceleration of transtensional opening of the modern gulf (Gastil and Krummenacher, 1977; Angelier and others, 1981; Henry, 1989). Since about 5 Ma, the surface of the eastern edge of the Baja continental block has been uplifted about 1 to 3 km, resulting in detachment and eastward translation of elevated continental crustal walls of the Gulf of California (Fenby and Gastil, 1991; Gastil and Fenby, 1991; Siem and Gastil, 1990, 1991). Pliocene to Holocene faulting appears to more active on the west side of the gulf, where it includes normal, dextral, and right-oblique faults with complex geometry and interaction (Stock, 1991; Mueller and Rockwell, 1991).

A paleomagnetic pole determined from the El Antimonio Formation in western Sonora overlaps two poles from coeval rocks of southwestern North America, including the correlative Luning Formation, at the 95% confidence level (Cohen and others, 1986). The clustering of the poles is significantly improved at the 95% confidence level by restoring 800 km of Late Jurassic sinistral displacement on the Mojave-Sonora Megashear, which separates the Serí terrane from North America (p. 79).

Paleomagnetic data suggest that the entire Baja California peninsula, including part of the Serí terrane and the Yuma and Cochimí terranes, was translated $10° ± 5°$ northward and rotated about 25° clockwise between mid-Cretaceous and early Miocene time (see Yuma terrane). However, no paleomagnetic data are available for metamorphosed rocks of the Serí terrane in northeastern Baja.

Tahué terrane

The Tahué terrane in western México is bounded on all sides by inferred contacts (Fig. 3). Its main components are mid-Paleozoic metasedimentary rocks of unknown origin that were accreted to North America by Late Jurassic time, and Jurassic to Cenozoic volcanic and plutonic rocks that formed within the magmatic arc along the western margin of North America. The Tahué terrane includes part of the Guerrero terrane of Campa-Uranga and Coney (1983) and Coney and Campa-Uranga (1987).

The oldest known rocks in the Tahué terrane are late Paleozoic chert, shale, micrite, quartzite, and overlying flysch in northern Sinaloa that have probable or possible correlatives throughout Sinaloa and southern Sonora (Fig. 17) (Carrillo-Martínez, 1971; Malpica, 1972). Fossils include Early Mississippian to Late Pennsylvanian fusulinids and mid-Pennsylvanian

to Early Permian conodonts (Carrillo-Martínez, 1971; Gastil and others, 1991). Probable correlatives include greenschist-facies metamorphic rocks in northern Sinaloa near El Fuerte derived from argillite, siliceous and andesitic flows and pyroclastic rocks, and rare limestone and chert (Mullan, 1978). Possible correlatives include greenschist- to amphibolite-facies metamorphic rocks in southern Sinaloa near Mazatlán, which were derived from flysch, conglomerate, thin limestone, and minor volcanic rocks (Henry and Frederikson, 1987) and which contain plant remains indicative of a maximum age of Carboniferous (R. Rodríguez, in Mullan, 1978).

The age and nature of basement to the Carboniferous rocks are problematic. Amphibolite-facies gneiss of probable Triassic age (Table 13) crops out near but not in contact with the probable Carboniferous rocks near El Fuerte (Fig. 17) (Mullan, 1978). The metasedimentary rocks near Mazatlán overlie and may have been deposited nonconformably on pre-Cretaceous quartz dioritic gneiss (Henry and Fredrikson, 1987). The tectonic significance and relation between the two gneiss outcrops are unknown. Low initial $^{87}Sr/^{86}Sr$ ratios in Cretaceous and Tertiary plutons (Table 13) imply that the Tahué terrane is not underlain by thick, old continental crust but instead by transitional crust, perhaps including gneissic rocks such as those near El Fuerte and Mazatlán (Fig. 17), or even by oceanic crust.

Sparsely distributed Late Jurassic to Early Cretaceous(?) magmatic arc rocks in northern and central Sinaloa probably were emplaced within an arc that developed on the western edge of North America. The carapace of the arc (Borahui Complex) consists of a thick (up to 8 km) sequence of metabasites derived from andesitic to basaltic flows, tuffs, and volcaniclastic rocks that locally unconformably overlie Carboniferous strata and metamorphosed correlatives (Mullan, 1978; Servais and others, 1982, 1986). Sparse outcrops of mafic and ultramafic plutonic rocks may represent the roots of the arc (Servais and others, 1982). Greenschist-facies metamorphism probably occurred in the Late Cretaceous (see below). The magmatic arc rocks are conformably overlain by and locally overthrust to the northeast or east-northeast by the Bacurato Formation, which consists of isoclinally folded greenschists derived from calcareous and tuffaceous sandstone, tuff, chert, and pelagic limestone grading upward to reefal limestones containing poorly preserved Aptian to Cenomanian fossils (Bonneau, 1969; Holguín-Q., 1978; Mullan, 1978; Servais and others, 1982, 1986).

The arc and its sedimentary cover are structurally overlain by thrust sheets consisting of weakly metamorphosed to unmetamorphosed dunite, serpentinite, pyroxenite, massive and layered gabbro, diabase dikes, pillow basalt, limestone, chert, and tuff that are interpreted as an ophiolitic assemblage (Ortega-Gutiérrez and others, 1979; Servais and others, 1982, 1986). The tectonic setting of the ophiolite may have been a backarc basin east of the Early Cretaceous magmatic arc (Fig. 17), as suggested by Ortega-Gutiérrez and others (1979), or a forearc basin west of the arc, as suggested by Servais and others (1982, 1986). The latter interpretation may be supported by limited structural data that imply

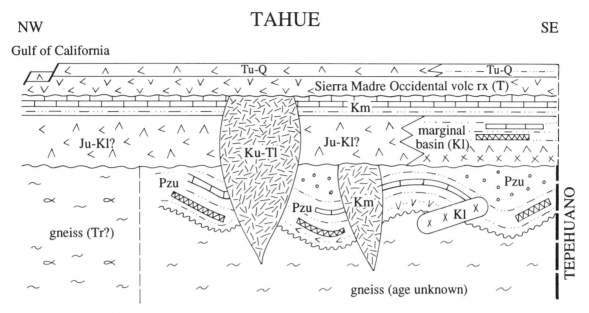

Figure 17. Schematic tectonostratigraphic section of Tahué terrane. Mesozoic and Cenozoic rocks are oriented west to east rather than northwest to southeast.

eastward tectonic transport (Servais and others, 1982). Contractional deformation and greenschist metamorphism of the Jurassic-Cretaceous(?) arc, thrusting of the partly conformable Bacurato Formation and the ophiolite assemblage eastward to northeastward onto the arc, and mylonitization along some thrusts are inferred to have occurred during Late Cretaceous to Eocene time (Servais and others, 1982, 1986).

Two classes of Late Cretaceous to Tertiary intrusive rocks have been identified in the Tahué terrane (Henry, 1975; Henry and Fredrikson, 1987). Weakly foliated and lineated tonalites and granodiorites older than about 85 Ma that crop out within 50 km of the coast in Sinaloa are interpreted as syntectonic rocks intruded during regional compression. Undeformed, more felsic granitoids younger than about 85 Ma that crop out throughout the terrane are interpreted as posttectonic intrusives. The ages of the granitoids decrease eastward (Henry, 1975). A similar range of Late Cretaceous to mid-Tertiary ages was reported from granitoids in southwestern Chihuahua near the Sinaloa border by Bagby (1979).

In southern Sinaloa and perhaps throughout the Tahué terrane, intrusion of Late Cretaceous–Paleogene granitic rocks was broadly synchronous with the accumulation of andesitic to rhyolitic tuffs, flows, volcaniclastic sedimentary rocks, and hypabyssal intrusives of the "lower volcanic complex" of McDowell and Keizer (1977) (Table 13) (Henry, 1975; Henry and Fredrikson, 1987). The Cretaceous to Eocene volcanic rocks of the lower volcanic complex, granitoids, and prebatholithic rocks are overlain unconformably by more than 1 km of mid-Tertiary silicic volcanic rocks of the "upper volcanic supergroup" of McDowell and Keizer (1977) and by mid-Tertiary basaltic andesites (Henry, 1975; McDowell and Clabaugh, 1979; Cam-

eron and others, 1989). Younger rocks include Miocene to Pliocene silicic to mafic dikes, domes, and volcanic rocks, Miocene to Quaternary alluvial and fluvial sedimentary rocks, and Quaternary basalt (Table 13).

On the west side of the Sierra Madre Occidental, fault geometry and stress orientation determined from fault-striae relations indicate that north-northwest–striking normal faults formed in response to east-northeast–west-southwest least principal stress by about 32 Ma (Henry, 1989; Henry and others, 1991). However, most faulting and tilting of Cenozoic strata in this region probably began after about 17 Ma. Total extension in southern Sinaloa is 20 to 50%, depending on the deep geometry of major normal faults. Many northwest-striking faults near the eastern margin of the Gulf of California are recently or currently active and probably accommodate transtensional Pacific–North America relative plate motion (Bryant and others, 1985; Henry, 1989; Stock and Hodges, 1989).

Paleomagnetic studies indicate that the upper volcanic sequence of the Sierra Madre Occidental probably has undergone negligible translation and rotation (Nairn and others, 1975; Hagstrum and others, 1987). Another study interpreted $10° ± 10°$ of northward translation and $25° ± 15°$ clockwise rotation of the lower volcanic sequence during the Late Cretaceous and Paleogene (Bobier and Robin, 1983), but this interpretation probably is invalid because the authors did not apply a structural correction to rocks that dip as much as 55° (C. Henry, personal communication, 1991).

Tarahumara terrane

The only outcrops of pre-Mesozoic rocks in the Tarahumara terrane are low-grade metasedimentary rocks near Boquillas,

TABLE 13. TAHUÉ TERRANE RADIOMETRIC DATA

Sample	System	Mineral*	Date (Ma)	References[†]	Comments
Pre-batholithic rocks					
Gneiss, El Fuerte	U-Pb	zr	Triassic	1	Upper intercept; unpublished data
Gabbros, Mazatlán	K-Ar	h	139 ± 3, 134 ± 3	2	Intrude Miss(?) or Jurassic(?) metasedimentary rocks
Amphibolite, southern Sinaloa	K-Ar	h	94 ± 1	3	Late reheating?
Cretaceous and Cenozoic magmatic rocks					
Quartz diorite (Recodo), southern Sinaloa	U-Pb	zr	102 ± 2	3	Syntectonic intrusion
Other quartz diorite, southern Sinaloa	K-Ar	h	98–87		6 samples; syntectonic
	K-Ar	b	99–61	3	7 samples
Granitoids, northern Sinaloa	K-Ar	h	93–90	3	3 samples; syntectonic
Granitic dike, northern Sinaloa	K-Ar	b	88 ± 2	4	Associated with Cu-W deposits;
	K-Ar	kf	77 ± 1	4	$^{87}Sr/^{86}Sr_i$: 0.7039
Quartz diorite, southern Sinaloa	K-Ar	h	83 ± 2	3	Posttectonic
	K-Ar	b	82 ± 1	3	
Granitic dike, northern Sinaloa	K-Ar	kf	78 ± 2	4	Associated with Cu-W deposits
Granodiorites, southern Sinaloa	K-Ar	h	75–73	3	3 samples;
	K-Ar	b	75–63	3	posttectonic
Granodiorite and quartz diorite, S Sinaloa	K-Ar	h	71–62	3	4 samples
	K-Ar	b	70–52	3	8 samples
Quartz monzonite, northern Sinaloa	K-Ar	h	68 ± 2	3	Post-tectonic
	K-Ar	b	66 ± 1	3	
Granodiorite, southern Sinaloa (San Ignacio)	U-Pb	zr	67 ± 1	3	
	K-Ar	h	65–63	3	6 samples
	K-Ar	b	65–52	3	8 samples
Rhyolite, southwestern Chihuahua	K-Ar	b	65 ± 1	5	
Mineralized breccia, N Sinaloa-NW Durango	K-Ar	wm	63 ± 1	4	Associated with Cu deposits; $^{87}Sr/^{86}Sr_i$: 0.7050
Granitoids, NC Sinaloa and SW Chihuahua	K-Ar	b	60–59	4	4 samples; associated with Cu; $^{87}Sr/^{86}Sr_i$: 0.7047-0.7050 (2 samples)
Granodiorite, southern Sinaloa	K-Ar	b	60 ± 1	4	Associated with Cu deposits;
	K-Ar	kf	53 ± 1	4	$^{87}Sr/^{86}Sr_i$: 0.7042
Granodiorite, central Sinaloa	K-Ar	b	59 ± 1	4	Associated with Cu deposits;
	K-Ar	h	58 ± 1	4	$^{87}Sr/^{86}Sr_i$: 0.7052
Granitoids, northern Sinaloa	K-Ar	b	58–51	6	3 samples
Granodiorite, northern Sinaloa	K-Ar	h	57 ± 1	4	Associated with Cu deposits;
	K-Ar	b	55 ± 1	4	$^{87}Sr/^{86}Sr_i$: 0.7063
Granitoids, northern Sinaloa	K-Ar	b	56–54	4	4 samples; associated with Cu; $^{87}Sr/^{86}Sr_i$: 0.7063-0.7048 (3 samples)
Granitoids, southern Sinaloa	K-Ar	h	56–50	3	5 samples
	K-Ar	b	55–47	3	9 samples
Quartz monzodiorite, southern Sonora	$^{40}Ar/^{39}Ar$	h	56 ± 1	7	Associated with W deposits
	K-Ar	h	53 ± 1	7	
	Rb-Sr		48 ± 1	7	6-pt isochron; $^{87}Sr/^{86}Sr_i$: 0.7057
	K-Ar	b	48 ± 1	7	
	$^{40}Ar/^{39}Ar$	b	47 ± 1	7	
Granodiorite, southern Sonora	K-Ar	b	54 ± 1	7	Associated with W deposits
Dacite, southwestern Chihuahua	K-Ar	b	52 ± 1	5	
Granodiorites, northern Sinaloa and southwestern Chihuahua	K-Ar	wm	52 ± 1	4	Hydrothermal Cu;
	K-Ar	wm	49 ± 1	4	$^{87}Sr/^{86}Sr_i$: 0.7036
Granodiorite, central Sinaloa	K-Ar		49	8	Source: Clark and others
Granodiorite, southwestern Chihuahua	K-Ar	b	48 ± 1	4	$^{87}Sr/^{86}Sr_i$: 0.7048
Granodiorite, southern Sinaloa (Candelero)	U-Pb	zr	48 ± 1	3	
	K-Ar	h	47–45	3	3 samples
	K-Ar	b	46–31	3	6 samples
Granitoids, southern Sinaloa	K-Ar	h	46, 44	3	
	K-Ar	b	48–34	3	6 samples
Granodiorite, western Durango	K-Ar	b	46 ± 1	9	
Mineralized breccia	K-Ar	wm	46 ± 1	4	Associated with Cu deposits
Andesite, western Durango	K-Ar	a, c, wr	44 ± 1	9	3 samples
Quartz diorite dike, southern Sinaloa	K-Ar	b	32 ± 1	3	

TABLE 13. TAHUÉ TERRANE RADIOMETRIC DATA (continued)

Sample	System	Mineral*	Date (Ma)	References†	Comments
Andesite dike, southern Sinaloa	K-Ar	h	30 ± 1	3	
Silicic tuffs and	K-Ar	b, kf, p	29–22	10	12 samples
flows, southern Sinaloa	K-Ar	b, p	28–17	3	5 samples
Andesite, northern Jalisco	K-Ar	p	24 ± 1	11	
Silicic tuff, northern Jalisco	K-Ar	b	21 ± 1	11	
Quartz diorite, southern Sinaloa	U-Pb	zr	20 ± 1	3	
(Colegio)	K-Ar	h	~19.5	3	
	K-Ar	b	~19	3	
Basalt, southern Sinaloa	K-Ar	wr	2 ± 1	12	
Basalt, northern Sinaloa	K-Ar		0.7	13	

*Mineral abbreviations: a = adularia; b = biotite; c = celadonite; h = hornblende; kf = potassium feldspar; p = plagioclase; wm = white mica; wr = whole rock; zr = zircon.

†1 = T. H. Anderson, personal communication; 2 = Henry and Fredrikson, 1987; 3 = Henry, 1975; 4 = Damon and others, 1983; 5 = Shafiqul-lah and others, 1983; 6 = Henry, 1974; 7 = Mead and others, 1988; 8 = Aguirre and McDowell, 1991; 9 = Loucks and others, 1988; 10 = McDowell and Keizer, 1977; 11 = Scheubel and others, 1988; 12 = Henry and Fredrikson, 1987; 13 = Clark, 1976.

northern Coahuila, that probably were metamorphosed in the late Paleozoic (Table 14) (Flawn and others, 1961). These rocks strongly resemble the Ouachita orogenic belt in the Marathon region of west Texas in terms of lithology, deformation style and intensity, and age (Flawn and Maxwell, 1958), and probably are the southern continuation of that belt. We interpret the Tarahumara terrane as deformed basinal sedimentary rocks that were obducted onto the North American shelf during the Pennsylvanian-Permian collision of North America and Gondwana; the Coahuiltecano terrane is inferred to be a remnant of Gondwana (Fig. 18).

Gravity studies indicate that a negative anomaly associated with the frontal zone of the Ouachita orogenic belt in west Texas continues about 200 km southward to south-southwestward into northeastern México (Handschy and others, 1987), and we have tentatively drawn the boundaries of the Tarahumara terrane near the boundaries of this anomaly. The Tarahumara–North America boundary corresponds very closely with the southeastern edge of buried Grenville lithosphere of North America as mapped using initial lead isotopic ratios of Tertiary igneous rocks in northern Chihuahua and western Texas (James and Henry, 1993). In terms of the findings of James and Henry, the Tarahumara terrane includes their central province and that part of the southeastern province not underlain by Phanerozoic continental crust accreted during the Ouachita orogeny. The isotopic data can be interpreted to indicate a minimum of 40 km of thrust displacement of the Tarahumara terrane northwestward onto the North America shelf.

Late Permian, Triassic, and Early Jurassic sedimentary rocks are absent from the Tarahumara terrane. Late Jurassic and Cretaceous siliciclastic sedimentary rocks, carbonates, and evaporites accumulated in the roughly north-south–trending Chihuahua Trough, overlapping the Tarahumara and Coahuiltecano terranes and North America (Navarro-G. and Tovar-R., 1974; Padilla y Sánchez, 1986). All Cretaceous and older rocks in the Tarahumara terrane were affected by the Laramide orogeny in the latest Cretaceous and Paleogene (Fig. 18) (Navarro-G. and Tovar-R., 1974; Padilla y Sánchez, 1986). Subduction-related calc-alkalic Cenozoic volcanic rocks in eastern Chihuahua, though not physically contiguous with the mid-Tertiary Sierra Madre Occidental province, probably are part of the province based on age and composition data (McDowell and Clabaugh, 1979). On average, they are more alkalic than rocks farther west in the province (Barker, 1979).

TABLE 14. TARAHUMARA TERRANE RADIOMETRIC DATA

Sample	System	Mineral*	Date (Ma)	References†	Comments
Greenschist, Sierra del Carmen,	K-Ar	wm	268 ± 5	1	
northern Coahuila	K-Ar	wm	271 ± 5	1	
	Rb-Sr	wr, wm	275 ± 20	1	Isochron age

*Mineral abbreviations: wm = white mica; wr = whole rock.
†1 = Denison and others, 1969

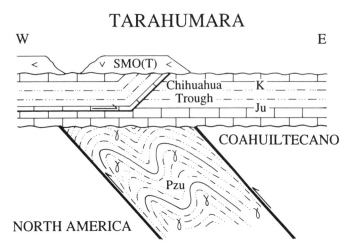

Figure 18. Schematic east-west structure section of Tarahumara terrane and adjacent terranes. SMO(T) indicates Tertiary volcanic rocks of Sierra Madre Occidental. Laramide deformation in Mesozoic rocks is schematic and simplified.

Tepehuano terrane

The pre–Late Jurassic geology of the Tepehuano terrane is poorly understood due to widespread Late Jurassic, Cretaceous, and Cenozoic cover. Isotopic data from xenoliths in Quaternary volcanic rocks imply that the terrane is partly underlain by Proterozoic continental crust. Sparse, widely separated outcrops beneath younger cover strata are deformed, poorly dated metaigneous and metasedimentary rocks that appear to reflect the development of part of the Jurassic Cordilleran magmatic arc, one or more subduction complexes(?) that include fragments of Paleozoic rocks, and one or more backarc basins, including mafic crust and siliciclastic fill. The internal structure of the Tepehuano terrane is simplified in Figure 19 because geologic and tectonic relations among many rock units still are not understood, but large Jurassic and Early Cretaceous horizontal contractions in all units probably are related to attachment of the terrane to the continent. The Tepehuano terrane includes the Parral and Sombrerete terranes and parts of the Cortes, Guerrero, and Sierra Madre terranes of Coney and Campa-Uranga (1987). The terrane roughly corresponds to the Mesa Central or Altiplano geomorphologic province.

Proterozoic xenoliths. Sm-Nd studies of granulite gneiss xenoliths from Quaternary volcanic rocks indicate that the Tepehuano terrane is underlain by Proterozoic continental crust (Fig. 19; Table 15). Samples with 1,720- to 1,520-Ma model ages probably represent mixing of roughly 10 to 30% recycled older crust and 70 to 90% newly derived mantle material of Grenville age (~1,000 Ma); samples with 1,050- to 660-Ma model ages probably represent mixing during Phanerozoic orogenesis of uncertain age (Ruiz and others, 1988b). Preliminary $^{207}Pb/^{206}Pb$ dates (Table 15), Pb isotopic ratios, and the geographic provinciality of fundamental differences in elemental abundances, particularly of

Sr and Nd, suggest that the lower crust comprises a heterogeneous assemblage of Proterozoic, Paleozoic, and perhaps Mesozoic rocks (J. Luhr, personal communication, 1992).

Pre-Oxfordian outcrops. Several sparsely distributed pre–Late Jurassic units crop out in northernmost Zacatacas (Córdoba, 1965; Anderson and others, 1991), southernmost Coahuila (Ledezma-Guerrero, 1967; Mayer-Pérez Rul, 1967), and eastern Durango (Pantoja-Alor, 1963). Structurally lowest (Fig. 19) is the Taray Formation, a strongly deformed assemblage of weakly metamorphosed deep-water flysch and olistostromal mélange containing blocks of chert, volcanic rocks, limestone with Paleozoic fossils, silicified serpentinite, and enigmatic dolomite-magnetite rock, which has been interpreted as a subduction complex (Ortega-Gutiérrez, 1984b; Anderson and others, 1990; Klein and others, 1990). Axial surfaces and bedding homoclines strike northeast-southwest, and matrix foliation dips southeast (Anderson and others, 1990). The depositional age of the Taray Formation must be younger than the Paleozoic age of some enclosed blocks, but the ages of most enclosed blocks and the age and cause of penetrative deformation are unknown (T. H. Anderson, personal communication, 1990).

The Taray Formation is overlain structurally or locally unconformably(?) by as much as 3 km of weakly metamorphosed, strongly deformed calc-alkalic volcanic and volcaniclastic rocks of the Rodeo Formation and Nazas Formation. Earlier workers interpreted partly nonmarine air-fall tuff, tuffaceous siltstone, and pyroclastic-flow tuff of the Nazas Formation to be younger than underlying feldspar porphyry and volcaniclastic rocks of the Rodeo Formation (e.g., Pantoja-Alor, 1963; Córdoba, 1965), but new data imply that the two may interfinger laterally as contemporaneous distal and proximal components of a Mesozoic volcanic arc (Jones and others, 1990). Deformed quartz porphyry of the Caopas Schist, which abuts and may be gradational with volcanic rocks of the Rodeo Formation, may have been a cogenetic, subvolcanic pluton in that arc (Jones and others, 1990; Anderson and others, 1991). The Nazas Formation probably includes volcanic and volcaniclastic rocks in northern San Luis Potosí that contain palmate fern fronds of probable Early or Middle Jurassic age (Maher and others, 1991). The Caopas Schist and Rodeo Formation have yielded Late Triassic and Jurassic radiometric dates, whereas the Nazas Formation yielded a single Middle Triassic date (Table 15). Most workers reconciled the inferred younger stratigraphic position of the Nazas Formation with available radiometric dates by rejecting dates from the Caopas Schist and Rodeo Formation, and by correlating the Nazas Formation with nonmarine strata of the Upper Triassic Huizachal Formation in the Guachichil terrane (Pantoja-Alor, 1963; López-Ramos, 1985; Salvador, 1987; Mitre-Salazar and others, 1991; Z. de Cserna, personal communication, 1990). However, in view of the obvious dissimilarities between the calc-alkalic volcanogenic rocks of the Nazas Formation and the dominantly nonmarine to shallow marine clastic rocks of the Huizachal Formation, it seems appropriate to discontinue the proposed correlation of the two units. We consider it more likely that the single radiometric date

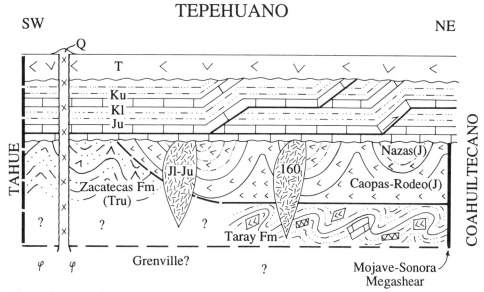

Figure 19. Composite tectonostratigraphic section and northeast-southwest structure section of Tepehuano terrane. Laramide deformation in Mesozoic rocks is depicted schematically. Relation of Zacatecas and Taray Formations is uncertain.

from the Nazas Formation is suspect, and that the Caopas, Rodeo, and Nazas Formations are parts of a Late Triassic to Jurassic volcanic arc (Anderson and others, 1991), hereafter termed the CRN arc, that was built on continental crust (Fig. 19). Jurassic plutonic rocks also crop out 400 km to the west-northwest in southern Chihuahua (Table 15).

The CRN arc was penetratively deformed (northwest-striking foliation and axial surfaces) and juxtaposed with the underlying Taray Formation prior to intrusion of a 160-Ma quartz porphyry (Table 15) (Anderson and others, 1990). The final stage of deformation may have been coeval with the deposition of Oxfordian limestone (Anderson and others, 1991). Laramide deformation of this region rooted into a detachment within the Oxfordian strata and did not affect the Taray Formation and CRN arc (Fig. 19) (Anderson and others, 1990, 1991).

In central and southeastern Zacatecas and western San Luis Potosí, the lower Carnian (Upper Triassic) Zacatecas Formation consists of isolated outcrops of strongly foliated, isoclinally folded marine sandstone, shale, conglomerate, and limestone that were metamorphosed under greenschist-facies conditions; intercalated greenstones are interpreted to derive from either interbedded volcanic rocks or younger intrusives (de Cserna, 1976; Mejía-Dautt and others, 1980; Ranson and others, 1982; Cuevas-Pérez, 1983; López-Ramos, 1985, p. 412–415; Servais and others, 1986). Adjacent outcrops of less-deformed pillow lavas, graywacke and shale, chert, tuff, and limestone previously were grouped with the Triassic Zacatecas Formation and interpreted as accreted oceanic crust (de Cserna, 1976). Later workers discovered Late Jurassic ammonites in the clastic rocks and pre-Valanginian radiolarians in the chert, and now these rocks generally are termed the Chilitos Formation or subdivided into

the Valdecañas, Plateros, and Chilitos Formations, and are inferred to overlie the Zacatecas Formation along an obscure contact (Cuevas-Pérez, 1983, 1985; Macdonald and others, 1986; Servais and others, 1986). The marine Zacatecas Formation may have been deposited on continental basement and graded eastward into, or been coeval with, nonmarine strata of the Huizachal Formation in the Guachichil terrane, but there is no field evidence for such an inference. Apparently, the Zacatecas Formation underwent an episode of penetrative deformation and metamorphism in the Early to Middle Jurassic, probably in a tectonically active environment that was dissimilar to the tectonic setting of both the coeval Huizachal Formation and the younger Chilitos Formation. We tentatively interpret the Zacatecas Formation as the upper levels of oceanic crust and overlying sedimentary cover that were deformed and metamorphosed during Jurassic accretion to North America, and suggest that they may be grossly correlative, in terms of depositional setting and tectonic environment, with the Taray Formation in northern Zacatecas (Fig. 19). The overlying Late Jurassic rocks probably were deposited in a Late Jurassic to Early Cretaceous backarc basin that may have continued south into Guanajuato.

Controversial, structurally complex metamorphic rocks of probable late Paleozoic to Jurassic age crop out in the vicinity of Santa María del Oro, northern Durango (Fig. 2; not shown in Fig. 19). Some workers recognize three units that they consider part of the upper Paleozoic Gran Tesoro Formation (Zaldívar-Ruiz and Garduño-M., 1984; Córdoba and Silva-Mora, 1989), whereas other workers recognize at least two fault-bounded units of pre–Late Jurassic age (Pacheco-G. and others, 1984; Aranda-García and others, 1988). The structurally lowest unit is muscovite-rich schist that contains greenschist- and amphibolite-

R. L. Sedlock and Others

TABLE 15. TEPEHUANO TERRANE RADIOMETRIC DATA

Sample	System	Mineral*	Date (Ma)	References[†]	Comments
Pre-Cretaceous rocks					
Lower crustal xenoliths	Sm-Nd		1,720–1,520	1	Model ages
Lower crustal xenoliths	Sm-Nd		1,050–660	1	Model ages
Lower crustal xenolith	$^{207}Pb/^{206}Pb$		1,100	2	Source: S. Bowring and others
Lower crustal xenoliths	$^{207}Pb/^{206}Pb$		1,071, 485,	3	
			39, 26	3	
Muscovite-amphibole schist,	K-Ar		326 ± 26	4	Metamorphic age
Santa María del Oro, Durango	K-Ar		311	5	Source: P. Damon
Nazas Formation			230 ± 20	6	
Metarhyolite, northern Zacatecas	Rb-Sr	wr	220 ± 30	7	Caopas (Rodeo?) Formation
Metarhyolite, northern Zacatecas	Rb-Sr	wr	200 ± 60	7	Caopas (Rodeo?) Formation
Volcanic-metamorphic rocks in well, SW Coahuila	Rb-Sr		199	8	
Quartz monzonite, southern Chihula	K-Ar	h	198 ± 7	9	
Metarhyolite, northern Zacatecas	Rb-Sr	wr	195 ± 20	7	Caopas (Roseo?) Formation
Metased rocks (Nazas?) in well in N Durango	K-Ar		168	8	
Quartz porphyry	U-Pb	zr	757 ±34;	10	Discordia intercepts;
			158 ± 4		supersedes 165-Ma date in Jones and others, 1990; intrudes Taray, Nazas
Tonalite, Guanajuato	K-Ar	wr	157 ± 9	11	
Metarhyolite, northern Zacatecas	Rb-Sr	wr	156 ± 40	7	Caopas (Roseo?) Formation
Quartz diorite, southern Chihuahua	K-Ar	h	155 ± 3	9	
Diorite, southern Chihuahua	K-Ar	h	149 ± 3	9	
Metarhyolite, northern Zacatecas	Rb-Sr	wr	141 ± 40	12	Caopas (Roseo?) Formation
Cretaceous-Cenozoic magmatic rocks					
Diorite, Guanajuato	K-Ar	h	122 ± 6	11	
Basalt, Guanajuato	K-Ar	wr	108 ± 6	11	
Granite stock, northern Zacatecas	K-Ar		90 ± 2	13	Source: P. Damon;
	$^{40}Ar/^{39}Ar$	b	38 ± 2	13	associated with F deposit
Diorite stock, east-central Durango	K-Ar	h	87 ± 2	14	
Andesite dikes, northern Durango	K-Ar		80, 70	4	
Diorite, northern Zacatecas	K-Ar	wm	75 ± 2,	15	2 samples
			74 ± 2		
Greenschist, northern Zacatecas	K-Ar	wm	75 ± 2	15	2 samples;
			73 ± 2		Metamorphic ages
Dacite stock, west-central Zacatecas	K-Ar	kf	54 ± 1	16	$^{87}Sr/^{86}Sr_i$: 0.7052
Andesite, central Durango	K-Ar	p	53 ± 1	17	
Silicic tuff, east-central Durango	K-Ar	b, kf	52 ± 2	14	2 samples
Silicic tuff, northern Durango	K-Ar		51	18	In Aguirre and McDowell, 1991
Andesite, east-central Durango	K-Ar	p	49 ± 3	14	
Silicic pyroclastic rocks, SC Zacatecas	K-Ar	b	47	19	Source: P. Damon
Granodiorite, east-central Durango	K-Ar	h	47 ± 1	18	In Aguirre and McDowell, 1991
Granodiorite, west-central Zacatecas	K-Ar	b	46 ± 1	16	Associated with Cu deposits
Andesite domes, east-central Durango	K-Ar	b, h	45 ± 1	14	2 samples
Silicic tuffs, east-central Durango	K-Ar	p, kf, b	43 ± 2	14	6 samples
Rhyolitic tuff, northern Durango	K-Ar	wr	43 ± 1	13	Associated with Hg deposits
	$^{40}Ar/^{39}Ar$	b	40 ± 1		
Silicic tuff, northern Durango	K-Ar		42	18	In Aguirre and McDowell, 1991
Granodiorite, northern Zacatecas	K-Ar	b	42 ± 1	7	
Rhyolite, southern Chihuahua	K-Ar	kf	42 ± 1	20	
Andesite, east-central Durango	K-Ar	p	41 ± 2	14	2 samples
Silicic tuff, south-central Zacatecas	K-Ar	kf	37	19	Source: P. Damon
Volcanic rocks, skarn	K-Ar	wr, wm	37–28	20	3 samples
Silicic tuff and lava, northern Zacatecas	K-Ar	wr, p, kf	38–28	13	3 samples; associated with Hg deposits
Rhyolite stock, east-central Durango	K-Ar	h	36 ± 1	18	
Rhyolitic tuff, northern Durango	K-Ar	wr	36 ± 1	13	Associated with Hg deposits
Granodiorite, San Luis Potosí	$^{40}Ar/^{39}Ar$	b	36 ± 1	13	Associated with Sb?
Rhyolite dome, east-central Durango	K-Ar	p	34 ± 1	14	
Quartz latite, east-central Durango	K-Ar	b	34 ± 1	21	Source: F. Felder
Rhyolite dikes, east-central Durango	K-Ar	b	34 ± 2	21	Source F. Felder
Andesite, southern Chihuahua	K-Ar	wr, kf	33, 31	22	

TABLE 15. TEPEHUANO TERRANE RADIOMETRIC DATA (continued)

Sample	System	Mineral*	Date (Ma)	References[†]	Comments
Silicic tuffs and lavas, central Durango	K-Ar	b, kf, p, wr	33–30	17	23 samples
Rhyolitic tuff, eastern Durango	K-Ar	wr, kf	31–30	13	2 samples; associated with Sn
Rhyolite, San Luis Potosí	K-Ar	wr, kf	31–27	13	4 samples; associated with F
Rhyolite, central Durango	K-Ar	wr	31 ± 1	13	
Granite stock, east-central Durango	K-Ar	b	30 ± 1	18	
Silicic dome/tuff, east-central Durango	K-Ar	p, kf	30 ± 2	14	8 samples
Silicic tuff, northern Zacatecas	K-Ar	wr, p, kf	30–24	13	3 samples; associated with Sn
	$^{40}Ar/^{39}Ar$	b	32 ± 1		
Silicic tuff and lava, central Zacatecas	K-Ar	wr, p, kf	30–21	13	4 samples; associated with Hg deposits
	$^{40}Ar/^{39}Ar$	wr	26	13	
Granite stock, northern San Luis Potosí	$^{40}Ar/^{39}Ar$	b, wm	29–28	13	3 samples; associated with Sn
Rhyolite, southern Chihuahua	K-Ar	wr	25	22	
Alkalic basalts, southwestern Chihuahua	K-Ar		24, 21	23	Contemporaneous with Basin and Range faulting
Alkalic mafic lavas, east-central Durango	K-Ar		22 ± 2	14	4 samples
Alkalic basalts, San Luis Potosí–Zacatecas	K-Ar		14–11	24	3 samples
Mafic volcanic rocks and dikes, C Durango	K-Ar	p, h	13–12	17	

*Mineral abbreviations: b = biotite; h = hornblende; kf = potassium feldspar; p = plagioclase; wm = white mica; wr = whole rock; zr = zircon.
[†]1 = Ruiz and others, 1988b; 2 = Ruiz and others, 1990; 3 = J. Luhr and others, unpublished data; 4 = Araujo and Arenas, 1986; 5 = Pacheco and others, 1984; 6 = Blickwede, 1981; 7 = Fries and Rincón-Orta, 1965; 8 = Wilson, 1990; 9 = Damon and others, 1981; 10 = T. H. Anderson, unpublished data, personal communication, 1990; 11 = Monod and others, 1990; 12 = Denison and others, 1969; 13 = Tuta and others, 1988; 14 = Aguirre and McDowell, 1991; 15 = Ranson and others, 1982; 16 = Damon and others, 1983; 17 = McDowell and Keizer, 1977; 18 = Clark and others, 1980; 19 = Ponce and Clark, 1988; 20 = Shafiqullah and others, 1983; 21 = Gilmer and others, 1988; 22 = Grant and Ruiz, 1988; 23 = Cameron and others, 1980; 24 = Luhr and others, 1991.

facies assemblages (chlorite, garnet, cordierite), displays a penetrative northwest-striking foliation, is locally blastomylonitic, and has yielded late Paleozoic K-Ar dates (Table 15). A structurally intermediate unit consists, according to conflicting reports, of either crenulated phyllite or metaandesite. The structurally highest unit consists of strongly deformed black slate containing early Mesozoic, probably Early Jurassic, palynomorphs and blocks of quartzite, Carboniferous limestone, tuff, and other sedimentary rocks. Isolated outcrops of basaltic to andesitic pillow lavas may be interbeds in the slate, olistoliths within the slate, or fault-bounded units unrelated to the Gran Tesoro Formation. Zaldívar-Ruiz and Garduño-M. (1984) and Córdoba and Silva-Mora (1989) considered all these rocks to be part of the Gran Tesoro Formation, which they inferred to be upper Paleozoic despite the presence of early Mesozoic fossils in the slate. Pacheco-G. and others (1984) and Aranda-García and others (1988) restricted use of the term Gran Tesoro Formation to the upper and perhaps middle units, which they considered a tectonized olistostrome, and recognized a fault between these units and the more strongly deformed and metamorphosed late Paleozoic schist. All units are intruded by quartz diorites that yielded Late Jurassic K-Ar dates in southern Chihuahua (Table 15), are overlain with uncertain relation by the Nazas Formation, and are overlain unconformably by Late Jurassic to Early Cretaceous marine strata correlative with rocks in eastern México (Aranda-García and others, 1987; Contreras y Montero and others, 1988).

A variety of Mesozoic rocks crop out near the cities of Guanajuato and León, Guanajuato (Fig. 2). The structurally

lowest rocks are strongly deformed greenschists of the Arperos Formation (also known in part or whole as Esperanza Formation or El Maguey Formation), which were derived from volcaniclastic flysch, tuff, chert, and limestone and shale that have been assigned probable Triassic or Jurassic ages and tentatively correlated with the Triassic Zacatecas Formation in the northern part of the terrane (e.g., Servais and others, 1982; de Cserna, 1989). However, recently reported radiolarians of probable Early Cretaceous age (Dávila-Alcocer and Martínez-Reyes, 1987) indicate that the entire assemblage probably is post-Jurassic. In adjacent Jalisco, 25 km north of León, the Arperos Formation is overlain unconformably by Albian carbonates including oolitic grainstone, reefstone, marly limestone, calcareous sandstone, and marl (Chiodi and others, 1988) interpreted as the western continuation of platform carbonates of the Gulf of Mexico sequence in eastern México (see below).

South and east of León, the Arperos Formation is structurally overlain by faulted slabs of mafic and ultramafic igneous rocks interpreted by Servais and others (1982) as an ophiolitic complex and by Monod and others (1990) as the disrupted remains of a once-complete oceanic island arc. Constituent rock types include massive and pillow basalts, at least partly of Early Cretaceous age; a diabase dike complex; distinct sheets of Early Cretaceous diorite and Jurassic(?) tonalite, each intruded by mafic dikes; partly serpentinized massive and layered gabbro; and serpentinized harzburgite, wehrlite, and pyroxenite (Table 15). The mafic and ultramafic rocks and the underlying Arperos Formation are strongly foliated and locally lineated, and contain kinematic indicators that record tectonic transport to the north to north-northeast. These rocks are inferred to be oceanic island arc tholeiites on the basis of trace element data (Ortiz and others, 1991; Lapierre and others, 1992), but we consider the island arc interpretation unlikely because the trace element data also show affinities with N-MORB (e.g., Sun, 1980) and because, in terms of lithology and structure, the igneous assemblage closely resembles Mesozoic ophiolites scattered throughout western México. We follow Servais and others (1982) in interpreting the mafic and ultramafic rocks as Late Jurassic(?)–Early Cretaceous oceanic crust of a backarc basin, presumably the same basin inferred to the north in Zacatecas (see above), that was shortened and thrusted north-northeastward onto volcaniclastic backarc basin fill (Arperos Formation) prior to deposition of Albian platform limestone. The thrust sheet of Jurassic(?) tonalite presumably was derived from the Jurassic magmatic arc in southwestern México, the original extent and present disposition of which are not well known.

In west-central Querétaro, strongly deformed, low-grade metasedimentary rocks of the Chilar Formation are lithologically and structurally similar to the Zacatecas Formation 300 km to the northwest. At least one episode of strong pre-Laramide deformation is documented. These rocks were provisionally inferred to be Paleozoic(?) or Triassic(?) by López-Ramos (1985, p. 443–446) and Late Jurassic(?) by Coney and Campa-Uranga (1987), who assigned them to the Toliman terrane. No direct radiometric or paleontologic data are available concerning the ages of protolith and metamorphism.

Late Jurassic to Recent rocks. Oxfordian to early Kimmeridgian carbonate and fine-grained clastic rocks that crop out throughout the northern Tepehuano terrane contain megafossils of both Tethyan and Boreal affinity (Buitron, 1984; Aranda-García and others, 1987; Salvador, 1987; Contreras y Montero and others, 1988). These strata are laterally continuous with coeval strata in the Guachichil and Coahuiltecano terranes; local formation names include Zuloaga, Novillo, Santiago, and La Gloria (Imlay, 1936; 1938; Heim, 1940; Cantú-Chapa, 1979). Coeval, correlative strata in the U.S. Gulf Coast region include the Smackover Formation (e.g., Imlay, 1943). Overlap by these strata indicates that the Tepehuano, Guachichil, and Coahuiltecano terranes were amalgamated within the continental platform of México by Late Jurassic time (Mitre-Salazar and others, 1991). Later fault displacements (Longoria, 1988) have been of insufficient magnitude to obscure regional facies patterns.

Late Jurassic to Late Cretaceous platform strata in the Tepehuano terrane are dominated by carbonates and local evaporites, but also include interbedded shale and flysch derived chiefly from the magmatic arc to the west (Fig. 19), with a marked increase in the abundance of clastic rocks in the late Cenomanian (Tardy and Maury, 1973; de Cserna, 1976; Cuevas-Pérez, 1983; Mitre-Salazar and others, 1991). Cretaceous platform limestone and shale crop out in the western part of the terrane (Chiodi and others, 1988).

Cretaceous and older rocks are intruded by widely distrib-

uted Paleogene granitoids (Table 15) and intermediate hypabyssal rocks (INEGI, 1980; Damon and others, 1983; López-Ramos, 1985, p. 426–427).

Rocks deformed during Laramide shortening are unconformably overlain by sparse outcrops of Paleogene nonmarine conglomerate and sandstone that probably were deposited in alluvial fans at the feet of fault blocks elevated during regional extension (Aranda-García and others, 1989). In northwestern Guanajuato, vertebrate fossils are late Eocene to early Oligocene, subordinate interbeds of basalt and andesite are present, and clasts were derived chiefly from silicic to intermediate volcanic rocks as well as from granitoids (increasing upsection) and Early Cretaceous limestone (Edwards, 1955). On the outskirts of Zacatecas city, conglomerate derived from mixed igneous and sedimentary sources unconformably overlies the Zacatecas Formation and Chilitos Formation (Edwards, 1955) and is overlain conformably by ignimbrite from an Oligocene caldera (Ponce and Clark, 1988). Near Durango city, conglomerate is derived chiefly from carbonates and clastic rocks, and is overlain by andesite that yielded a K-Ar date of ~52 Ma (Table 15), implying a Paleocene–early Eocene age for the conglomerate (Córdoba, 1988).

Tertiary magmatic rocks (Table 15) include Paleogene granitoids, Paleogene andesitic lava, tuff, and hypabyssal intrusives of the "lower volcanic complex" of McDowell and Keizer (1977), and mid-Tertiary silicic ignimbrites and basaltic andesites of the "upper volcanic supergroup" of McDowell and Keizer (1977) (Swanson and others, 1978; Córdoba, 1988; Cameron and others, 1989; McDowell and others, 1990; Wark and others, 1990). Late Cenozoic rocks include Miocene-Pliocene alkali basalts and basanites that were erupted during Basin and Range extension (Table 15), Neogene and Quaternary fluvial sedimentary rocks, and Quaternary basalts containing crustal and mantle xenoliths (Luhr and others, 1990; Heinrich and Besch, 1992).

Structural and geophysical data. The Tepehuano terrane underwent an unresolved amount of north- to northeast-vergent Laramide shortening in the latest Cretaceous and Paleogene (Pantoja-Alor, 1963; Ledezma-Guerrero, 1967, 1981; Enciso de la Vega, 1968; INEGI, 1980). Laramide deformation affected rocks at least as far south as 24°N (Ledezma-Guerrero, 1981) and at least as far west as 104°30′W, where pre-Tertiary rocks are covered by Tertiary volcanics of the Sierra Madre Occidental (Pantoja-Alor, 1963).

Regional stresses changed from east-northeast–west-southwest compression to east-northeast–west-southwest tension about 30 Ma, initiating block tilting, diking, and veining. High-angle normal faulting generally ascribed to Basin and Range deformation and associated basaltic volcanism were underway by 24 Ma and continue at present (Table 15) (Aguirre-Díaz and McDowell, 1988; Grant and Ruiz, 1988; Córdoba and Silva-Mora, 1989; Henry and others, 1991; Henry and Aranda-Gómez, 1992). On the basis of structural analysis of faults, fractures, and lineaments, Aranda-Gómez and others (1989) concluded that the southern part of the Mesa Central of México (i.e., southern Tepehuano terrane) experienced distinct episodes of northeast-southwest and northwest-southeast tension during the Cenozoic. The relative ages of these episodes are not yet understood.

A paleomagnetic pole determined from the Nazas Formation south of Torreón, Coahuila overlaps two poles from coeval rocks of southwestern North America at the 95% confidence level (Nairn, 1976; Cohen and others, 1986). The clustering of the poles is significantly improved at the 95% confidence level by restoring 800 km of Late Jurassic sinistral displacement on the Mojave-Sonora Megashear, which separates the Tepehuano terrane from North America (p. 79).

Yuma composite terrane

The Yuma composite terrane of Baja California (and southern California) consists of a Jurassic-Cretaceous volcanic arc subterrane to the west and a Triassic-Jurassic basinal subterrane to the east (Fig. 20). The subterranes were amalgamated during the Early Cretaceous and intruded by the Peninsular Ranges batholith during the Early and Late Cretaceous. The Yuma composite terrane corresponds to the Alisitos terrane of Campa-Uranga and Coney (1983) and to the Santa Ana terrane of southern California (Howell and others, 1985; Coney and Campa-Uranga, 1987).

The western (arc) subterrane. The western (arc) subterrane is known by different names on opposite sides of the east-west Agua Blanca fault near latitude 31°N (Fig. 3). To the north, volcanic rocks between the Agua Blanca fault and southern California generally have been correlated with the Upper Jurassic–Lower Cretaceous Santiago Peak Formation of southern California (Larsen, 1948; Hawkins, 1970). To the south, the western subterrane is known as the Alisitos Formation, which is

at least partly of Early Cretaceous age. Recent studies seem to show that a perceived age difference between the Santiago Peak and Alisitos Formations is imaginary, that the Agua Blanca fault does not mark a major division of the arc subterrane, and that the two formations are correlative parts of the same Late Jurassic to chiefly Early Cretaceous volcanic arc that was the extrusive equivalent of the older part of the Peninsular Ranges batholith (see below). In the following paragraphs, we summarize characteristics of the Santiago Peak and Alisitos Formations and describe other volcanic rocks from the western (arc) subterrane that are not clearly part of the Santiago Peak–Alisitos association.

The Santiago Peak Formation consists predominantly of calc-alkalic and tholeiitic, andesitic to rhyolitic massive flows, tuff, agglomerate, and breccia, with minor amounts of basalt and volcaniclastic rocks (Larsen, 1948; Adams, 1979; Balch and others, 1984; Buesch, 1984). On the basis of Tithonian bivalves, belemnites, radiolarians, and trace fossils in the volcaniclastic rocks (Fife and others, 1967; Jones and others, 1983), the Santiago Peak Formation is widely inferred to be Upper Jurassic. However, recent U-Pb zircon studies of massive flows and breccias have yielded Early Cretaceous dates (Table 16), implying that the more abundant volcanic rocks in the Santiago Peak Formation are Early Cretaceous and thus coeval with the older part of the Peninsular Ranges batholith (Kimbrough and others, 1990; Herzig and Kimbrough, 1991). A 122-Ma date from several granite clasts is a maximum age for part of the volcanic breccia in the Santiago Peak Formation (D. Kimbrough, personal communication, 1990).

The Alisitos Formation is a thick (>6 km) sequence of calc-alkalic rhyolitic to andesitic flows, pyroclastic rocks, and breccias with thin interbeds of volcaniclastic sandstone and shale

YUMA

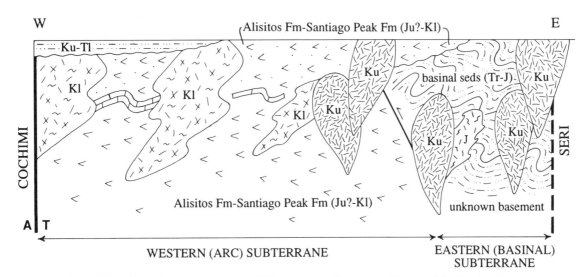

Figure 20. Schematic structure section of Yuma composite terrane. Plutons of Peninsular Ranges batholith locally intrude boundary between Yuma and Serí terranes and reverse fault boundary between eastern (basinal) and western (arc) subterranes. Plutons are younger to east and display eastward increases in $^{87}Sr/^{86}Sr_i$, $\delta^{18}O$, and REE fractionation.

TABLE 16. YUMA TERRANE RADIOMETRIC DATA

Sample	System	Mineral*	Date (Ma)	References[†]	Comments
Volcanic arc (western) subterrane					
Andesite beneath Alisitos Formation at Arroyo Calamajué	U-Pb	zr	155	1	Unpublished data
Santiago Peak Formation	$^{207}Pb/^{206}Pb$	zr	130 ± 5	2	
Santiago Peak Formation	U-Pb	zr	120	2	
Santiago Peak Formation	U-Pb	zr	114	2	
Santiago Peak Formation	U-Pb	zr	107	2	
Siliceous volcanic rocks (Alisitos Formation?) at Arroyo Calamajué	Rb-Sr	wr	103 ± 4	3, 1	
	U-Pb	zr	~122		
Greenschist-facies andesite near Loreto	K-Ar	h	92 ± 2	1	Minimum age
Peninsular Ranges Batholith					
S-type granitoids in eastern subterrane	U-Pb	zr	1,600; 156 ± 12	4	Discordia intercepts
	Rb-Sr	wr	168 ± 12	4	5 samples
Tonalite near Loreto	K-Ar	h	144 ± 9	5	
Western part of batholith	U-Pb	zr	140–105	6, 7	Unpublished data
Eastern part of batholith	U-Pb	zr	105–80	6	
Gabbro	K-Ar	h	126 ± 4	5	
Hypabyssal volcanic rock	U-Pb	zr	127 ± 5	5	Source: L. Silver
Granodiorite near El Arco	K-Ar	b	117 ± 4	8	
	K-Ar	h	110 ± 3	8	
Granodiorite near El Arco	K-Ar	h	115 ± 3	8	
	K-Ar	b	111 ± 3	8	
Copper porphyry	K-Ar	kf	107 ± 3	8	
Granitoids near El Arco	K-Ar	wr	107–99	9	3 samples
Quartz monzonite near 27°30'N	K-Ar		93 ± 2	10	
Granodiorite near Loreto	K-Ar	b	87 ± 2	11	
Upper Cenozoic volcanic rocks					
Rhyolite tuff, Baja California Sur	K-Ar	b, kf, wr	28–23	12, 13	7 samples
Basalt near 25°N	K-Ar	p	28 ± 1	12	
Volcanic rocks, Baja California	K-Ar	p, wr, h, kf, b	25–10	14, 12 13, 15	78 samples; mainly calc-alkalic
Volcanic rocks, Baja California	K-Ar	wr, p, kf	12–1	12, 13, 15	30 samples; alkaline and tholeiitic

*Mineral abbreviations: b = biotite; h = hornblende; kf = potassium feldspar; p = plagioclase; wr = whole rock; zr = zircon.
[†]1 = Gastil and others, 1991; 2 = Kimbrough and others, 1990; 3 = Griffith, 1987; 4 = Todd and others, 1991; 5 = Gastil and others, 1978; 6 = Silver and others, 1979; 7 = Silver and Chappell, 1988; 8 = Barthelmy, 1979; 9 = Damon and others, 1983; 10 = Schmidt, 1975; 11 = McLean, 1988; 12 = Gastil and others, 1979; 13 = Hausback, 1984; 14 = Gastil and Krummenacher, 1977; 15 = Sawlan and Smith, 1984.

and reef limestone (Santillán and Barrera, 1930; Allison, 1955; Gastil and others, 1975, 1981; Almazán-Vázquez, 1988a, b). Some limestones contain Aptian-Albian fauna, but much of the formation is unfossiliferous and of unknown age. The Alisitos Formation crops out as far south as latitude 28°N (Rangin, 1978; Barthelmy, 1979) and is known from the subsurface beneath the eastern margin of the Vizcaíno desert (~27°50'N); similar rocks of unknown age crop out near Loreto (McLean, 1988). The Alisitos Formation probably is correlative with the Early Cretaceous part of the Santiago Peak Formation (Kimbrough and others, 1990).

Other Mesozoic igneous rocks reported from the western (arc) subterrane of the Yuma terrane have uncertain origins and relation to the Santiago Peak and Alisitos Formations. The Alisitos Formation overlies Late Jurassic andesite and basalt (Table 16) at Arroyo Calamajué (Fig. 2). Initial reports of Late Triassic or Early Jurassic clastic and volcanic rocks beneath the

Alisitos Formation near Rancho San José (Fig. 2) (Minch, 1969) were based on incorrect fossil identification; the rocks are Cretaceous and probably part of the Alisitos (J. Minch, personal communication, 1991; Strand and others, 1991). Weakly metamorphosed Late Cretaceous andesite and volcaniclastic rocks northwest of Loreto (Fig. 2) probably are a younger part of the Alisitos Formation, although they also resemble Late Cretaceous rocks in the Tahué terrane (Gastil and others, 1981). North of El Arco (Fig. 2), gabbro, diorite, serpentinite, pyroxenite, and pillow basalt of unknown age are faulted against metavolcanic and metasedimentary rocks of possible Cretaceous age (Rangin, 1978; Barthelmy, 1979). It has been speculated that these rocks represent a disrupted ophiolite and adjacent or overlying sedimentary rocks that were intercalated within numerous thrust sheets during mid-Cretaceous contraction (see below) (Rangin, 1978; Radelli, 1989), but few data are available concerning their age, origin, and structural evolution.

The eastern (basinal) subterrane. The eastern (basinal) subterrane is considered by many workers to be divisible into an older unit north of the Agua Blanca fault and a younger unit south of the fault. To the north, undated siliciclastic flysch between the Agua Blanca fault and southern California generally has been correlated with the Triassic(?)-Jurassic Bedford Canyon Formation, the Late Triassic(?) Frenchman Valley Formation, and the Triassic(?) Julian Schist of southern California (Gastil and others, 1975, 1981; Criscione and others, 1978; Todd and others, 1988; Reed, 1989). The Triassic-Jurassic age of the siliciclastic Bedford Canyon Formation is based on olistostromal Middle to Late Jurassic limestone clasts (Moran, 1976) and Rb-Sr studies of sedimentary rocks that yield isochron ages of about 230 and 175 Ma (Criscione and others, 1978). The Jurassic fauna include several ammonite species that have affinities southward rather than northward (Imlay, 1963, 1964). On the basis of lithologic similarity to part of the Franciscan Complex, the protoliths of the Bedford Canyon Formation and, by analogy, of the Frenchman Valley Formation and the Julian Schist, have been interpreted as trench or trench-slope deposits (Moran, 1976; Criscione and others, 1978); forearc basin and even backarc basin environments also are possible (Todd and others, 1988). Recent work has shown that the eastern subterrane is intruded by strongly deformed S-type granitoids as old as Middle Jurassic (Fig. 20), but the regional extent of these plutons is not yet certain (Todd and others, 1991). The following geologic history can be deduced for the Julian Schist at its type locality near Julian: deposition of a fine- to coarse-grained siliciclastic protolith in uncertain tectonic setting; intrusion by thin pegmatite veins, perhaps in association with intrusion of Jurassic S-type granitoids; Early Cretaceous penetrative deformation and metamorphism; and Early Cretaceous intrusion by the Peninsular Ranges batholith.

To the south, the eastern (basinal) subterrane includes unnamed flysch units at several isolated localities. Near 30°00′N, a 6-km-thick greenschist- and amphibolite-facies flysch unit that includes rare quartzite conglomerate and andesite contains probable Aptian-Albian fauna (Gastil and others, 1981; Phillips, 1984) and has been interpreted as a backarc sequence (Gastil and others, 1986b). Near 30°30′N, flysch of unknown age contains rare interbeds of rhyolite and andesite (Gastil and Miller, 1984). At Arroyo Calamajué, broken formation derived from metamorphosed flysch of unknown age may be correlative with the Cretaceous flysch farther north (Gastil and Miller, 1983; Griffith, 1987). North of El Arco, prebatholithic metamorphosed clastic rocks crop out to the northeast of probable Alisitos rocks (Barthelmy, 1979; Radelli, 1989).

Relation of subterranes. Contact relations between the eastern and western subterranes are complex. Locally, volcanic rocks of the Santiago Peak Formation (western subterrane) unconformably overlie deformed, metamorphosed, and uplifted flysch of the eastern subterrane. However, field relations and geochemical and geophysical studies indicate that in most areas the eastern (basinal) subterrane overlies the western (arc) subterrane along a steeply east-dipping suture that probably formed during late Early Cretaceous collision of the arc subterrane with the marginal basin subterrane and continental rocks of the Serí terrane farther east. In southern California and much of northern Baja California, this suture is obscured by the younger (Late Cretaceous) part of the Peninsular Ranges batholith (Fig. 20), as well as by Late Cretaceous and Cenozoic cover. North of the Agua Blanca fault, the position of the suture is inferred from geochemical and geophysical studies, and U-Pb dating of tectonic and posttectonic plutons indicates collision during the Early Cretaceous (Todd and others, 1988). South of the Agua Blanca fault, the suture crops out in several locations; locally, the western subterrane is juxtaposed directly with the Serí terrane, i.e., the eastern Yuma subterrane is absent. Steeply plunging lineations, asymmetric kinematic indicators, and the inferred depth of crustal exposure indicate thrust or reverse displacement; U-Pb dating of syntectonic and posttectonic plutons indicates that collision, southwest-northeast contraction, and upper greenschist- to lower amphibolite–facies metamorphism occurred between about 105 and 95 Ma (Griffith, 1987; Griffith and Goetz, 1987; Goetz and others, 1988; Radelli, 1989; Windh and others, 1989). Younger episodes of west-vergent thrusting and mylonitization (~80 Ma) and uplift along normal faults (~62 Ma) that affected the eastern (marginal basin) subterrane and adjacent Serí terrane have been recognized along the East Peninsular Ranges fault zone in Alta California (Dokka, 1984; Todd and others, 1988; Goodwin and Renne, 1991), but similar deformation in Baja California has not been recognized.

Peninsular Ranges batholith. Plutons of the Peninsular Ranges batholith intruded the Yuma terrane during Cretaceous time. This batholith is a composite of a 140- to 105-Ma (U-Pb) western part and a 105- to 80-Ma (U-Pb) eastern part that youngs to the east (Fig. 20; Table 16). The older part of the batholith is cogenetic with, and intrudes, Santiago Peak-Alisitos volcanic rocks of the western subterrane. The younger part of the batholith intrudes both subterranes of the Yuma terrane, the suture between them, and the older part of the batholith. Near the U.S.-México border, the older part of the batholith is more petrologically diverse and has lower initial $^{87}Sr/^{86}Sr$ ratios, lower $\delta^{18}O$ values, and less fractionated rare earth element (REE) patterns than the younger part (Silver and others, 1979; Hill and others, 1986; Gromet and Silver, 1987). Sr, Nd, and Pb isotopic systems have mantle values in the west but are progressively more evolved to the east (Silver and Chappell, 1988). The western part of the batholith is typical of primitive island arcs built on oceanic lithosphere; the entire batholith appears to be an example of the formation of continental crust in an area previously devoid of continental lithosphere (Silver and Chappell, 1988).

The boundary between the older and younger parts of the batholith is nearly coincident with regional features including a magnetite-ilmenite line, a boundary between I-type and S-type granitoids, a sharp gravity gradient, a linear magnetic anomaly (Todd and Shaw, 1985; Gastil and others, 1986a; Jachens and others, 1986, 1991), and the suture between the eastern and western subterranes (Fig. 20). Petrographic and structural observations in penetratively deformed plutons as young as 100 Ma are interpreted to indicate syntectonic intrusion of the older part of the batholith (Todd and others, 1988). Continuous exposures of the Peninsular Ranges batholith are found as far south as El Arco (28°N), but isolated outcrops have been discovered on the east coast of Baja California Sur as far south as 26°N (Table 16), and diorite and gneissic xenoliths have been collected in Tertiary basalt at latitude 26°20′N (Demant, 1981; Hausback, 1984; López-Ramos, 1985, p. 28; McLean and others, 1987; McLean, 1988). Positive gravity anomalies associated with the batholith continue at least as far south as 26°N (see below).

Postbatholithic rocks. Along the west coast of northern Baja California, Late Cretaceous to Eocene postbatholithic marine clastic strata rest nonconformably on, and probably were derived from, the Alisitos Formation and Peninsular Ranges batholith (Gastil and others, 1975; Bottjer and Link, 1984). Cretaceous strata contain abundant *Coralliochama Orcutti*, a warm-water rudist absent from Cretaceous rocks of the Great Valley Group in California (D. Bottjer, personal communication, 1992). Some postbatholithic rocks in northern Baja California were deposited in tectonically active basins of uncertain tectonic setting (Boehlke and Abbott, 1986; Cunningham and Abbott, 1986). Eocene sedimentary rocks in the northern Yuma terrane indicate bypassing of the eroded Peninsular Ranges batholith and volcanic arc and tapping of source terranes in Sonora (Bartling and Abbott, 1983). Middle Miocene basalts, tuffs, and fluvial and shallow marine clastic rocks along the coast between Ensenada and San Diego were derived from western volcanic and Franciscan sources that subsequently were submerged or displaced (Ashby, 1989).

Postbatholithic Late Cretaceous and Paleogene clastic rocks in Baja California Sur probably were derived from the southward continuation of the Peninsular Ranges batholith, which crops out locally on the eastern coast. Latest Cretaceous and Paleogene marine clastic rocks containing benthic and planktonic forams and minor tuff (e.g., Bateque and Tepetate Formations) crop out in the eastern Vizcaíno Peninsula and along the Pacific coast east of Isla Magdalena (Heim, 1922; Mina, 1957; Fulwider, 1976; Hausback, 1984; López-Ramos, 1985; McLean and others, 1987; Squires and Demetrion, 1989, 1990a, b). Wells in these regions and in the northern Bahía Sebastián Vizcaíno have penetrated Late Cretaceous to Eocene marine clastic rocks and minor tuff (Mina, 1957; López-Ramos, 1985, p. 30–32, 50–52). The Late Cretaceous strata are similar to the Valle Formation in the western part of the Vizcaíno Peninsula (Cochimí terrane), implying Late Cretaceous overlap of a buried northwest-trending Yuma-Cochimí fault boundary (**p. xx**). Eocene nonmarine rocks near Loreto probably are more proximal lateral equivalents of the shallow marine strata (McLean, 1988).

Late Cenozoic volcanism, sedimentation, and faulting. Late Oligocene to early Miocene(?) (Table 16) shallow marine clastic rocks and subordinate tuffs and flows of the San Gregorio Formation (also called El Cien Formation or Monterrey Formation) in central Baja California Sur contain commercially viable

phosphorite deposits (Hausback, 1984; Alatorre, 1988; Grimm and others, 1991). The Middle to upper Miocene Salada Formation, which crops out in the Magdalena plain in southwestern Baja California Sur, is a thin (<100 m) sequence of richly fossiliferous marine sandstone, conglomerate, siltstone, and mudstone deposited in a shallow embayment (Smith, 1992). Miocene calc-alkalic andesites and volcaniclastic rocks (Comondú Formation) in Baja California Sur and southern Baja California, and correlative rocks in the Serí terrane in northeastern Baja California, are remnants of a chain of coalescing stratovolcanoes that was active 24 to 11 Ma (Gastil and others, 1979; Sawlan and Smith, 1984; Sawlan, 1991). Vent-facies rocks of the axial core of this calc-alkalic arc are exposed on the eastern margin of Baja from 29° to 25°N (Hausback, 1984). The waning stage of orogenic magmatism was contemporaneous with alkalic and tholeiitic volcanism that started by about 13 Ma in eastern Baja California and within the developing Gulf of California rift (Hausback, 1984; Sawlan and Smith, 1984; Sawlan, 1991). The geologic-tectonic map of Baja California compiled by Fenby and Gastil (1991) depicts all sedimentary and volcanic rocks of Oligocene and younger age.

The eastern Baja California peninsula, including the eastern Yuma, western Serí, and eastern Pericú terranes, has undergone extension with or without dextral slip since at least the Middle Miocene (Angelier and others, 1981; Dokka and Merriam, 1982; Hausback, 1984; Stock and Hodges, 1989, 1990). The least principal stress and extension directions changed from east-northeast–west-southwest to northwest-southeast about 6 Ma, roughly coeval with the initiation or acceleration of transtensional opening of the modern Gulf of California. Since about 5 Ma, the surface of the eastern edge of the Baja continental block has been uplifted about 1 to 3 km, resulting in detachment and eastward translation of elevated continental crustal walls of the Gulf of California (Fenby and Gastil, 1991; Gastil and Fenby, 1991). Pliocene to Holocene faulting in eastern Baja California includes dextral slip on northwest-striking strike-slip faults, down-to-the-gulf normal faulting, and oblique-slip faulting (e.g., Umhoefer and others, 1991). The geologic-tectonic map of Baja California compiled by Fenby and Gastil (1991) depicts all faults on which displacement is interpreted to be Oligocene and younger.

Geophysical data. A positive low-pass filtered gravity anomaly trends south-southeast from the peninsular ranges in Baja California to at least 26°N in Baja California Sur (Couch and others, 1991). Gravity models interpret this anomaly to indicate southeastward continuation of the Cretaceous batholith in the subsurface.

Paleomagnetic investigations of rocks of different age, lithology, and magnetic character in the Yuma terrane have yielded Cretaceous and Paleogene paleolatitudes that are significantly shallower than expected for stable North America. Calculated relative paleolatitudes, i.e., the implied northward latitudinal displacement of specific localities relative to North America since rocks at those localities were magnetized, include 12.3° ± 7.4°, 13.2° ± 6.8°, and 4.6° ± 6.0° for Cretaceous plutons (Teissere and Beck, 1973; Hagstrum and others, 1985); 15.0° ± 3.8° and

18.2° ± 6.7° for Late Cretaceous sedimentary rocks (Fry and others, 1985; Filmer and Kirschvink, 1989); and about 5° ± 5° for Paleogene sedimentary rocks (Flynn and others, 1989). These and other results from correlative rocks in southern California (summarized in Lund and Bottjer, 1991 and Lund and others, 1991) imply at least 10° and perhaps as much as 20° of northward translation and 25° to 45° of clockwise rotation of Baja with respect to stable North America between about 90 and 40 Ma. These data are discussed more fully on pages 80–81.

Zapoteco terrane

The Zapoteco terrane is a fragment of Proterozoic continental crust consisting mainly of crystalline basement rocks of Grenville age overlain nonconformably by rare cratonal Paleozoic strata. On the basis of petrologic, geochronologic, and paleomagnetic data we infer that the Precambrian and early Paleozoic rocks of the Zapoteco terrane formed part of the Grenville province of southeastern Canada; however, other interpretations are permissible. The Zapoteco terrane probably was displaced to the south of the southern margin of North America during the Paleozoic, and by the latest Paleozoic it hosted a magmatic arc within or west of the western margin of Pangea. The Zapoteco terrane corresponds to the Oaxaca terrane of Campa-Uranga and Coney (1983) and Coney and Campa-Uranga (1987).

Oaxacan Complex. The oldest unit in the Zapoteco terrane is the Oaxacan Complex (Fig. 21), an assemblage of meta-anorthosite, quartzofeldspathic orthogneiss, paragneiss, calcsilicate metasedimentary rocks, and charnockite that was formed by Grenvillian metamorphism of miogeoclinal or continental rift deposits and plutonic rocks (Ortega-Gutiérrez, 1981a, b, 1984a). The protoliths experienced peak granulite-facies metamorphic

temperatures of 710° ± 50°C and pressures of approximately 7 kbar (Mora and others, 1986). Metamorphism probably occurred between 1,100 and 1,000 Ma, and slightly younger Sm-Nd cooling ages (Table 17) probably postdate peak metamorphism (Patchett and Ruiz, 1987). There are no radiometric data to support speculative Archean to Middle Proterozoic ages (e.g., Bazan, 1987). Fold axes and lineations in the Oaxacan Complex plunge gently to the north-northwest, and mesoscopic folds record east-west shortening (Ortega-Gutiérrez, 1981b).

Paleozoic to Cenozoic rocks. The Oaxacan Complex is overlain nonconformably by thin-bedded shale, sandstone, limestone, and conglomerate of the Tiñu Formation, which contains early Tremadocian (earliest Ordovician) trilobite taxa that resemble those in Early Ordovician rocks in South America, southeastern Canada, and northwestern Europe, and that are dissimilar to those of southwestern North America (Pantoja-Alor and Robison, 1967). Younger Paleozoic strata include Carboniferous marine sandstone and shale of the Santiago and Ixtaltepec Formations, and continental sandstone, siltstone, and conglomerate of the Matzitzi Formation of Pennsylvanian and probable Permian age; depositional environments of these rocks are not yet certain (Pantoja-Alor and Robison, 1967; Robison and Pantoja-Alor, 1968). The Matzitzi Formation is the oldest unmetamorphosed stratigraphic unit that physically overlaps the fault contact between Zapoteco and Mixteco terranes.

Near Caltepec, Puebla, the Oaxacan Complex is intruded by cataclastic granitoids that are correlated with the Devonian Esperanza granitoids in the Mixteco terrane (F. Ortega-Gutiérrez, unpublished data). The granitoids become more strongly mylonitic closer to the contact of the Zapoteco terrane with the Mixteco terrane, implying syntectonic intrusion of the two terranes during the Early to Middle Devonian. The Oaxacan Complex also

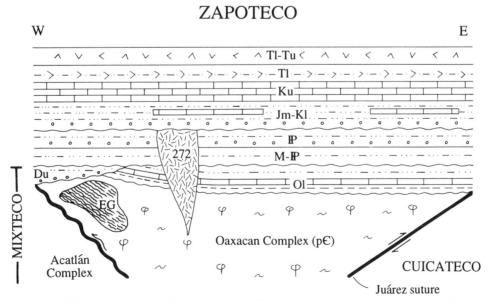

Figure 21. Schematic tectonostratigraphic section of Zapoteco terrane. Upper Devonian rocks at western margin overlap Mixteco and Zapoteco terranes.

TABLE 17. ZAPOTECO TERRANE RADIOMETRIC DATA

Sample	System	Mineral*	Date (Ma)	References†	Comments
Oaxacan Complex					
Paragneiss	U-Pb	zr	1,168 ±180; 747 ±330	1	Discordia intercepts
Gneiss and syntectonic pegmatite	U-Pb	zr	1,080 ± 10	2	
Gneiss	U-Pb	zr	1,050 ± 20	3	
Paragneiss	U-Pb	zr	1,020	4	Slightly discordant
Posttectonic pegmatites	U-Pb	zr	975 ± 10	2	
	U-Pb	zr	960 ± 15	3	Concordant age
Gneiss	Sm-Nd	g	960, 940	5	Cooling ages
Pegmatite	K-Ar	b	950 ± 30	6	
	K-Ar	b	930 ± 30	6	
	K-Ar	b	906 ± 30	7	
	Rb-Sr	b	870 ± 35	7	
	Rb-Sr	kf	770 ± 35	7	
	Rb-Sr	b	770 ± 20	7	
	K-Ar	b	680 ± 20	7	
Other rocks					
Granite	Rb-Sr		272 ± 8	8	$^{87}Sr/^{86}Sr_i$: 0.7047
Ignimbrite	K-Ar	b	17 ± 1	9	

*Mineral abbreviations: b = biotite; g = garnet; kf = potassium feldspar; zr = zircon.
†1 = Robinson, 1991; 2 = Anderson and Silver, 1971; 3 = Ortega-Gutiérrez and others, 1977; 4 = Robinson and others, 1989; 5 = Patchett and Ruiz, 1987; 6 = Fries and others, 1962a; 7 = Fries and Rincón-Orta, 1965; 8 = Ruiz-Castellanos, 1979; 9 = Ferrusquía-Villafranca, 1976.

is intruded by a small, undeformed Early Permian granitoid (Table 17).

Much of the Zapoteco terrane is overlain by Mesozoic and Cenozoic rocks that are very similar to those in the Mixteco terrane. These rocks include Late Jurassic to Early Cretaceous shallow-water and nonmarine clastic rocks with minor limestone and coal, mid-Cretaceous carbonates, Campanian-Maastrichtian conglomerate and sandstone derived from the Juchatengo subterrane of the Mixteco terrane, Paleogene red beds and volcanic rocks, mid-Tertiary andesite, and Neogene calc-alkalic volcanogenic rocks (Fig. 21, Table 17) (Carfantan, 1986; Ortega-Gutiérrez and others, 1990). Neogene nonmarine strata and volcanic rocks were deposited in elongate north-north-west–striking grabens that formed from about 19 to 12 Ma along the Oaxaca fault at the eastern margin of the Zapoteco terrane (Ferrusquía-Villafranca and McDowell, 1988; Centeno-García and others, 1990). The geometry and timing of extension may indicate that the Basin and Range province continued south of the TMVB in the Middle Miocene (Henry and Aranda-Gómez, 1992).

Geophysical data. The paleopole of primary magnetization in the Oaxacan Complex is >40° from the Grenville Loop of the North America polar wander path; this has been intepreted to indicate a position near Quebec during Grenvillian time (Ballard and others, 1989). Paleozoic cover of the Zapoteco terrane was remagnetized between Late Permian and Jurassic time, possibly by Permo-Triassic intrusions, and may have rotated counterclockwise as much as 28° with respect to the adjacent Mixteco terrane (McCabe and others, 1988). Paleomagnetic arguments for Jurassic to Early Cretaceous southward displacement of the Mix-

teco terrane with respect to cratonal North America (Urrutia-Fucugauchi and others, 1987; Ortega-Guerrero and Urrutia-Fucugauchi, 1989) also must apply to the attached Zapoteco terrane.

TERRANE BOUNDARIES

In this section, we discuss the rationale for distinction of terranes and for the delineation of terrane boundaries shown in Figure 3, and we summarize available data concerning the orientation, nature, and kinematic history of terrane boundaries. Because much of México is covered by Cretaceous and Cenozoic rocks, and because major displacement on most terrane-bounding faults is Jurassic or older, few terrane-bounding faults crop out at the surface, particularly in northern and central México. Nevertheless, the surface trace of most faults can be determined to within 10 to 200 km based on scattered exposures of basement rocks and geophysical and isotopic data. The straight dashed lines shown in Figure 3 are not meant to imply vertical faults; as discussed in this section, the subsurface orientation of most faults is unknown. As noted elsewhere, many of these terranes probably are composite, and future work may lead to partitioning into smaller terranes or subterranes.

Boundaries of Tarahumara terrane

The eastern and western boundaries of the Tarahumara terrane are not exposed. The western boundary is inferred to be a major southeast-dipping suture, probably of Pennsylvanian-

Permian (Ouachitan) age, at which deformed basinal sedimentary rocks of the Tarahumara terrane (i.e., Ouachita orogenic belt) were thrust at least 40 km onto continental margin sedimentary rocks of the Pedregosa Basin and North American shelf (Walper and Rowett, 1972; King, 1975; Armin, 1987; Handschy and others, 1987; James and Henry, 1993). On the east, the Tarahumara terrane is bounded by the Coahuiltecano terrane, which consists of metamorphic rocks that probably are part of stranded South American (Gondwanan) continental crust. Because of the dearth of pre-Jurassic outcrops in this region, it is impossible to ascertain the location and nature of the Tarahumara-Coahuiltecano contact, but we speculate that it is a southeast-dipping fault at which the Tarahumara terrane was overthrust by forearc and/or arc rocks of Gondwana. The southern boundary of the Tarahumara terrane is the Mojave-Sonora Megashear (see below), which may have truncated a southward continuation of the Tarahumara/Ouachita orogenic belt into central México. In Part 2 we propose an alternate view in which the Tarahumara terrane never extended farther south than it does currently, and in which the megashear separates it from unrelated accreted terranes to the south.

Southern boundary of North America, Tarahumara, and Coahuiltecano terranes (Mojave-Sonora Megashear)

The Mojave-Sonora Megashear (MSM) of the southwestern United States and northwestern México is a controversial sinistral fault system inferred to be the site of 700 to 800 km of Late Jurassic displacement, based on the truncation of Precambrian basement rocks, the offset of latest Precambrian to early Paleozoic sedimentary rocks, the offset of Triassic sedimentary rocks, and the offset of a Jurassic volcanic arc in northern México (Fig. 22) (Silver and Anderson, 1974, 1983; Stewart and others, 1984; 1990; Anderson and others, 1990). The MSM probably is a vertical fault or fault system, but it is difficult to prove this

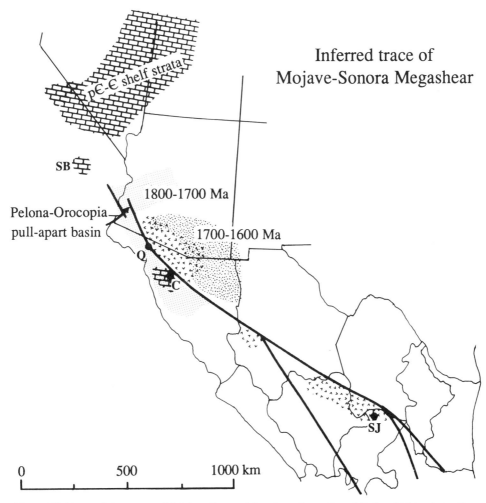

Figure 22. Trace of the proposed Mojave-Sonora Megashear in northern México. Lighter dot pattern indicates 1,800- to 1,700-Ma basement province; heavier dot pattern, 1,700- to 1,600-Ma basement province (both from Anderson and Silver, 1979); brick pattern, uppermost Proterozoic to Cambrian shelf strata (from Stewart and others, 1984); v pattern, Jurassic volcanic arc (from Anderson and Silver, 1979, and other sources as in text). Abbreviations: C, Caborca; Q, Quitovac; SB, San Bernardino Mountains area; SJ, San Julian uplift.

because along most of its length it is obscured by younger sedimentary rocks and modified by younger structural features. In Part 2, we discuss at length the case against and, in our opinion, the stronger case for large displacement on the MSM (p. 78).

Eastern and southern boundaries of Serí terrane

The boundary between the Serí and Tahué terranes at latitude 28°N separates outcrops and inferred subcrops of Precambrian basement in the Serí terrane to the north from Mississippian and inferred late Paleozoic basement rocks of the Tahué terrane to the south. Initial $^{87}Sr/^{86}Sr$ values of Late Cretaceous–Paleogene volcanic rocks are 0.7064 to 0.7080 north of the boundary but only 0.7036 to 0.7063 south of the boundary (Tables 10, 11), suggesting that sialic Precambrian crust underlies the Serí terrane but not the Tahué terrane (Damon and others, 1983). The inferred boundary also marks the southern edge of the outcrop area of the Upper Triassic–Lower Jurassic Barranca Group in the Serí terrane. The orientation of and history of displacement on the inferred fault boundary are unknown, but displacement apparently has been negligible since the emplacement of overlapping Early Cretaceous volcanic rocks.

In southern Baja California, the Serí terrane is tectonically interleaved with the western (arc) subterrane of the Yuma terrane (eastern Yuma subterrane is absent) within a 5-km-wide, steeply east-dipping suture zone (Griffith, 1987; Griffith and Goetz, 1987; Goetz and others, 1988). Juxtaposition was latest Early Cretaceous (106 to 97 Ma), based on U-Pb ages of deformed and undeformed plutons in and near the suture.

The nature of the Serí-Yuma contact in southern California and northern Baja California, where Mesozoic basinal rocks of the eastern Yuma subterrane intervene between the Serí terrane and the western (arc) Yuma subterrane, is enigmatic. The Serí-eastern Yuma boundary is intruded by numerous plutons and has not been studied in detail, although some screens between plutons appear to be transitional in lithology and age between the Paleozoic miogeoclinal strata of the Serí terrane and the Mesozoic basinal strata of the eastern Yuma subterrane (Todd and others, 1988). We provisionally infer a prebatholithic fault boundary between the two rock units, but if further work demonstrates that the contact is depositional, then the Mesozoic basinal strata should be considered a subterrane of the Serí terrane rather than of the Yuma terrane. The boundary between the eastern and western Yuma subterranes is inferred to be a steeply east-dipping suture zone of Early Cretaceous age on the basis of geologic, geophysical, and geochemical discontinuities (Todd and Shaw, 1985; Todd and others, 1988) and thus is geometrically similar to and may be related to the Serí-Yuma suture in southern Baja California.

Cochimí-Yuma boundary

The boundary between Cretaceous arc and forearc basin rocks of the Yuma terrane and arc rocks and blueschists of the Cochimí terrane is not exposed, but its position in the subsurface is known within a few tens of kilometers from the distribution of outcrops and subcrops and from distinctive gravity anomalies. The boundary probably lies west of PEMEX well Totoaba-1 in Bahía Sebastián Vizcaíno (Fig. 2), which penetrated mid-Cretaceous flysch of the Valle Formation and bottomed in andesite that may be correlative with the Alisitos Formation of the Yuma terrane (López-Ramos, 1985, p. 52). The boundary must be east of exposures of the Cochimí terrane on the Vizcaíno Peninsula at 28° to 27°N and on Isla Santa Margarita and Isla Magdalena at 25° to 24°N (Fig. 2). The terrane boundary probably strikes north-northwest between these two outcrop areas but is buried beneath late Cenozoic sedimentary rocks of the Pacific continental shelf. Gravity studies have measured a north-northwest-trending, low-pass filtered, positive gravity anomaly above the Pacific shelf south of about 30°N, indicating that submerged areas of the terrane probably are underlain by comparatively dense rocks similar to the ophiolitic, arc, and blueschist rocks known from outcrops of the terrane (Couch and others, 1991).

The current proximity of basement rocks of marked differences in lithology and metamorphic and structural history implies that the boundary between the Yuma and Cochimí terranes must be a major fault zone. The nature and the orientation of the fault or faults are unknown, but possibilities include east- or west-vergent thrust faults related to accretion, and younger strike-slip faults. The age of juxtaposition of the two terranes is post-Jurassic and pre–latest Cretaceous. By the earliest Cretaceous, arc subterranes of the Cochimí terrane had collided with continental North America, but it is not certain that they were originally juxtaposed with the central Yuma terrane because paleomagnetic studies infer that both the Yuma and Cochimí terranes were south of their present latitude prior to the Eocene (p. 80). Available data are too meager to assess possible postcollisional strike-slip displacement between the two terranes. Major displacement may have ceased by the Late Cretaceous because the boundary apparently is overlapped by widespread Late Cretaceous to Eocene marine strata that have been mapped and drilled in the eastern Vizcaíno region and the Iray-Magdalena basin (López-Ramos, 1985, p. 50–52). Alternatively, the boundary may have undergone significant Cenozoic displacement that juxtaposed strata deposited in similar, although distant, depositional environments.

Yuma-Tahué boundary (pre–Gulf of California)

Prior to the opening of the Gulf of California, the Yuma terrane in southern Baja California was adjacent to the Tahué terrane. Relations between the Yuma terrane and Tahué terrane were obliterated by late Cenozoic magmatism and the opening of the Gulf of California. We infer that the pre-gulf boundary consists of or included the fault or faults at which Baja California, including the Yuma terrane and the western part of the Serí terrane, was translated northward with respect to North America in the Late Cretaceous and Cenozoic.

Boundaries of Pericú terrane

The "La Paz fault" at the western margin of the Pericú terrane probably includes structures related to several episodes of displacement. Over the past several million years, this fault has been characterized by down-to-the-west normal displacement, probably associated with the opening of the Gulf of California (Curray and Moore, 1984). Holocene displacement is indicated by visible scarps along the fault trace (A. Carrillo-Chávez, unpublished data). Normal slip during the Miocene may have been accompanied by a sinistral (Hausback, 1984) or dextral (Sedlock and Hamilton, 1991) component of strike-slip displacement. Basement rocks of the Pericú terrane have uncertain relation to buried basement rocks west of the fault that provisionally are assigned to the Yuma terrane, but early Tertiary strike-slip displacement inferred between the two terranes implies that the buried boundary between them is subvertical.

Prior to the opening of the Gulf of California, the Pericú, western Nahuatl, and southern Tahué terranes were mutually adjacent. The original boundaries among these terranes have been obscured by late Cenozoic strike-slip faulting, opening of the Gulf of California, and volcanism in the Trans-Mexican Volcanic Belt. Nevertheless, the protoliths of prebatholithic rocks in the three terranes are quite similar and may indicate common parentage (Henry, 1986). Metamorphic rocks of the Pericú terrane were derived from clastic rocks and carbonates at least as old as Triassic, and probably Paleozoic. The lithology and age of these rocks are very similar to those of the Paleozoic rocks of the Tahué terrane. The host rocks for granitoids in the western Nahuatl terrane are interbedded andesite, metagraywacke, and marble of possible Jurassic and Cretaceous age. These rocks are similar to the Cretaceous marble and volcanic rocks of the Tahué terrane and were correlated with these rocks by Henry and Fredrikson (1987). These relations may indicate that the Pericú and Tahué terranes were part of the same parent Paleozoic terrane, and that the western Nahuatl, Tahué, and (presumably) Pericú terranes were part of the same composite terrane by the late Mesozoic.

Tahué-Tepehuano boundary

Thick volcanic rocks of the Sierra Madre Occidental completely obscure the location, nature, and orientation of the contact between the Tahué and Tepehuano terranes. Mitre-Salazar and others (1991) inferred that the two terranes were juxtaposed in the Late Cretaceous, based on differences between arc-derived Cretaceous rocks in the Tahué terrane and basinal (carbonates and marine clastic) rocks in the Tepehuano terrane. An alternate explanation is that Jurassic closure of an intervening ocean basin resulted in collision of the two terranes and the cessation of magmatism in the Tepehuano terrane. In other words, the stratigraphic differences between the Tahué and Tepehuano terranes in Cretaceous time may simply reflect differences between adjacent arc and backarc environments.

Tepehuano-Guachichil boundary

The Tepehuano-Guachichil terrane boundary is an inferred fault, called the San Tiburcio lineament by Mitre-Salazar (1989) and Mitre-Salazar and others (1991), that is overlapped by late

Mesozoic strata. The boundary is defined by the eastern outcrop limit of Triassic(?)-Jurassic magmatic arc rocks (Caopas Schist, Rodeo and Nazas Formations) and Late Triassic marine strata (Zacatecas Formation), and by the western outcrop and subcrop limit of nonmarine strata of the Upper Triassic to Lower Jurassic Huizachal Formation (López-Ramos, 1985). We do not accept proposed correlations of the Huizachal Formation with either the Nazas Formation or Zacatecas Formation. We suggest that the San Tiburcio lineament is a southeastward splay of the Mojave-Sonora Megashear (p. 79) with Late Jurassic dextral slip of unknown magnitude. The San Tiburcio lineament may have been reactivated during Laramide orogenesis (Mitre-Salazar, 1989).

Guachichil-Maya boundary

Basement rocks along this boundary are obscured by late Mesozoic-Cenozoic sedimentation, Laramide folding and thrusting, and Cenozoic volcanism, but abundant well logs and samples indicate that a subvertical boundary separates widespread Permo-Triassic plutonic rocks to the east (Maya terrane) from Precambrian gneiss and Paleozoic schist and sedimentary rocks to the west (Guachichil terrane) (López-Ramos, 1972). This steeply dipping boundary is a fault or fault zone that strikes about N20W and appears to be offset by east-northeast–striking high-angle(?) faults east of Tuxpán, Veracruz. Schist and gneiss in some wells in the Maya terrane probably represent country rocks metamorphosed by the Permo-Triassic intrusions, as indicated by cooling ages from some samples (Table 7). The boundary is placed at the western limit of Permo-Triassic plutonic rocks, which do not crop out in the Guachichil terrane. We propose that reported tonalite or diorite in a few wells in the Guachichil terrane instead may be gneissic basement; the ambiguity inherent in identifying basement on the basis of small samples of drill core was discussed by López-Ramos (1972).

Several interpretations of the nature, timing, and kinematics of displacement on the buried fault boundary between the Guachichil and Maya terranes have been proposed. First, the boundary may be a high-angle dextral fault system that accommodated southward displacement of the Yucatán block of the Maya terrane during the Jurassic opening of the Gulf of Mexico (Padilla y Sánchez, 1986). Second, it may consist of thrust and reverse faults that were produced during Triassic collision events, with Yucatán displacement accommodated on the offshore (intra-Maya terrane) Golden Lane-Tamaulipas fault zone (e.g., Pindell, 1985). Third, as discussed in Part 2, it may have been a high-angle sinistral fault system that was a strand of the Jurassic Mojave-Sonora Megashear (p. 79).

Trans-Mexican Volcanic Belt

Late Cenozoic volcanic rocks mask the identity of and relations among older rocks in the Trans-Mexican Volcanic Belt (TMVB) in southern México (Figs. 1, 3). Boreholes in the México basin ("la cuenca de México") between Cuernavaca and Pachuca have penetrated Tertiary volcanic and continental clastic rocks, Late Cretaceous limestone, shale, and sandstone, Early Cretaceous marine limestone, and underlying anhydrite (Mooser and others, 1974; de Cserna and others, 1987). Spectral analysis of regional aeromagnetic data in the western part of the TMVB has been used to infer that the Cenozoic volcanic rocks are underlain by Cretaceous-Tertiary granitic basement west of Guadalajara and by older(?) crystalline basement east of Guadalajara (Campos-Enríques and others, 1990). Although insufficient data are available to map terrane boundaries through the volcanic cover, it is likely that the TMVB coincides with a major fault or fault zone that forms the southern boundary of the Tahué, Tepehuano, and Guachichil terranes, and the northern boundary of the Nahuatl, Mixteco, Zapoteco, and Cuicateco terranes (Fig. 3). Anderson and Schmidt (1983) invoked several hundred kilometers of Jurassic sinistral slip on this inferred fault, whereas Gastil and Jensky (1973) inferred several hundred kilometers of dextral slip of latest Cretaceous and earliest Tertiary age. We are unable to document geologic evidence for the older, sinistral event, but restoration of the postulated younger, dextral event aligns the Cretaceous batholiths of the Tahué, Pericú, and Nahuatl terranes and various mineralization belts in central and southern Mexico (Gastil and Jensky, 1973; Clark and others, 1982).

Eastern boundary of Nahuatl terrane

The location of the fault boundary between the Nahuatl terrane and the Mixteco terrane is controversial. At its westernmost exposure, the Acatlán Complex of the Mixteco terrane is faulted westward above Cretaceous carbonates along the Papalutla reverse fault. The basement and parent terrane of these carbonates are unknown. If the carbonates are underlain by Acatlán Complex, they are part of the Mixteco terrane and the Mixteco-Nahuatl terrane boundary is west of the carbonates. However, suitable major faults are not exposed in this area. Alternatively, the carbonates may mark an eastward facies change from the Cretaceous marine siliciclastic and volcanic rocks that comprise most of the Nahuatl terrane, in which case the carbonates are part of the Nahuatl terrane and the Papalutla fault is the Mixteco-Nahuatl terrane boundary. A third possibility is that the Cretaceous carbonates overlap and completely obscure the original Mixteco-Nahuatl terrane boundary. Currently, we cannot determine which of these possibilities is the most likely.

The fault boundary between the Nahuatl terrane and the Chatino terrane has been completely obliterated by intrusion of Tertiary granitoids east of Zihuatanejo and Petatlán, Guerrero. The granitoids have yielded a 33 ± 8-Ma three-point Rb-Sr isochron (González-Partida and others, 1989), and U-Pb studies of the plutons are in progress (K. Robinson, personal communication, 1991).

Zapoteco-Mixteco boundary

Basement rocks of the Zapoteco and Mixteco terranes (Oaxacan Complex and Acatlán Complex, respectively) are directly juxtaposed at the subvertical to northeast-dipping Caltepec fault zone 140 km north of the city of Oaxaca (Ortega-Gutiérrez, 1980). The 300- to 400-m-wide fault zone contains cataclastic and mylonitic rocks derived from Precambrian granulitic gneiss of the Oaxacan Complex and gneissic granitoids of the Acatlán Complex and is overlain unconformably by the Pennsylvanian-Permian Matzitzi Formation (Fig. 12). Basement rocks of the two terranes probably were sutured by the Middle Devonian, based on the syntectonic(?) intrusion of both terranes by the Early to Middle Devonian Esperanza granitoids, the Early to Middle Devonian age of penetrative deformation and high-temperature metamorphism of the Acatlán Complex, and the presence of clasts derived from basement rocks of both terranes in the Upper Devonian Tecomate Formation of the Mixteco terrane (Ortega-Gutiérrez, 1978b, 1981a, b; Ortega-Gutiérrez and others, 1990; Yañez and others, 1991).

Northern boundary of Chatino terrane

From west to east along its northern boundary, the Chatino terrane is in fault contact with the Nahuatl, Mixteco, Zapoteco, and Cuicateco terranes. This contrasts with the relations shown by Campa-Uranga and Coney (1983) and Coney and Campa-Uranga (1987), who placed a narrow western arm of the Cuicateco (their Juárez) terrane between the Chatino and Zapoteco (their Xolapa and Oaxaca) terranes. The Chatino-Nahuatl contact and many reaches of the Chatino-Mixteco and Chatino-Zapoteco contacts are obliterated by Cenozoic plutons. Where the contact has escaped later intrusion, it crops out as a major mylonitic fault zone known by different fault names on opposite sides of a cross-cutting Cenozoic high-angle fault that causes the sharp bend in the boundary (Fig. 3). The contact with the Cuicateco terrane and the eastern part of the Zapoteco terrane is called the Chacalapa fault zone, and the contact with the Mixteco terrane and the western part of the Zapoteco terrane is called the Juchatengo fault zone (Ortega-Gutiérrez and others, 1990).

The south-dipping, 2- to 5-km-thick Chacalapa fault zone consists of mylonite and ultramylonite derived from both the Zapoteco and Chatino terranes, and metamorphosed mid-Cretaceous limestone from the Zapoteco terrane (Ortega-Gutiérrez, 1978b; Ortega-Gutiérrez and others, 1990; F. Ortega-Gutiérrez and R. Corona-Esquivel, unpublished data). Blastomylonitic granite with a shallowly east-southeast–plunging lineation yielded a poorly constrained Rb-Sr whole-rock isochron age of about 110 Ma (Ortega-Gutiérrez and others, 1990). Within several hundred meters of the Chacalapa fault zone, Precambrian granulites of the Zapoteco terrane contain retrograde amphibolite-facies assemblages similar to prograde amphibolite-facies assemblages in the Chatino terrane, implying synkinematic metamorphism. The Chacalapa fault zone is interpreted as the site of Late Cretaceous thrusting of the Chatino terrane over the Zapoteco terrane, based on structural analysis, the presence near the fault zone of fault-bounded synorogenic strata of Campanian-Santonian age, and the Tertiary age of nearby posttectonic strata (Carfantan, 1981; Grajales-Nishimura, 1988; Ortega-Gutiérrez and others, 1990).

Near Juchatengo (97°W), the Juchatengo fault zone is about 100 to 200 m thick and consists of north- to northeast-dipping mylonite, ultramylonite, and cataclasite (Ratschbacher and others, 1991) probably derived from the Chatino terrane and the Juchatengo subterrane of the Mixteco terrane. Shear criteria and quartz c-axes indicate top-to-the-north motion during low-grade metamorphism; the interior of the mylonite zone is hydrothermally altered. Footwall gneiss and migmatite of the Chatino terrane display evidence for coaxial north-south stretching. Hanging-wall Cretaceous(?) sedimentary and volcanic rocks, assigned to the Juárez terrane by Ratschbacher and others (1991) but here interpreted to overlie the Juchatengo subterrane of the Mixteco terrane, display evidence for brittle north-south extension.

At Tierra Colorada, 50 km northeast of Acapulco, the Juchatengo fault zone is about 1 km thick and consists of mylonite, ultramylonite, and cataclasite derived from migmatite of the Chatino terrane and mid-Cretaceous carbonates of the Mixteco terrane (Ratschbacher and others, 1991). Shear criteria and fault-striae solutions indicate top-to-the-northwest ductile flow in response to northwest-southeast extension. Mylonitic fabrics and faults in the hanging wall display a component of left-lateral strike-slip motion, indicating that displacement probably was left-normal. Hanging-wall rocks are intruded by the Tierra Colorada pluton, which yielded a 60-Ma Rb-Sr date (Table 8). According to Ratschbacher and others (1991), a contact aureole indicates that this pluton also intrudes the mylonites and thus constrains ductile deformation to prior to 60 Ma. However, these workers also noted (p. 1235) that the contact aureole is faulted "with a deformation geometry consistent with that of the mylonite zone," indicating "continued uplift in the Tertiary." An alternate interpretation of this area, based on unpublished mapping, kinematic analysis of structures in the fault zone, and a synthesis of regional radiometric data, is that the Tierra Colorada pluton intruded the hanging wall about 60 Ma and was juxtaposed with mylonites of the Juchatengo fault zone during early Tertiary tectonic exhumation, extension, and mylonitization of the Chatino terrane from beneath the Mixteco and Zapoteco terranes (Robinson and others, 1989, 1990; Robinson, 1991; G. Gastil and K. Robinson, personal communication, 1991).

Clearly, the kinematics and age of the Juchatengo and Chacalapa fault zones are unresolved problems in need of additional study. There is disagreement on the dip of the mylonite zone (S or N), the type of faulting (thrusting or left-oblique extension), and the timing of deformation (Cretaceous or early Tertiary).

Perhaps all observations can be reconciled by a two-stage history involving Mesozoic thrusting and Paleogene left-oblique extension. During stage one, northward thrusting of the Chatino terrane on a south-dipping fault or faults (shown in Fig. 5) caused amphibolite-facies metamorphism of both hanging wall and footwall and produced a thick south-dipping zone of cataclastic rocks. Thrusting may have started during the Jurassic(?) and Early Cretaceous, as indicated by Rb-Sr and U-Pb dates from gneissic granitoids of the Xolapa Complex in the Chatino terrane

(Table 1), and may have continued into the Late Cretaceous, as indicated by the Campanian-Santonian age of synorogenic strata (Ortega-Gutiérrez and others, 1990). During stage two (not shown in Fig. 5), left-oblique transtension produced north-dipping mylonites with normal fault geometry, tectonic thinning of the hanging-wall section, exhumation of the Xolapa Complex footwall, and an undetermined amount of left-lateral displacement along the northern boundary of the Chatino terrane (Robinson and others, 1989; 1990; Ratschbacher and others, 1991; Robinson, 1991). Left-oblique transtension probably occurred during the Paleogene or possibly the late Late Cretaceous. It began later than about 80 Ma, the youngest metamorphic age from deformed orthogneiss and migmatite, and may have been active until about 30 Ma, the oldest intrusion age of undeformed granitoids that cut mylonites east of Zihuatanejo and Petatlán, Guerrero (Table 9). Complex relations like those at Tierra Colorada may indicate protracted, possibly episodic, extension, mylonitization, and tectonic exhumation of the Chatino terrane.

Boundaries of Cuicateco terrane

The Mesozoic oceanic rocks of the Cuicateco terrane are easily distinguished from Precambrian and Paleozoic continental rocks of the bounding Maya and Zapoteco terranes. Along its western margin, the Cuicateco terrane is overthrust by the Zapoteco terrane along the shallowly southwest-dipping Juárez suture, which contains mylonitic rocks derived from the Zapoteco terrane and from granitoids inferred to be a subterrane of the Cuicateco terrane (Ortega-Gutiérrez and others, 1990). The Cuicateco-Zapoteco boundary has been modified by Cenozoic normal and right-lateral displacement on several strands of the high-angle Oaxaca fault (Centeno-García and others, 1990). Along its eastern margin, the Cuicateco terrane is thrust eastward over Paleozoic(?) metamorphic rocks and Jurassic red beds of the Maya terrane along the Vista Hermosa fault. Thrusting on these bounding faults, internal deformation of the Cuicateco terrane, and internal deformation of the adjacent Maya terrane probably occurred more or less synchronously during Late Cretaceous time and terminated prior to the deposition of a Cenozoic overlap assemblage, although structural and stratigraphic relations near the city of Oaxaca indicate pre–Early Cretaceous displacement on the Cuicateco-Zapoteco boundary (F. Ortega-Gutiérrez and others, unpublished data).

Maya-Chortis boundary

We have placed the boundary between the southern Maya terrane and Chortis terrane along the Motagua fault in central Guatemala, one of three east-west–striking faults that probably form the boundary between the North America and Caribbean plates (Molnar and Sykes, 1969; Dengo, 1972; Malfait and Dinkelman, 1972; Muehlberger and Ritchie, 1975; Schwartz and others, 1979). As discussed in Part 2, the southern boundary of the Maya terrane was the site of latest Cretaceous arc collision, northward obduction of Early Cretaceous oceanic crust, and hundreds to thousands of kilometers of postcollisional sinistral displacement. Cumulative displacement on the Polochic fault is controversial (Deaton and Burkart, 1984a; T. Anderson and others, 1985, 1986; Dengo, 1986; Burkart and others, 1987), but carefully documented offset of major anticlinoria, belts of Pb-Zn mineralization, conglomerate clasts and their source, the Miocene volcanic belt, stratigraphic contacts, and granitoids indicates about 130 km of Cenozoic sinistral slip (Deaton and Burkart, 1984a; Burkart and others, 1987). Cumulative sinistral displacement on the Motagua fault is unknown, but interpretations of geophysical data from the Cayman Trough and Yucatán basin and plate tectonic reconstructions imply a minimum of 1,100 km and perhaps more than 2,500 km of latest Cretaceous and Cenozoic slip (Macdonald and Holcombe, 1978; Pindell and Dewey, 1982; Burke and others, 1984; Rosencrantz and Sclater, 1986; Pindell and others, 1988; Rosencrantz and others, 1988; Pindell and Barrett, 1990; Rosencrantz, 1990).

PART 2: TECTONIC EVOLUTION OF MÉXICO

Richard L. Sedlock, Robert C. Speed, and Fernando Ortega-Gutiérrez

INTRODUCTION

In Part 2 of this volume, we develop a comprehensive model of the tectonic evolution of México since the mid-Proterozoic. In Part 1, we delineated and described the terranes that comprise México and northern Central America; here, we propose a plausible tectonic evolution of México that accounts for the geologic history of each terrane and for the displacement, attachment, and redistribution of the terranes. Our fundamental premise is that most of México is a conglomeration of terranes accreted to the southern margin of North America during Phanerozoic time. Some aspects of our model of the paleogeographic reconstruction of México are modified after ideas in the H-1 and H-3 Ocean-Continent Transects completed for the DNAG program (Ortega-Gutiérrez and others, 1990; Mitre-Salazar and others, 1991).

Previous models divided the country into a few simplified continental blocks and focused chiefly on Mesozoic displacements during the breakup of Pangea and on the evolution of the Gulf of México and Caribbean plate (Walper, 1980; Pindell and Dewey, 1982; Anderson and Schmidt, 1983; Coney, 1983; Pindell and others, 1988). Papers of regional scope by Morán-Zenteno (INEGI, 1985) and de Cserna (1989) summarized the geology of México in the context of geographic and morpho-tectonic provinces but eschewed terrane analysis. Campa-Uranga and Coney (1983) and Coney and Campa-Uranga (1987) divided México into tectonostratigraphic terranes but did not address their tectonic evolution. The concept of terranes is absent from the schematic syntheses of the Jurassic to Tertiary tectonic

evolution of México by Tardy and others (1986) and Servais and others (1986), which instead infer large-scale continuity of arcs, backarc basins, and forearc basins beneath Cenozoic and late Mesozoic cover.

PREMISES AND OTHER CONSTRAINTS

Investigations of the geology of México are handicapped by sparse basement outcrop, limited availability of many studies, and insufficient radiometric dating. Kinematic analysis of terranes also is complicated by the gaping hole in our knowledge of Proterozoic, Paleozoic, and early Mesozoic plate motions and plate boundaries in the region, and by poorly constrained late Mesozoic motions of all plate pairs except North America–South America. These obstacles inhibit the testing of the many, to some extent irreconcilable, hypotheses that have been proposed for different aspects of the tectonic evolution of México. To maintain internal consistency in our tectonic model, we base our reconstruction on numerous formal premises and less formal constraints as discussed below. Many premises and constraints are straightforward and conventional, but others address pivotal but poorly understood aspects of the geology of México.

Reference frame and time scale

In this study, we adopt the North American continent as the kinematic reference frame to which we will relate motions within México. Because the position and orientation of North America are not completely established for the Phanerozoic, we have omitted north arrows and latitude references on our paleogeographic reconstruction. The term Proterozoic North America denotes contiguous Proterozoic and older crystalline continental crust at the onset of the Phanerozoic. We use the DNAG time scale (Palmer, 1983).

Southern margin of Proterozoic North America

Premise 1: The southern margin of Proterozoic North America, which is well established in the southern Appalachians and Great Basin, extends no more than a few hundred kilometers south of the frontal trace of the Ouachita orogenic belt.

The gross shape of Proterozoic North America was created in the latest Proterozoic and early Paleozoic by rifting and drifting of an originally larger continent (Fig. 23), producing well-documented passive continental margins at the eastern margin north of Georgia (Rankin, 1975; Thomas, 1977, 1991) and at the western margin of the continent north of southern California (Stewart, 1972; Stewart and Poole, 1974; Bond and Kominz, 1984).

It is difficult to pinpoint the location and nature of the southern margin of Proterozoic North America between the southern Appalachians and southern California due to virtually continuous sedimentary and volcanic cover strata of Mesozoic and Cenozoic age. Indirect evidence is recognized in the Ouachita orogenic belt (Fig. 24), where allochthonous off-shelf early Pa-

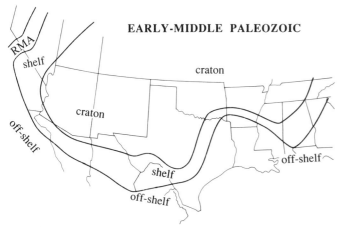

EARLY-MIDDLE PALEOZOIC

Figure 23. Inferred distribution of craton, shelf, and off-shelf (basinal) depositional environments at southern margin of North America during early to middle Paleozoic time. Modern political boundaries shown for reference.

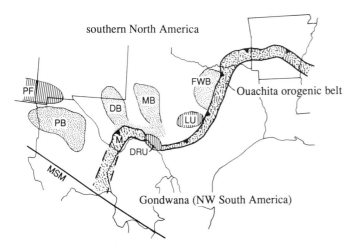

Figure 24. Ouachita orogenic belt in south-central United States and its probable continuation into north-central México (Tarahumara terrane). Also shown are location of foreland basins (dot pattern) and uplifts and Pedregosa forebulge (vertically lined) caused by the collision of Gondwana and North America. Modified after Ross (1986), Armin (1987), and Viele and Thomas (1989). Abbreviations: DB, Delaware basin; DRU, Devils River uplift; FWB, Fort Worth basin; LU, Llano uplift; M, Marathon region; MB, Midland basin; PB, Pedregosa basin; PF, Pedregosa forebulge.

leozoic strata were thrust onto early Paleozoic platformal cover of North America in the late Paleozoic (Premise 5). Passive-margin facies that might represent the precollision southern margin of Proterozoic North America are not exposed in the Ouachita orogenic belt, but the existence of such rocks is inferred on the basis of Cambrian carbonate boulders in Pennsylvanian conglomerate in the Haymond Formation of the Marathon Mountains (Fig. 23) (Palmer and others, 1984). The passive-margin sequence may have been completely buried by far-traveled Ouachitan thrust

sheets (Viele, 1979a, b; Lillie and others, 1983) or may have been anomalously narrow due to transform-dominated, as opposed to rift-dominated, rifting of long reaches of the southern margin of Proterozoic North America (Cebull and others, 1976; Thomas, 1977, 1985, 1989, 1991). In any event, it is highly likely that the Ouachita orogenic belt structurally overlies the southern margin of Proterozoic North America, based on seismic profiling, the large dimensions of the belt, and the emplacement of the belt onto continental platform strata (Keller and others, 1989a, b).

Premise 2: The Ouachita orogenic belt, and thus the southern margin of Proterozoic North America, continues south along the Chihuahua-Coahuila border but cannot be demonstrated south of those states (Fig. 24).

Premise 3: The southern passive margin of Proterozoic North America extended from the Chihuahua-Coahuila border region westward through central Chihuahua and Sonora (Fig. 23).

The Ouachita orogenic belt does not crop out in México west or south of the Marathon region. Most workers have projected the belt westward into central Chihuahua on the basis of lithologic similarities in the two regions (Flawn and others, 1961), but recent gravity and lead isotope studies indicate that the belt probably continues southward in the subsurface from the Marathon region along the Chihuahua-Coahuila border (Handschy and others, 1987; Aiken and others, 1988; James and Henry, 1993). Basinal strata in central Chihuahua that have been interpreted as part of the Ouachita orogenic belt probably were deposited on the subsiding late Paleozoic North America continent shelf (Bridges, 1964a; Mellor and Breyer, 1981; Handschy and others, 1987). The original southwestward extent of the Ouachita orogenic belt is difficult to ascertain (Premise 5), but geologic and gravity data indicate that it now extends no farther south than the northern border of Durango (Fig. 24).

Two end-member alternatives have been proposed for the location of the southern margin of Proterozoic North America west of the Chihuahua-Coahuila border. According to the first alternative, the margin continues westward across central Chihuahua to central Sonora, based on facies gradients in Paleozoic platformal cover in the southwestern United States and Sonora that indicate a nearly straight, east-west–trending, south-facing passive margin across northwest México (Stevens, 1982; Palmer and others, 1984; Stewart, 1988). This model implies that oceanic lithosphere developed south of the continental margin in the early Paleozoic. Steep facies gradients indicate that the continent-ocean transition zone may have been narrow, little disrupted, and nonsubsiding relative to typical rift margins (Thomas, 1985). The cratonal-platformal aspect of early Paleozoic North American strata in northern México suggests that the motion of an outboard oceanic plate or plates relative to North America was nearly pure strike slip, probably not transtensional, and certainly not transpressional. According to the second alternative, North American basement continues southward into central México (Guzmán and de Cserna, 1963; Shurbet and Cebull, 1980, 1987). Proponents of this alternative also infer late Precambrian to early Paleozoic northwest-striking transform offset of the continental margin (Cebull and others, 1976), as has been inferred in the Ouachita and Appalachian orogens (Thomas, 1977, 1991). The second alternative is unattractive because the 70- to 80-km thickness of the lithosphere in central México is less than in cratonal areas of North America, and because, in the late Paleozoic and early Mesozoic, part and perhaps most of what is now central México was an oceanic realm that lay west of the western margin of Pangea, with a clear history of arc magmatism and accretionary tectonics.

Premise 4: In the latest Proterozoic, the passive margin of southwestern North America stretched northward from Sonora into southeastern California and south-central Nevada (Fig. 23).

The late Proterozoic southern margin of North America between central Chihuahua and Sonora (Premise 3) and east-central California and south-central Nevada (Stewart and Poole, 1974) is difficult to determine because of superposed sinistral and dextral strike-slip faulting of late Proterozoic(?), late Paleozoic, Mesozoic, and Cenozoic age (Premises 11 through 13). North American basement may have formed a southwest-trending promontory between Chihuahua and central California in the late Proterozoic (Dickinson, 1981). Following the reasoning in Premise 3, such a promontory must have been disrupted prior to the late Paleozoic, producing (1) displaced terranes correlative with southwestern North America or (2) a large continental mass with a paleomagnetic pole path that indicates past proximity to southwestern North America. We find no evidence for either product, although it is conceivable that the disrupted terranes or continental mass may have been attached to northern South America in the late Paleozoic or Mesozoic, given the existence there of basement rocks of Grenville age (Rowley and Pindell, 1989).

Facies distribution indicates that the trend of the western margin of North America at the latitude of central and southern California changed from north-south or northeast-southwest to northwest-southeast by the Pennsylvanian and perhaps earlier (Stevens and Stone, 1988; Stone and Stevens, 1988). Late Paleozoic to possible early Mesozoic southeastward translation of a narrow tract of latest Proterozoic to Cambrian platformal rocks from southeastern California to northern Sonora (Premise 10) does not require pronounced southwestward bulging of the continental margin and is consistent with the roughly linear northwest trend shown in Figure 23.

Paleogeography of Pangea in the vicinity of México

Premise 5: The supercontinent Pangea was formed by diachronous Pennsylvanian–Early Permian collision of Gondwana (Africa and South America) with North America during the Alleghany orogeny in the Appalachian region and the Ouachita orogeny in the southern United States.

Late Paleozoic deformation, metamorphism, and synorogenic sedimentation in the Appalachian and Ouachitan orogens are widely ascribed to diachronous collision between Gondwana and North America (Hatcher, 1972; King, 1975; Walper, 1980). Reconstructions based on paleomagnetic, biostratigraphic, and other geologic and geophysical evidence show that North America and South America were in close proximity as part of the Pangea supercontinent during the Pennsylvanian, Permian, and Triassic (Ross, 1979; Van der Voo and others, 1984; Scotese and McKerrow, 1990).

Premise 6: Diachronous collision, or "zippering," of North America and South America terminated in the Permian near the Chihuahua-Coahuila border due to the unfavorable orientation of the margins of the colliding plates and perhaps to the cessation of convergence.

The termination of the Ouachita orogenic belt (Tarahumara terrane) near the Chihuahua-Coahuila border can be explained in several ways. First, the original southwestern continuation of the belt may have been offset southeastward to Ciudad Victoria, Tamaulipas (northern Guachichil terrane) by sinistral slip on the Mojave-Sonora Megashear or other faults (Flawn and others, 1961). This possibility is unlikely in light of recently documented differences between Ouachitan rocks in west Texas and contemporaneous rocks near Ciudad Victoria (Stewart, 1988). Second, the Ouachita orogenic belt may have continued westward from the Chihuahua-Coahuila border region to southern California, subparallel to the inferred southern margin of Proterozoic North America (Premise 3), but later was disrupted or obscured by tectonism and magmatism. This alternative is awkward because it places Gondwanan crystalline basement rocks directly west of a Permo-Triassic arc in eastern México that probably records eastward subduction of oceanic lithosphere beneath the western margin of Pangea. A third possibility, which we adopt here, is that collision did not occur south or west of the Chihuahua-Coahuila border, for either or both of two reasons: (1) the orientation of the two continental margins may have been unsuitable for collision, e.g., a north-south western boundary of South America; and (2) the convergence rate between North America and South America may have slowed dramatically during the ongoing collision between the two continents.

Low-grade metasedimentary rocks underlying the Coahuiltecano terrane, metamorphic country rocks of the Permo-Triassic batholith in the Coahuiltecano and northern Maya terranes, and metamorphic basement rocks underlying the Yucatán platform are interpreted here as remnants of Gondwanan continental crust.

Origin of Grenville basement in México

Premise 7: Precambrian rocks of Grenville age in México and central Cuba were derived from and are allochthonous with respect to the North American Grenville province.

High-grade metasedimentary and metaigneous rocks of the Grenville province rim eastern and southern Proterozoic North America, probably underlie the Ouachita orogenic belt, and are inferred to extend southwestward into northern México (Fig. 25). Ages of intrusion and metamorphism of rocks in the Llano uplift of central Texas range from 1,305 to 1,091 Ma (Walker, 1992), and rocks in this age range crop out or have been penetrated by wells in central and northern Chihuahua. Outcrops and xenoliths of basement rocks that have been correlated with the Grenville province are known from several disjunct terranes in central and southern México and from central Cuba (Renne and others, 1989).

Basement rocks with lithology, history, and age similar to the Grenville province have been recognized in northern South America (Kroonenberg, 1982; Priem and others, 1989), but to date there are no strong arguments in favor of their possible original contiguity with and subsequent separation from the Grenville province in North America. We adopt Premise 7 because evidence is lacking for pre–late Paleozoic contiguity of North America and South America, and thus for original contiguity and perhaps continuity of the Grenville province and Grenville-like rocks in South America.

The outcrops of Grenville rocks in eastern and southern México are commonly interpreted as culminations of continuous Grenvillian basement in eastern and central México (de Cserna, 1971; López-Infanzón, 1986; Ruiz and others, 1988b). We suggest that such continuity of basement is very unlikely because the western edge of the Maya terrane probably was a major kinematic boundary in late Paleozoic to mid-Jurassic time, separating relatively rigid continental crust of the Maya terrane to the east from tectonically active, kinematically distinct terranes to the west (pp. 94–103). Our preferred alternate interpretation is that exposures of Grenville rocks are discrete, fault-bounded fragments of basement derived from the North American Grenville province, that much or most of México is underlain by basement blocks of not only Grenville but also younger (early and middle Paleozoic) age, and that the basement blocks are allochthonous with respect to North America and to one another. The interpretation of the basement of central México as a heterogeneous assemblage of rocks of diverse age and composition is supported by recent Pb, Sr, and Nd isotopic studies.

Permian-Triassic arc in eastern México

Premise 8: In the Late Permian and Triassic, a continental magmatic arc developed at the western margin of central Pangea near what is now eastern México.

Late Permian and Triassic magmatic arc rocks are present in a roughly linear swath that extends from Coahuila (Coahuiltecano terrane) to Chiapas (Maya terrane) in eastern México; these rocks probably were emplaced into continental crust (López-Ramos, 1972, 1985; Damon and others, 1981; López-Infanzón,

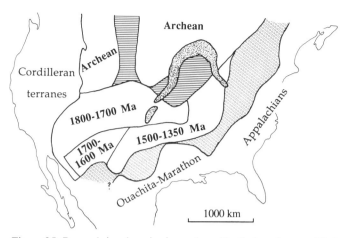

Figure 25. Precambrian domains in southern North America, modified after Anderson and Silver (1979), Bickford and others (1986), and Hoffman (1989). Horizontal lines indicate 2,000- to 1,800-Ma provinces; circle pattern, 1,300- to 1,000-Ma Grenville province; irregular dot pattern, 1,100-Ma rift.

1986; Wilson, 1990). Remnants of the arc crop out near Valle San Marcos (242 ± 2 Ma granodiorite, Rb-Sr), Potrero de la Mula (213 ± 14 Ma I-type granite, Rb-Sr), and Las Delicias (210 ± 4 Ma granodiorite, K-Ar) in the Coahuiltecano terrane; they crop out in the Chiapas Massif (~260 to 220 Ma, K-Ar), Maya Mountains of Belize (237 to 226 Ma, K-Ar), and Guatemala (227 Ma, Rb-Sr, and 238, 213, and 212 Ma, ^{40}Ar/^{39}Ar cooling ages) in the southern part of the Maya terrane; and they have been penetrated by numerous petroleum wells in the states of Veracruz, Nuevo León, and Tamaulipas in the Gulf coastal plain (~275 to 210 Ma, K-Ar).

Global and regional reconstructions indicate that this arc probably formed on the western margin of Pangea above an east-dipping subduction zone that consumed oceanic lithosphere of a plate or plates west of Pangea, and that the Mexican reach of the arc was roughly coeval with adjacent reaches of continental magmatic arc in the southwestern United States and northwestern South America (Dickinson, 1981; Scotese and McKerrow, 1990). In the southwestern United States, Permo-Triassic arc rocks include Permian volcanic rocks as old as 283 Ma (U-Pb), Late Permian to Early Triassic granitoids, and Early Triassic volcaniclastic rocks that contain clasts of the granitoids in the northwestern Mojave Desert and Death Valley (Carr and others, 1984; Walker, 1988; Snow and others, 1991; Barth and others, 1992). In northwestern South America, Permo-Triassic magmatic rocks crop out in the Cordillera de Mérida in northwestern Venezuela, the Sierra Nevada de Santa Marta in northern Colombia, and the Cordillera Central and Cordillera Oriental in central Colombia (Shagam and others, 1984; Restrepo and Toussaint, 1988; Case and others, 1990); in the eastern Cordillera of Peru (Cobbing and Pitcher, 1983); and in Chile north of about 28°S (Farrar and others, 1970; McBride and others, 1976; Halpern, 1978; Aguirre, 1983).

The Mexican reach of the Permo-Triassic continental arc may have been partially disrupted by later tectonism but is presumed to initially have been laterally continuous with the reach in northwestern South America (p. 99). The discontinuity between the northern end of the Mexican reach of the arc and the southern end of the U.S. reach in the Mojave–Death Valley region probably reflects the initial geometry of the plate boundary rather than tectonic offset (p. 95).

Late Paleozoic–Cenozoic strike-slip faulting

Premise 9: Since late Paleozoic time, the western margin of North America has been cut by sinistral and dextral strike-slip fault systems that accommodated the margin-parallel transport of tectonostratigraphic terranes. These fault systems probably extended south into México and perhaps into northwestern South America.

Paleomagnetic, biostratigraphic, and geologic data support the widely held view that the western margin of North America, at least at the latitude of the United States and Canada, was the site of margin-parallel transport of tectonostratigraphic terranes during much of Cenozoic, Mesozoic, and perhaps Paleozoic time (e.g., Coney and others, 1980; Howell and others, 1985). Below, we outline evidence for episodes of sinistral and dextral displacement within and along the western margin of México at different times since the late Paleozoic.

Late Paleozoic–Early Triassic truncation of southwestern United States

Premise 10: A fragment of southwestern North America was tectonically removed from the continental margin in eastern California and translated to the southeast during Pennsylvanian to Early Triassic time.

The distribution of cratonal, shelfal, and basinal rocks in southeastern and east-central California and western Nevada is interpreted by many workers to indicate Pennsylvanian to Early Triassic truncation of the southwestern margin of Proterozoic North America at an enigmatic northwest-striking structure or structures (Fig. 26). Displaced fragments include part of the Roberts Mountain allochthon, which was accreted to western North America during the Antler orogeny, and latest Proterozoic to early Paleozoic passive margin strata of North America (Stevens and others, 1992). Proposed ages of displacement are Late Permian to earliest Triassic (Hamilton, 1969; Burchfiel and Davis, 1981), Pennsylvanian to Early Permian (Stevens and Stone, 1988; Stone and Stevens, 1988; Stevens and others, 1992), and both late Paleozoic and Early Triassic (Walker, 1988). We propose that the displaced fragment of the Roberts Mountain allochthon was translated southeastward about 400 km to the Mojave region during the Pennsylvanian to Early Permian at a left-lateral fault or faults (Walker, 1988; Stevens and others, 1992). To the southeast, this left-lateral fault

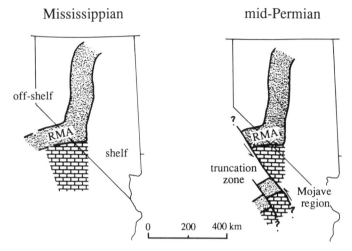

Figure 26. Diagrams showing proposed late Paleozoic truncation of southwestern margin of North America. Brick pattern indicates passive margin facies; irregular dot pattern, accreted Roberts Mountain Allochthon. Modified from Walker (1988).

or faults may be the Mojave-Sonora Megashear, in which case about half of the inferred postulated displacement on the megashear is Paleozoic (Stevens and others, 1992). Alternatively, the fault or faults may have been outboard of the future trace of a Jurassic megashear (see discussion of Premise 11).

Late Jurassic Mojave-Sonora Megashear

Premise 11: The Mojave-Sonora Megashear in the southwestern United States and northwestern México accommodated about 700 to 800 km of sinistral slip from 160 to 145 Ma, and was roughly coincident with the Jurassic magmatic arc.

In Part 1 we identified the southern boundary of the North America, Tarahumara, and Coahuiltecano terranes as the Mojave-Sonora Megashear (MSM), a major left-lateral fault postulated to have been active in the Late Jurassic (Silver and Anderson, 1974, 1983; Anderson and Silver, 1979). The MSM was first recognized in Sonora on the basis of the truncation of the northeast-southwest trend of 1,700- to 1,600-Ma basement rocks by 1,800- to 1,700-Ma basement rocks, the juxtaposition of dissimilar Mesozoic volcanic and sedimentary rocks as young as Oxfordian, and penetrative deformation features in Proterozoic and Mesozoic rocks near the proposed fault trace (Fig. 22). Other evidence for displacement along the MSM (Fig. 22) includes: (1) strong similarity between latest Proterozoic and Cambrian shelfal rocks southwest of the fault near Caborca, Sonora (Serí terrane) and rocks northeast of the fault in southern California and Nevada (Stewart and others, 1984); (2) strong stratigraphic and sedimentologic similarity between Middle Ordovician quartzite in central Sonora (Serí terrane) and the Eureka quartzite in southeastern California (Ketner, 1986); (3) about 800 km sinistral offset of the western margin of the continent, as inferred from the distribution of Paleozoic shelfal rocks between the Mojave region and Caborca (Stewart and others, 1984, 1990); (4) about 800 km of sinistral offset of Jurassic volcanic rocks in northern Sonora (North America) and northern Zacatecas (Tepehuano terrane) (Anderson and others, 1990); (5) probable transpressive Late Jurassic deformation in Jurassic rocks in northern Sonora and northern Zacatecas (Connors and Anderson, 1989; Anderson and others, 1991) and in northern Sonora (Rodríguez-Castañeda, 1990); (6) outcrop of northwest-striking mylonitic rocks with horizontal lineation near Quitovac in northwestern Sonora (Connors and Anderson, 1989; Tosdal and others, 1990b); (7) significant improvement of the clustering of paleomagnetic poles of Late Triassic–Early Jurassic sedimentary rocks on both sides of the MSM after restoration of 800 km of sinistral slip (Cohen and others, 1986); and (8) strong stratigraphic and faunal similarities between Late Triassic rocks in Sonora and southwestern Nevada (Stanley and others, 1991). Isotopic changes in Cenozoic basaltic andesites across the inferred trace of the MSM are similar to, but more subtle than, changes across other lithospheric boundaries in western North America; sample coverage is insufficient to determine whether the boundary is abrupt or gradational (Cameron and others, 1989). Bouguer gravity anomalies in eastern Sonora and western Chihuahua show an east-west disturbance of a generally north-south trend along the proposed trace of the megashear (Aiken and others, 1988; Schellhorn and others, 1991).

There are two main arguments against major sinistral displacement on the MSM. Differences in some lithologic characteristics between shelfal and off-shelf rocks in Sonora and time-equivalent rocks in southeastern California and southern Nevada have been interpreted by some workers as sufficient evidence that the two regions were not contiguous during deposition, that they formed more or less in place, and that major displacement on the MSM is not likely (Poole and Madrid, 1988; Stewart and others, 1990). However, some of these differences may be due to depositional lateral variation. The second argument against the megashear is based on the presence of cratonal-platformal Paleozoic rocks at El Capitán, northwestern Sonora, and in the Gila Mountains, southwestern Arizona. Because these areas are southwest of the proposed megashear and northwest of the inferred continental margin near Caborca, it has been argued that either the megashear lies to the south, passing westward from Sonora to the Gulf of California, and that slip on the megashear is Paleozoic (Stevens and others, 1992), or that the passive margin was and is continuous and unbroken from Sonora to central California (Leveille and Frost, 1984; Hamilton, 1987). Interpretations stemming from the El Capitán–Gila Mountains outcrops may be invalid if these cratonal-platformal rocks are allochthonous and emplaced in their present position after displacement on the MSM (L. T. Silver, oral presentation at U.S. Geological Survey workshop, January 1990).

Another point of contention concerning the megashear is whether the age of the proposed displacement is Late Jurassic or late Paleozoic. A Late Jurassic age is widely cited, but Stevens and others (1992) suggested that major sinistral displacement on the MSM occurred in the late Paleozoic, synchronous with truncation of the southwestern margin of the continent. The MSM may have formed at or near the inherited Proterozoic edge of North America. Reactivation of a Paleozoic fault may account for Late Jurassic, possibly transpressional deformation along the trace of the fault, so the argument for a Late Jurassic age of displacement must rest on the improved clustering of paleomagnetic poles and on the apparent offset of Jurassic volcanic rocks and Triassic sedimentary rocks. The proposed offset of the Jurassic arc does not specify a distinctive piercing point, and the near coincidence of the trace of the MSM and the apparent axis of the arc make it difficult to accurately determine the offset. We consider sinistral displacement on the MSM to be of Late Jurassic age, but available data also are consistent with several hundred kilometers of late Paleozoic sinistral displacement.

Along most of its proposed length, the MSM not only has been covered by post-Jurassic strata but also has been modified by post-Jurassic structures. In southern and central California, Jurassic basins that may have formed at releasing steps or bends in the megashear were strongly deformed in the Late Cretaceous and Tertiary, obscuring the original nature of the MSM (Harper and others, 1985; Tosdal and others, 1990a). In Sonora, the megashear appears to have been overthrust in the Late Cretaceous (DeJong and others, 1990). In the southwestern United States and northwestern México, the megashear may have been within or near the Jurassic magmatic arc at the western margin of North America, but alternate models of its position in these regions cannot be discounted (Stewart and others, 1984, 1990).

The trace of the MSM is fairly well defined in Sonora and Chihuahua, but poorly defined in eastern México. Elsewhere in Part 2, we advance the argument that the MSM branches into several strands eastward from southeastern Chihuahua (Fig. 22). From north to south, these branches include but are not limited to a fault along the southern boundary of the Coahuiltecano terrane, a northwest-striking fault separating the northern and southern Guachichil terranes, and the San Tiburcio lineament between the Guachichil and Tepehuano terranes (Mitre-Salazar, 1989; Mitre-Salazar and others, 1991). The San Marcos fault in the Coahuiltecano terrane (Fig. 3), which parallels the MSM and was active in the Late Jurassic, may connect with the MSM to the west, in Chihuahua. We emphasize that northward to eastward overthrusting of all these strands by Jurassic and Cretaceous strata during Laramide orogenesis has obscured the boundaries between basement rocks of each pair of terranes. For example, Upper Triassic to Lower Jurassic Huizachal Formation, characteristic of the Guachichil terrane, is mapped as far north as Galeana, Nuevo León (~24°30′N). Nevertheless, the likely position of the Guachichil-Coahuiltecano boundary, i.e., the megashear, is several tens of kilometers south of Galeana, beneath nappes that have transported the Huizachal Formation and other rocks of the Guachichil terrane to the north and east.

Cretaceous translation of Baja California

Premise 12: A crustal fragment comprising most of the present Baja California peninsula has been translated about 15° northward since about 90 Ma.

One of the most controversial aspects of the tectonic evolution of México is whether rocks in Baja California underwent large amounts of margin-parallel transport during Mesozoic and Tertiary oblique convergence. Paleomagnetic studies of rocks of different age, lithology, and magnetic character have been interpreted to indicate that much or all of Baja California and adjacent southern California has been translated northward a minimum of 10° and perhaps as much as 20° since about 90 Ma (Hagstrum and others, 1985, 1987; Lund and Bottjer, 1991; Lund and others, 1991b). Most studies also have recognized 25° to 45° of concomitant clockwise rotation. The rocks that are affected include the Yuma terrane, that part of the Serí terrane in Baja California, the Cochimí terrane excepting blueschist-facies subduction complex rocks, and rocks in southern California south of the Transverse Ranges that appear to correlate with the Yuma and Serí terranes. In the following discussion we refer to this region as Baja or the Baja block; it is similar to the Peninsular Ranges composite terrane of Bottjer and Link (1984) but does not include the Pericú terrane at the southern tip of the peninsula, for which no paleomagnetic data are available. Although anomalously low paleolatitudes were reported in a study of volcanic rocks on the eastern margin of the gulf (Bobier and Robin, 1983), we do not consider this region part of the Baja block because these workers did not correct for structural tilt.

Paleomagnetic investigations of rocks of different age, lithology, and magnetic character throughout the Baja block have yielded Cretaceous and Paleogene paleolatitudes that are significantly shallower than would be expected if they had been part of stable North America. Below we list some typical relative paleolatitudes, i.e., the northward latitudinal displacement of specific localities relative to North America since rocks at those localities were magnetized (Lund and Bottjer, 1991). Cretaceous plutons: $12.3° \pm 7.4°$ (Teissere and Beck, 1973), $13.2° \pm 6.8°$, and $4.6° \pm 6.0°$ (Hagstrum and others, 1985); Triassic chert, limestone, pillow basalt, and sandstone: $18.0° \pm 11.3°$ (Hagstrum and others, 1985); Jurassic clastic sedimentary rocks: $15.1° \pm 6.7°$ (Morris and others, 1986); Late Cretaceous sedimentary rocks: $18.2° \pm 6.7°$, $17.5° \pm 6.7°$, and $14.1° \pm 9.6°$ (Fry and others, 1985; Morris and others, 1986), $15.0° \pm 3.8°$ (Filmer and Kirschvink, 1989), and $17.7° \pm 6.9°$ (Smith and Busby-Spera, 1991); and Paleogene sedimentary rocks: $15.0° \pm 9.8°$ (Morris and others, 1986). Studies of Tertiary sedimentary and volcanic rocks indicate no resolvable motion with respect to North America since

about 40 Ma (Hagstrum and others, 1987; Lund and others, 1991a, b). Included within the values listed above is the widely accepted 2° to 3° of northward displacement of Baja California during Late Cenozoic opening of the Gulf of California.

Paleomagnetic data from Baja, including studies listed above and others summarized by Lund and Bottjer (1991) and Lund and others (1991b), display an average range of relative paleolatitude values of about 12° to 15°. The dispersion of values is not peculiar to Baja, and thus does not undermine the viability of the paleomagnetic data base. For example, such dispersion is characteristic of the paleomagnetic data base for cratonal North America, where an individual site may yield a relative paleolatitude as much as 12° different from the average of many sites on the craton (Lund and Bottjer, 1991). The importance of the Baja data base is that the *average* Cretaceous and Paleogene relative paleolatitudes of Baja are consistently lower than those of coeval North America (Lund and Bottjer, 1991; Lund and others, 1991b), and it is on this basis that we premise 15° of northward displacement of Baja since about 90 Ma.

The interpretation of Cretaceous and Paleogene northward displacement of Baja has been countered by numerous paleomagnetic and geologic arguments (Beck, 1991; Butler and others, 1991; Gastil, 1991; Lund and others, 1991b): (1) relative paleolatitudes from Baja are not statistically different from those of North America; (2) results may be due to experimental error; (3) results from the Peninsular Ranges batholith are due to tilting of the batholith about a subhorizontal axis; (4) in sedimentary rocks, the inclination may be flattened by compaction of incompletely lithified sediment or by very strong anisotropy in strongly magnetized rocks; (5) results are due to an irregularity in the geomagnetic field during the Cretaceous and early Paleogene; (6) results may reflect remagnetization; (7) no fault or faults have been identified along which Baja may have been displaced; and (8) correlations of geologic units between Baja and Sonora, such as Paleozoic sedimentary rocks and Cretaceous plutonic rocks, allow little net displacement. Below, we sequentially address each argument.

1. Recent studies of Late Cretaceous and early Paleogene strata on North America and in the Baja block offer very strong evidence that there are significant differences between data sets from the two areas (Lund and others, 1991b).

2. In modern paleomagnetic studies, experimental error ought to be random and less than about 20°, yet no Cretaceous paleomagnetic poles from Baja fall within a circle of radius 20° centered on the Cretaceous reference pole (Beck, 1991). Also, modern magnetic cleaning methods ensure that large amounts of present-day overprint are not present.

3. En masse tilting of the Peninsular Ranges batholith as a means of deriving anomalously low paleolatitudes is not supported by regional geologic relations or recent geobarometric studies of the batholith itself. The western part of the batholith apparently has not been tilted since at least the Turonian, based on the ages of flat-lying overlying strata (Gastil, 1991). Using amphibole barometry, Ague and Brandon (1992) defined pa-

leohorizontal in the northern part of the batholith that differed from a rather conjectural one used by Butler and others (1991). On the basis of the newly determined orientation of this surface, Ague and Brandon calculated about 1,000 ± 450 km of northward displacement of Baja since the emplacement of the batholith.

3. and 4. In order to explain all anomalous poles from Baja by batholith tilting and sediment compaction, the direction and amount of tilting must have produced paleomagnetic poles that converged with those produced by flattening and perhaps rotation in the sediments. Such a coincidence is possible but the probability is very small (Beck, 1991).

4. Inclination flattening of as much as 20° does not account for the distribution of Cretaceous poles from Baja (Beck, 1991). Likewise, studies of grain size versus magnetization in Cretaceous siltstones and sandstones indicate that compaction is not the cause of anomalously low inclinations (Smith and Busby-Spera, 1991).

5. It is very unlikely that the anomalous paleomagnetic poles from Baja are due to an irregularity of the Late Cretaceous paleomagnetic field (Beck, 1991). No irregularities are apparent in the Cretaceous paleomagnetic field over cratonal North America or over the world as a whole. Any irregularity somehow must have been focused on the western edge of North America, maintaining itself for at least 30 m.y. and tracking the leading edge of the continent as it moved westward over hotspots.

6. Almost all paleomagnetic poles from Baja are interpreted as carrying a primary magnetization. Only a few of these poles fall near the post-Cretaceous apparent polar wander path for North America, indicating that undetected remagnetization of all Baja poles has not occurred (Beck, 1991).

7. Possible locations of faults that may have accommodated displacement of Baja are discussed by Gastil (1991). In this volume, we adopt Gastil's Oaxaca-Baja Megashear II, a hypothetical fault or fault system that roughly coincides with the Gulf of California. The trace of this fault system apparently has been obliterated by late Cenozoic stretching, extension, subsidence, and sedimentation in the Gulf of California.

8. Large displacement of Baja along a fault system in the Gulf of California has been challenged on the basis of several geologic correlations across the gulf. However, regional geologic relations and correlations, such as reefal limestones in Cretaceous arc rocks, high-grade metamorphism associated with arc magmatism (Gastil, 1991), and compositional and isotopic trends within the Cretaceous batholith (Silver and others, 1979; Silver, 1986), have proven to be inconclusive tests of the proposed displacements. For example, we cannot dismiss the possibility that Cretaceous arc rocks in Baja correlate equally well with coeval arc rocks farther south in México, for which few comparable geochemical data are available.

A stronger correlation across the Gulf of California is the boundary between Paleozoic shelfal and Paleozoic basinal strata (Stewart and others, 1990; Gastil and others, 1991). This correlation may be reconciled with Cretaceous-Paleogene northward displacement of Baja only if the Paleozoic rocks in Baja earlier

had been displaced southward from correlative rocks in Sonora. In our reconstruction, we postulate a two-stage displacement history of Baja California involving Permian-Jurassic southward displacement and Cretaceous-Paleogene northward displacement. Some critics may charge that such a scenario relies too much on coincidence, but we counter that the perceived coincidence is a function of the timing of evolution relative to plate tectonics: if humans as earth scientists had evolved 10 m.y. in the future (assuming constant Pacific–North America relative motion), the coincidence would not be an issue inasmuch as the Paleozoic rocks in Baja and Sonora would not be adjacent. In fact, much of western North America appears to have been affected by southward displacement of terranes in the Triassic to the mid-Cretaceous, followed by northward displacement in the Late Cretaceous and Tertiary (Beck, 1989).

Other geologic evidence is consistent with Baja at the latitude of southern México in the mid-Mesozoic. Late Jurassic granite clasts in Late Jurassic conglomerate of the Cochimí terrane that contain mid-Proterozoic inherited or xenocrystic zircon may have been derived from either mid-Proterozoic to Mesozoic continental crust in southern México or early Proterozoic crust intruded by abundant mid-Proterozoic granitoids in northwestern México (Anderson and Silver, 1977a, b; Kimbrough and others, 1987). A southern source is tentatively inferred because most granitoids in northwestern México are slightly older and more alkalic than the conglomerate clasts in the Cochimí terrane (Kimbrough and others, 1987). Limestone blocks in flysch of the eastern subterrane of the Yuma terrane contain Jurassic ammonites that have stronger affinities to the south than to the north (Imlay, 1963, 1964). Late Cretaceous sedimentary rocks in northern Baja contain abundant *Coralliochama Orcutti* rudists that are absent from cooler water assemblages in the Great Valley of California (D. Bottjer, personal communication, 1992) and are consistent with, although not indicative of, a more southerly position.

The view that the Baja block and other Cordilleran blocks or terranes were subjected to large displacements requires that convergence between North America and subducting oceanic lithosphere to the west was oblique, at times highly oblique, for tens of millions of years (Beck, 1991). Plate motion calculations, although rather speculative for times as old as the Mesozoic, seem to support this conclusion. Another implication of the large displacements is that terrane dispersion was not obstructed by a continental buttress (Beck, 1991).

Cretaceous slip in the TMVB

Premise 13: A fault zone in central México now concealed by thick volcanic rocks of the Trans-Mexican Volcanic Belt accommodated about 435 km of dextral slip since the mid-Cretaceous.

The late Cenozoic Trans-Mexican Volcanic Belt (TMVB) in central México may conceal a preexisting fault zone of uncertain displacement history (Demant, 1978; Nixon, 1982). Gastil and

Jensky (1973) inferred about 435 km of post–mid-Cretaceous slip on a dextral fault system in order to realign batholithic and mineralization belts. Anderson and Schmidt (1983) proposed about 300 km of Jurassic slip on a sinistral fault system as a geometric requirement of their kinematic model. We are unable to assess the speculative sinistral offset in the Jurassic, but we accept the proposed post–mid-Cretaceous dextral slip.

Cenozoic translation of the Chortis block/terrane

Premise 14: The Chortis block was translated at least 1,000 km eastward on a sinistral fault system near the southern margin of México since the Eocene.

Proterozoic and Phanerozoic metamorphic basement rocks with a predominant north-south structural grain appear to have been truncated near the roughly east-west–trending Middle America trench in southern México (Figs. 3, 27) (de Cserna, 1967, 1971). Karig and others (1978) noted that this truncation may have occurred during Tertiary sinistral displacement of the Chortis block (Chortis terrane). Estimates of the amount and timing of sinistral displacement of Chortis range from 1,000 to 2,000 km since the middle Eocene to 150 km during the late Miocene (Burkart and others, 1987; Rosencrantz and others, 1988). The alternative of greater magnitude displacement is supported by the length of the truncated margin, based on length-slip magnitudes of modern faults, and the presence of a wide belt of highly deformed, locally mylonitic rocks in the Motagua fault zone (Erikson, 1990). West-northwest–trending structures in the Chatino terrane may have been produced by Paleogene transtension during eastward displacement of the Chortis block (Robinson and others, 1989, 1990; Robinson, 1991). The Cenozoic evolution of the southern margin of México is discussed further on pages 109–112.

Opening of the proto-Caribbean and Gulf of Mexico

Premise 15: The Gulf of Mexico formed when the Yucatán block was rifted from southern North America in the late Middle Jurassic to Late Jurassic.

Premise 16: An ocean basin formed near what is now the Caribbean Sea when South America was rifted from southern North America during Early Cretaceous time.

Western Pangea underwent intracontinental extension and rifting beginning in the Late Triassic. Late Triassic to Early Jurassic red beds and volcanic rocks at the western and northern margins of the Gulf of Mexico probably were deposited in grabens produced during an early phase of brittle extension. In Middle Jurassic time, extension was accommodated by ductile stretching of continental lithosphere in the U.S. Gulf Coast region and the northern margin of the Yucatán block. Continued rifting of Pangea caused latest Middle to Late Jurassic southward drifting of Yucatán from the U.S. Gulf Coast, with contemporaneous formation of oceanic lithosphere in the Gulf of Mexico, and Early Cretaceous opening of a "proto-Caribbean" ocean basin between the drifting North American and South American continents (Schlager and others, 1984; Buffler and Sawyer, 1985; Pindell, 1985; Dunbar and Sawyer, 1987; Klitgord and Schouten, 1987; Salvador, 1987; Pindell and others, 1988; Pindell and Barrett, 1990). Rifting and drifting of western Pangea is discussed further on pages 96–98.

MESOZOIC AND CENOZOIC EVOLUTION OF OCEANIC PLATES BORDERING MÉXICO

In this section we summarize the constraints imposed on the tectonic evolution of the Mexican region by studies of the oceanic plates that once lay to the west and south.

Western México

Using linear velocities predicted by plate motion studies based on global plate–hotspot circuits (Engebretson and others, 1985; DeMets and others, 1990) and on marine magnetic anomalies (Stock and Molnar, 1988; Stock and Hodges, 1989), we have calculated the normal and tangential components of relative motion between the North American plate and oceanic plates to the west over the last 180 Ma (Table 18). Normal and tangential components of relative velocity are shown for five sites on the North American plate. On the basis of global reconstructions of the orientation of the North American plate (e.g., Scotese and others, 1988), we base our calculations on a north-south trend of the plate boundary at the latitude of México and southern California prior to 161 Ma, a N20W trend between 161 and 85 Ma (roughly parallel to the Sierra Nevada batholith), and a N40W trend since 85 Ma (parallel to the modern San Andreas fault system in California). Geographic coordinates refer to a plate

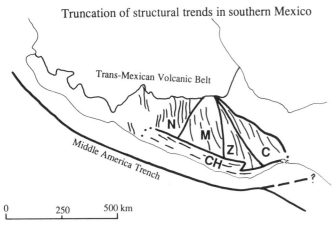

Figure 27. Apparent truncation of structural trends in metamorphic basement in southern México. Data from many sources, including INEGI (1980) and de Cserna (1989). Abbreviations for terranes: C, Cuicateco; Ch, Chatino; M, Mixteco; N, Nahuatl; Z, Zapoteco.

TABLE 18. CALCULATED MOTIONS OF OCEANIC PLATES RELATIVE TO NORTH AMERICA SINCE 180 Ma*

Coordinates Pre–5 Ma / Post–5 Ma	38°N, 123°W / 38°N, 123°W		33°N, 118°W / 35°N, 120°W		30°N, 114°W / 32°N, 116°W		26°N, 112°W / 28°N, 114°W		21°N, 108°W / 23°N, 110°W		Subducted Plate
Time (Ma)	Norm	Tang	Norm	Tang	Norm	Tang	Norm	Tang	Norm	Tang	
0–5	10 c	50 r	5 c	50 r	2 c	49 r	3 e	49 r	7 e	49 r	Pacific
5–17	45 c	1 l	27 e	48 r	24 e	50 r	24 e	50 r	19 e	50 r	
17	40 c	15 r	3 e	25 f	3 e	25 r	3 e	25 r	3 e	25 r	Cocos
			40 c	12 r			90 c	27 r	102 c	25 r	
28	45 c	20 r	45 c	18 r	10 e	50 r	73 c	5 r	74 c	3 r	
36	70 c	8 r	75 c	4 r	80 c	0					
42	70 c	20 l	80 c	25 l	95 c	30 l	105 c	34 l	105 c	18 l	Farallon
50	100 c	27 l	105 c	30 l	120 c	38 l	135 c	43 l	135 c	32 l	
59	140 c	10 l	145 c	12 l	145 c	8 l	135 c	2 l	125 c	4 r	
68	110 c	90 r	105 c	87 r	120 c	26 r	120 c	26 r	115 c	20 r	
74	120 c	58 r	115 c	56 r	110 c	54 r	110 c	21 l	100 c	24 l	Kula
85	80 c	110 r	75 c	110 r	75 c	105 r	75 c	96 r	75 c	35 r	
100	90 c	65 r	90 c	60 r	90 c	55 r	85 c	52 r	80 c	48 r	
119	56 c	17 r	59 c	15 r	62 c	13 r	62 c	13 r	65 c	10 r	
135	80 c	17 l	75 c	21 l	65 c	25 l	55 c	30 l	45 c	35 l	Farallon
145	21 c	32 l	19 c	33 l	17 c	34 l	15 c	34 l	15 c	35 l	
161	100 c	26 l	100 c	28 l	100 c	34 l	100 c	37 l	95 c	40 l	
180	125 c	23 l	125 c	26 l	125 c	30 l	125 c	35 l	120 c	40 l	

Left axis: Assumed orientation of plate boundary — N40°W / N20°W (at 85 Ma); N20°W / N-S (at 161 Ma). Vertical axis: Time before Present (Ma).

*Rates in millimeters/year = kilometers/Ma; errors not shown. Norm, Tang = normal and tangential components of convergence; c = extension; r = right-lateral; l = left-lateral. Velocities calculated using linear velocities and stage poles of Engebretson and others (1985), Pindell and others (1988), Stock and Molnar (1988), Stock and Hodges (1989), and DeMets and others (1990).

margin for which 300 km of post–5.5-Ma displacement has been restored in the Gulf of California (Stock and Hodges, 1989; Sedlock and Hamilton, 1991). We premise that the Kula plate was subducted beneath western México with a large component of right obliquity in the Late Cretaceous, following the "southern option" of Engebretson and others (1985). Alternatively, a different, unnamed plate moving with a large component of dextral tangential motion relative to North America may have lain adjacent to México in the Late Cretaceous (T. Atwater, D. Engebretson, presentations at U.S. Geological Survey, January 1990). Uncertainties associated with the calculated rates in Table 18 are large, particularly for the Mesozoic, so we do not premise velocities for specific times or sites in our reconstruction. Rather, we regard the rates as a guide to the expected sense of obliquity and the times of changes in relative motion, as has proven valuable in geologic studies of the western United States, Canada, and Alaska (e.g., Page and Engebretson, 1984). It should be noted that the amount of tangential relative motion that was accommodated in the forearc and arc of a particular subduction zone probably was significantly less than the values shown in Table 18; in modern obliquely convergent margins the ratio is less than 0.5 (Jarrard, 1986).

The following general history at Mexican latitudes (south of 32°N) may be inferred from Table 18. From 180 until 145 Ma, rapid convergence of the Farallon and North America plates included a normal component of ⩽100 mm/yr and a sinistral component of about 30 to 40 mm/yr. The component of sinistral slip may have increased more markedly than shown during the Late Jurassic (161 to 145 Ma) due to an increase in the northward absolute motion of the North America plate (May and Butler, 1986). Convergence slowed dramatically about 145 Ma, although the sinistral component remained at about 35 mm/yr. From 135 until 100 Ma, moderate convergence included a normal component of 60 mm/yr and a tangential component that changed from sinistral to dextral about 119 Ma. From 100 to 74 Ma at latitudes south of about 23°N and until 85 Ma at latitudes north of 23°N, the Farallon plate was subducted moderately rapidly with a dextral component of 35 to 55 mm/yr. About 85 Ma, spreading was initiated between the Farallon plate and the newly formed Kula plate (and perhaps another, unidentified plate) at the latitude of western México (Kula plate labeled in right column and outlined by heavy lines in Table 18). The Kula plate was subducted beneath North America rapidly and with a large dextral component until about 59 Ma (Woods and Davies, 1982; Engebretson and others, 1985). Throughout the Late Cretaceous and Paleogene, the Farallon plate was subducted nearly orthogonally beneath México, with a minor component (0 to 40 mm/yr) of either dextral or sinistral motion. Ward (1991) suggested that Farallon–North America motion 36 to 20 Ma at the latitude of northern México (32° to 26°N) slowed greatly or ceased due to the unsubductibility of very young Farallon lithosphere.

Subduction beneath western North America has been superseded by dextral shear along the lengthening Pacific–North America transform plate boundary since the intersection of the Pacific-Farallon spreading ridge with the trench about 25 Ma (Atwater, 1970, 1989). Neogene and Quaternary Pacific–North America motion at Mexican latitudes has been transtensional, with large components of boundary–normal extension in early Miocene and late Miocene time (Pacific plate labeled in right column and outlined in heavy lines in Table 18). The dextral tangential component of relative motion at Mexican latitudes was markedly slower in the interval 20 to 11 Ma than at other times. South of the transform margin, subduction of the southern part of the old Farallon plate (now called the Cocos plate, labeled in right column and outlined in heavy lines in Table 18) continues. Models of the tectonic evolution of parts of the transform margin are presented by Atwater (1970, 1989), Dickinson and Snyder (1979), Lonsdale (1989, 1991), Lyle and Ness (1991), and Sedlock and Hamilton (1991).

Caribbean region

The Caribbean basin and the Gulf of Mexico formed in the larger context of southeastward drift of South America relative to North America. Relative motion of South America was to the southeast at about 30 mm/yr between the early Middle Jurassic (about 180 to 175 Ma) and early Late Cretaceous (100 to 84 Ma) and has been small and nonuniform since that time (Klitgord and Schouten, 1987; Pindell and others, 1988; Pindell and Barrett, 1990). Although Mesozoic growth of the gap between the continents is well established, the processes of gap-filling by horizontal stretching, sea-floor spreading, and terrane migration are not fully understood. The Yucatán platform withdrew to the southeast as part of South America until the Late Jurassic. The kinematics of Yucatán displacement and, thus, of the evolution of the Gulf of Mexico are uncertain, but the following sequence of events is likely (Buffler and Sawyer, 1985; Dunbar and Sawyer, 1987; Salvador, 1987; Winker and Buffler, 1988; R. Buffler, personal communication, 1988): (1) Late Triassic(?) to Early Jurassic brittle extension and graben formation, with unknown but probably negligible cumulative displacement; (2) Middle Jurassic ductile horizontal stretching of continental lithosphere in the gulf region, with perhaps 600 km cumulative stretching in the direction of North America–South America drifting; and (3) Late Jurassic sea-floor spreading, with about 450 km of cumulative displacement in the direction of North America–South America drifting. As a result of this deformation, the continental crust at the northern margin of the southern Maya terrane (Yucatán platform) was strongly stretched and intruded by dikes. When Yucatán–North America relative motion ceased near the end of the Jurassic, relict South American crust of the Maya terrane became part of the North American plate. South America began to drift away from the southern margin of North America (Maya terrane) in the earliest Cretaceous, creating an intervening Caribbean ocean basin. The Caribbean basin attained its current (maximum) dimensions by early Late Cretaceous time. A Jurassic seaway between North America and South America, inferred on the basis of faunal distributions, probably covered submerged pre-drift Pangean continental crust.

At least three hypotheses have been advanced to explain the origin of the oceanic lithosphere that currently occupies the gap between North America and South America. First, Caribbean lithosphere may have formed in place during the Cretaceous drifting of South America away from North America (Freeland and Dietz, 1972; Salvador and Green, 1980; Klitgord and Schouten, 1986). Second, the Caribbean lithosphere may be Jurassic Farallon plate that was inserted eastward into the growing gap between the continents during the Early and early Late Cretaceous and displaced progressively eastward during the Cenozoic (Malfait and Dinkelman, 1972). Third, in-place Caribbean lithosphere that formed during Cretaceous drifting was consumed beneath an eastward-migrating salient of the Farallon plate that became independent of the Farallon (Cocos) plate in the Tertiary (Pindell and Dewey, 1982; Sykes and others, 1982; Burke and others, 1984; Pindell and others, 1988; Pindell and Barrett, 1990).

We employ the third hypothesis because it is most consistent with up to 1,100 km of late Eocene(?) and younger sinistral displacement on the Caribbean–North America plate boundary in the Cayman Trough (Macdonald and Holcombe, 1978; Rosencrantz and Sclater, 1986; Rosencrantz and others, 1988) and with early Paleogene opening of the Yucatán Basin (Rosencrantz, 1990). In most versions of the third hypothesis, an island arc terrane, speculatively identified as Cuba, Jamaica, or the Nicaragua Rise, collided with the southern Maya terrane in the latest Cretaceous and subsequently was translated eastward to northeastward into the Caribbean basin (Pindell and Dewey, 1982; Burke and others, 1984; Pindell and others, 1988; Pindell and Barrett, 1990). In an alternate model, the southern Maya terrane collided with the northern, continental arc, margin of the Chortis terrane, which subsequently underwent no more than a few hundred kilometers of sinistral displacement in Guatemala and several hundred kilometers of net east-west extension during the Cenozoic (Donnelly, 1989). Many important aspects of these models may prove to be untestable because it is so difficult to determine the location of plate boundaries and the timing and kinematics of displacement during the Jurassic to early Tertiary, particularly in and near the western margin of the proto-Caribbean basin—that is, south of México.

RECONSTRUCTION OF THE TECTONIC EVOLUTION OF MÉXICO

Precambrian to Devonian

Early to middle Proterozoic basement. The oldest rocks in México crop out in northwestern Sonora, south of the inferred trace of the Mojave-Sonora Megashear, and are inferred to underlie the entire Serí terrane. Metasedimentary and metavolcanic rocks were intruded by calc-alkalic plutons about 1,750 to 1,710 Ma, and younger layered gneisses were deformed and metamorphosed about 1,685 to 1,645 Ma (Anderson and Silver, 1977b, 1981). These rocks were probably displaced southeastward by the megashear from a northeast-trending belt of similar rocks in the southwestern United States (Figs. 22, 25) (Anderson and Silver, 1979). Slightly younger (1,650-Ma metamorphic age) Proterozoic rocks crop out in northern Sonora north of the megashear and are probably correlative with a northeast-trending belt of similar rocks that includes the Pinal Schist in southern Arizona (Anderson and Silver, 1977b, 1979, 1981). Both suites of metamorphic rocks were intruded by plutons that probably are part of a belt of anorogenic granitoids extending from southern California to the midcontinent (e.g., Anderson, 1983) and by much less abundant 1,100-Ma granites (Anderson and Silver, 1977a, b, 1981).

Grenville basement. The early to middle Proterozoic metamorphic rocks described above are bounded to the southeast by a northeast-trending belt of rocks that have yielded Grenville (1,300 to 1,000 Ma) radiometric dates (Fig. 24) (Bickford, 1988). The Grenville belt apparently continues southwestward from central Texas to northern and central Chihuahua, where ~1,000-Ma amphibolite dikes cut ~1,300-Ma granite. Although some outcrops are fault-bounded, we infer that the Grenville rocks in Chihuahua are autochthonous or little displaced with respect to the main Grenville belt. The original southwestward extent of Grenville and older Proterozoic rocks is uncertain, but the apparent termination of contiguous Precambrian basement at the latitude of central or southern Chihuahua implies truncation of the southern margin of North America in this area by rifting or strike-slip faulting (Stewart, 1988). The timing of truncation is bracketed by the youngest ages of Grenville rocks (about 1,000 Ma) and by the nonconformable deposition of latest Precambrian to Cambrian passive-margin facies on eroded crystalline basement in the Serí terrane and southern North America (about 600 Ma). We infer that these truncated margins evolved into the latest Precambrian to early Paleozoic west-trending (present coordinates) passive margin in Chihuahua and Coahuila and the northwest-trending passive margin between Sonora and eastern California (Premises 3, 4; Fig. 23).

Basement rocks elsewhere in México and in central Cuba that have yielded Grenville radiometric dates include gneiss in the northern and southern Guachichil terrane, at least part of the Acatlán Complex in the Mixteco terrane, the Oaxacan Complex in the Zapoteco terrane, at least part of the Chuacús Group in the southern Maya terrane, and allochthonous metasedimentary rocks in central Cuba that have yielded K-Ar and $^{40}Ar/^{39}Ar$ cooling ages on phlogopite of 950 to 900 Ma (Renne and others, 1989). Also, lower crustal xenoliths in the central Tepehuano terrane contain some zircons of Grenville age and have yielded $^{207}Pb/^{206}Pb$ ages of Grenville and younger age. Unambiguous correlations between these rocks and those of the autochthonous North American Grenville province have not been demonstrated, but, like Stewart (1988), we infer that the former are allochthonous fragments that were separated from the latter in the latest Proterozoic (1,000 to 600 Ma). Some allochthons may have been tectonically removed from the southwestern end of the North American Grenville belt during latest Proterozoic truncation of unknown sense and displacement. Other allochthons may be parts of the Grenville belt in eastern North America that were displaced into the Iapetus ocean basin during latest Precambrian to Cambrian rifting and then southwestward during Paleozoic (Acadian) strike-slip faulting. Few distinctive features tie the orphaned Grenville blocks to specific parts of the Grenville province. We provisionally infer that most of the allochthonous Grenville rocks, which are lithologically similar to Grenville basement in Chihuahua and Texas, were derived from the southern North American Grenville province.

Paleontologic and paleomagnetic data indicate that the Oaxacan Complex of the Zapoteco terrane, although grossly similar to coeval rocks in Chihuahua and western Texas, did not originate in the southern North American Grenville province. Early Ordovician strata that unconformably overlie the Oaxacan Complex contain trilobites that are dissimilar to those of the southwestern United States but have strong affinities with fauna in Argentina, southeastern Canada, Scandinavia, and Great Britain (Robison and Pantoja-Alor, 1968; Whittington and Hughes, 1972). The paleopole to primary magnetization in the Oaxacan Complex is >40° from the Grenville loop of the North American polar wander path, which can be interpreted to show that these rocks were adjacent to Grenville rocks near southeastern Canada about 1,000 Ma (Ballard and others, 1989). According to the PALEOMAP global plate reconstruction, southeastern Canada was contiguous with Scandinavia until the latest Proterozoic to Cambrian opening of the Iapetus Ocean (Scotese and McKerrow, 1990; C. Scotese and others, unpublished data). The paleontologic and paleomagnetic data are consistent with the following tectonic history. The Oaxacan Complex formed adjacent to Grenville rocks in southeastern Canada and southern Scandinavia about 1,300 to 1,000 Ma (Fig. 28A). In the latest Proterozoic to Cambrian, the Oaxacan Complex was rifted away from southeastern Canada and either remained attached to Scandinavia or was stranded as a microcontinent within the growing proto-Tethys Ocean (Fig. 28B). Faunal similarities imply that the Oaxacan Complex was not far from either Scandinavia or southeastern Canada in the earliest Ordovician.

Alternatively, the evolution of the Zapoteco terrane can be modeled by emphasizing the similarity of Early Ordovician fauna above the Oaxacan Complex to coeval fauna in Argentina. The global plate reconstruction proposed in the SWEAT hypothesis (Dalziel, 1991) places the North American Grenville province adjacent to southern South America at the end of the Proterozoic. The exact location of the Oaxacan Complex in this reconstruction is impossible to determine, but the dissimilarity of the Early Ordovician trilobite fauna of the Zapoteco terrane to coeval fauna in the southwestern United States suggests that by the early Paleozoic the Zapoteco terrane lay east of the Patagonia assemblage, i.e., in a position relative to North America similar to that predicted by the southeastern Canada–Scandinavia link (Fig. 28; cf. Fig. 3 of Dalziel, 1991). Thus, according to either model, the Zapoteco terrane probably lay east of the North American Grenville belt in the early Paleozoic.

We propose that during Ordovician-Silurian diachronous

closure of the proto-Tethys Ocean, the Zapoteco microcontinent was translated westward and southward through proto-Tethys by margin-parallel right-slip during late Taconian and/or Acadian orogenesis. Platformal sedimentation probably occurred in the open ocean southeast and perhaps well outboard of eastern North America. In the Early to Middle Devonian, the Zapoteco terrane collided with the Acatlán Complex (basement of the Mixteco terrane), which consists of an ophiolite of unknown age that was obducted in the early or middle Paleozoic onto a subduction complex containing sedimentary rocks derived from a Grenville-aged source (Ortega-Gutiérrez, 1981a, b; Robinson and others, 1989; Yañez and others, 1991). An Early to Middle Devonian age of collision is indicated by the ages of metamorphism and deformation of the Acatlán Complex, the ages of granitic intrusion of basement in both terranes, and the presence of clasts of both basement terranes in Late Devonian marine strata in the Acatlán Complex (Ortega-Gutiérrez, 1978b, 1981a, b; Ortega-Gutiérrez and others, 1990; Yañez and others, 1991). The collision probably occurred at a boundary with a convergent component within the oceanic realm south of eastern North America (Fig. 28C), but Paleozoic plate configurations, boundaries, and relative motions are insufficiently known to specify the location, nature, and orientation of that boundary. The collision has been interpreted to be a result of Acadian orogenesis (Yañez and others, 1991). Late Paleozoic translation of the amalgamated terranes to a position west of southwestern North America (Fig. 28D) is discussed below.

As postulated in Premise 7, we consider the exposures of Grenville rocks in Chihuahua, eastern México, and Oaxaca to be discrete, fault-bounded fragments derived from the North American Grenville province, rather than culminations of continuous Grenvillian basement in eastern and central México. Continuity of Grenville basement into central México is considered unlikely because the western edge of the Maya terrane probably was a major kinematic boundary in the late Paleozoic to mid-Jurassic (pp. 94–103). Our hypothesis is that much or most of México is underlain by Grenville and Paleozoic basement blocks of diverse origin, including those described above, and that the basement blocks have been displaced with respect to North America and to one another during the complex younger tectonic history of the region.

Latest Proterozoic to early Paleozoic basement. Outcrops or subcrops of igneous and metamorphic continental basement in several parts of southern and eastern México are inferred to be of latest Proterozoic to early or middle Paleozoic age (Fig. 29). The Las Ovejas Complex in the Chortis terrane contains amphibolite, gneiss, and migmatite that yielded a poor Rb-Sr isochron of 720 ± 260 Ma (Horne and others, 1976). Metasedimentary and metaplutonic rocks in the Chiapas Massif in the southern Maya terrane were derived from protoliths of probable Late Proterozoic to early Paleozoic age, and are overlain nonconformably by undeformed late Paleozoic strata. Deep Sea Drilling Project cores obtained from the Catoche Knolls in the Gulf of México (Fig. 29) contain gneiss, amphibolite, and phyllite that yielded early Paleozoic (~500 Ma) ages (Schlager

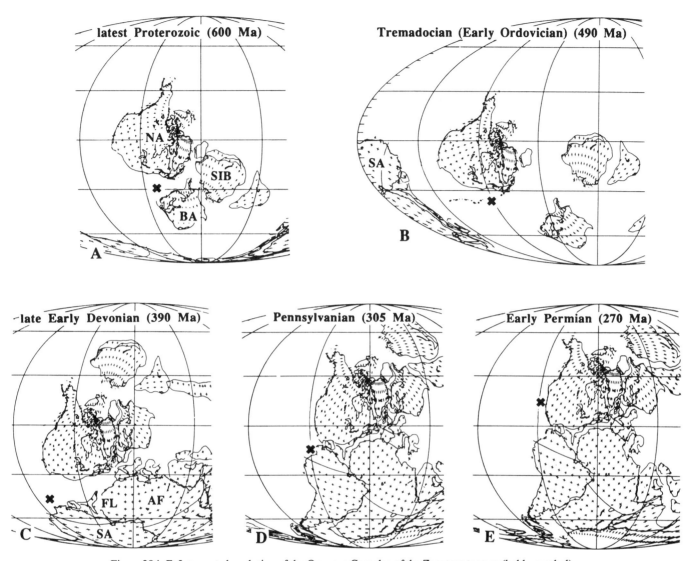

Figure 28A-E. Interpreted evolution of the Oaxacan Complex of the Zapoteco terrane (bold x symbol), in framework of reconstruction of Scotese and McKerrow (1990). Abbreviations: AF, Africa; BA, Baltica; FL, Florida; NA, North America; SA, South America; SIB, Siberia. Amalgamation with Acatlán Complex of the Mixteco terrane probably occurred in the Early to Middle Devonian. Location and orientation of strike-slip faults (not shown) responsible for translation of Zapoteco terrane are uncertain.

and others, 1984). Scant radiometric dates from schist, granite, gneiss, quartzite, and rhyolite in wells in the Coahuiltecano terrane and the Yucatán Peninsula (Maya terrane) imply middle Paleozoic magmatism and metamorphism. We infer that all of these rocks are relics of South American continental crust that were stranded by Cretaceous drifting of South America away from North America. We discuss the poorly constrained kinematics of these rocks in a section below.

Unmetamorphosed Silurian strata in the northern Guachichil terrane contain schist clasts similar to the nearby Granjeno Schist, implying a pre-Silurian age for the schist. However, radiometric dates from the Granjeno Schist imply late Paleozoic (330

to 260 Ma) metamorphism, suggesting to us that the schist clasts were derived from a different source and that the Granjeno Schist may be no older than Mississippian (p. 27).

Early Paleozoic sedimentary rocks. The passive margin at the southern and southwestern edges of North America persisted from latest Proterozoic to middle or late Paleozoic time and probably was the site of deposition of cratonal and shelfal ("miogeoclinal") strata. Basinal ("eugeoclinal") rocks probably were deposited in deeper water farther offshore to the south and west.

Cratonal strata were deposited on Proterozoic North American basement in northern Chihuahua and Sonora and the southwestern United States from at least the Ordovician until the

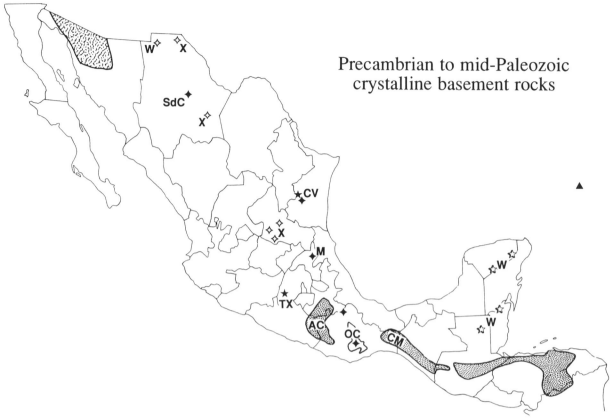

Figure 29. Outcrops and subcrops of upper Proterozoic to middle Paleozoic crystalline basement rocks in México and northern Central America. Irregular pattern indicates mid-Proterozoic rocks in Sonora; filled diamonds, outcrops of Grenville rocks at Sierra del Cuervo (SdC), Ciudad Victoria (CV), Molango (M), and in Oaxacan Complex (OC); open diamonds, subcrops of Grenville rocks inferred from wells (W) and xenoliths (X); dot pattern, outcrop of uppermost Proterozoic(?) to lower Paleozoic basement rocks, including Acatlán Complex (AC) and Chiapas Massif (CM); filled triangle, lower Paleozoic rocks at Catoche Knoll, DSDP Leg 77, Hole 538A; filled stars, outcrops of crystalline basement rocks of probable mid-Paleozoic age at Ciudad Victoria (CV) and in Taxco region (TX); open stars, subcrops from wells (W) of crystalline basement of possible but unproven mid-Paleozoic age.

Mississippian (Imlay, 1939; Greenwood and others, 1977). These strata were mantled to the south and west by latest Proterozoic-Devonian shallow-water shelfal strata in Sonora and Baja California (Serí terrane) that strongly resemble coeval rocks in the Death Valley and San Bernardino–western Mojave regions of California. The shelfal rocks thicken away from the craton and are overthrust by allochthonous siliceous and detrital basinal rocks of Ordovician-Mississippian age (Serí terrane) that probably were deposited on Paleozoic proto-Pacific ocean floor an unknown distance south or west of the edge of North America. These early and middle Paleozoic deep-water rocks are similar in some respects to those in the Roberts Mountain allochthon in Nevada and, to a lesser degree, rocks in the Ouachita orogenic belt (Stewart, 1988; Stewart and others, 1990), but there are significant differences in the timing of deformation and emplacement over shelfal rocks (see below).

Early and middle Paleozoic shelfal rocks in the northern Guachichil terrane are now faulted against Grenville gneiss in a structural position analogous to that of the Talladega slate in the southern Appalachians (Tull and others, 1988), but initially they may have been deposited on the gneiss in or outboard of an unidentified sector of the passive southern margin of North America.

Devonian and Carboniferous

Deposition and deformation of shelfal and basinal strata. Shelfal and basinal rocks were deposited in shelf, slope, and ocean-floor environments south and west of the passive margin of southwestern North America during the Devonian and Carboniferous (Fig. 23). We infer that late Paleozoic basinal strata in the northern Tahué terrane, eastern Nahuatl terrane, and possibly the Pericú terrane were deposited on oceanic or transitional crust at an indeterminate distance from North America.

The Tahué terrane contains Carboniferous to Early Permian clastic rocks, siliceous to intermediate volcanic rocks, chert, and thin carbonates that are grossly similar to Paleozoic rocks in northeastern Baja California (López-Ramos, 1985, p. 9). Basement beneath these strata is not exposed, but isotopic ratios of younger plutons imply a lack of thick sialic continental crust (Damon and others, 1983). In the Tierra Caliente Complex of the eastern Nahuatl terrane, protoliths of metasedimentary and meta-igneous rocks appear to be Paleozoic, at least in part, and may have been deposited on transitional crust (Elías-Herrera, 1989). In the Pericú terrane, strongly deformed and metamorphosed sedimentary protoliths deposited on unknown basement have yielded late Paleozoic metamorphic dates. Because these terranes probably arrived at the active margin of western México later than most terranes, we infer that they were in the northern proto-Pacific basin in the Paleozoic (pp. 99–100).

The evolution of early to middle Paleozoic basinal strata differed with position around the southern and western margins of North America. In the Great Basin, basinal strata were deformed prior to and during the Devonian and thrust eastward onto shelfal strata during two distinct events: the Late Devonian–Early Mississippian Antler orogeny (Roberts Mountain allochthon) and the Late Permian or Early Triassic Sonoma orogeny (Golconda allochthon) (Roberts and others, 1958). In the western Ouachitan orogen, basinal strata were deformed and incorporated into a growing subduction complex at the southern margin of North America during the Mississippian and thrust northward onto the North American shelf during Pennsylvanian and earliest Permian time (Viele and Thomas, 1989). In Sonora, they were internally deformed at an unknown location during the early Late Mississippian and thrust northward(?) onto shelfal strata in Late Permian to Middle Triassic time (Poole and Madrid, 1988; Stewart and others, 1990). Because deformation in Sonora was west of the Permo-Triassic arc at the western margin of Pangea, we infer that it was not caused by collision of Gondwana but rather by accretion of an unidentified island arc or microcontinent.

Magmatism, metamorphism, and tectonism. Several fragments that contain continental basement of Grenville or early to middle Paleozoic age experienced magmatism and regional metamorphism during the Devonian and Carboniferous. The Granjeno Schist probably was metamorphosed to greenschist facies in Mississippian time, prior to its faulting against Grenville gneiss of the northern Guachichil terrane. Early Paleozoic gneiss at Catoche Knolls in the Gulf of Mexico yielded a $^{40}Ar/^{39}Ar$ plateau age (biotite) of 350 Ma that probably reflects a Mississippian thermal overprint (Schlager and others, 1984). Wells in the Coahuiltecano terrane and Maya terrane penetrated Carboniferous granitic gneiss and metaandesite, respectively (Denison and others, 1969; Marshall, 1984). Gneiss, schist, and low-grade metamorphic rocks that are inferred to underlie the Coahuiltecano terrane probably were metamorphosed in the Devonian to Carboniferous. Late Proterozoic(?) to mid-Paleozoic basement rocks of the Maya terrane in Guatemala and Belize were intruded by granitoids in the Mississippian and were locally overlain by

silicic lavas and pyroclastic rocks in the Late Pennsylvanian and Early Permian. The Acatlán Complex in the Mixteco terrane was intruded by the Esperanza granitoids, penetratively deformed and metamorphosed in the Late Silurian to Middle Devonian, and deformed and metamorphosed during the latest Devonian to Carboniferous (Ortega-Gutiérrez, 1978a, b; Yañez and others, 1991). Late Proterozoic to mid-Paleozoic basement rocks in the Chortis terrane were intruded by granitoids in the Pennsylvanian.

We infer that Paleozoic magmatism and metamorphism in Mexican terranes probably were related to consumption of oceanic lithosphere at one or more subduction zones between Gondwana and the passive southern margin of North America (Fig. 30). Throughout early and middle Paleozoic time, Gondwana was an indeterminate distance south and east of North America. Plate reconstructions indicate that northward motion of Gondwana with respect to North America and the consumption of intervening oceanic lithosphere was underway at least as early as Devonian time (Scotese and McKerrow, 1990). Deep-water strata deposited in the ocean basin between Gondwana and southern North America were incorporated into a subduction complex by Mississippian time, implying southward subduction of oceanic lithosphere at a trench at the northern edge of Gondwana (Fig. 30) (Viele and Thomas, 1989).

The positions and displacements of Mexican basement blocks with respect to Gondwana and North America cannot be pinpointed with available data. Below we outline some schematic tectonic histories that pertain only to those basement blocks that were affected by middle to late Paleozoic magmatism or metamorphism and thus, presumably, were not part of the passive southern margin of North America. The few Mexican basement blocks unaffected by Devonian-Carboniferous magmatism or metamorphism, e.g., southern Guachichil terrane and northern Guachichil terrane, may have been attached to North America.

1. Basement blocks were part of the forearc or arc along the northern edge of Gondwana (label B in Fig. 30), where they were intruded and metamorphosed during southward subduction of oceanic lithosphere beneath Gondwana. As integral parts of the northern edge of Gondwana, the blocks were penetratively deformed and accumulated thick synorogenic sediments during late Paleozoic collision with southern North America.

2. Basement blocks were intruded and metamorphosed in the Gondwana forearc or arc, as above, but were stripped from northern Gondwana prior to late Paleozoic Ouachitan collision and translated westward or southwestward (present coordinates) to an ocean basin south of the southwestern United States (west of label A in Fig. 30), presumably due to diachronous oblique convergence between Gondwana and North America.

3. Basement blocks were intruded or metamorphosed at one or more cryptic subduction zones or collision zones north of the major subduction zone on the northern margin of Gondwana (label C in Fig. 30). In the middle Paleozoic, the blocks were isolated continental fragments embedded in oceanic or transitional crust an unknown distance from each other and from

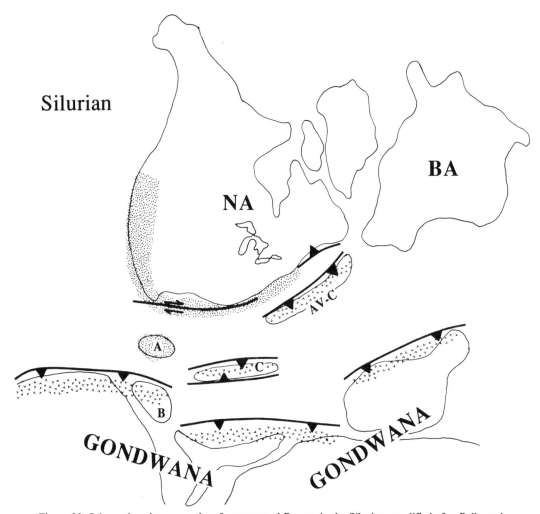

Figure 30. Schematic paleogeography of west-central Pangea in the Silurian, modified after Bally and others (1989). Blocks A, B, and C are schematic and not directly representative of existing blocks (see text). Stipple pattern indicates deposition of passive margin facies; v pattern, magmatic arcs at northern margin of South America, in Avalon-Carolina (AV-C) terrane, and in block C; BA, Baltica; NA, North America. Collision of Avalonia-Carolina terrane may have been completed by Silurian time.

North America and Gondwana (labels A, C, AV-C in Fig. 30). Trapped between the converging North America and Gondwana plates, the blocks were deformed during accretion to Gondwana or late Paleozoic Gondwana–North America collision. The tectonic evolution of these blocks may resemble that of the Sabine block (terrane) in the south-central United States (Viele and Thomas, 1989; Mickus and Keller, 1992).

4. Basement blocks were intruded or metamorphosed at one or more cryptic subduction zones or collision zones, as above, but they escaped Ouachitan collision by southwestward translation to the ocean basin south of the southwestern United States, perhaps due to diachronous oblique convergence between North America and Gondwana.

Interpreting the tectonic history of Mexican basement blocks in terms of the above alternatives is frustrated by limited outcrop, incomplete radiometric dating and structural analysis, and complex Mesozoic tectonic overprints. Paleozoic plate kinematics cannot be determined because Paleozoic oceanic lithosphere, which must have bounded the basement blocks, has been completely subducted or is masked by Mesozoic and Cenozoic cover. Here, we speculate briefly on the Paleozoic tectonic history of individual basement blocks.

The events leading to the Devonian amalgamation of the Zapoteco and Mixteco terranes were discussed above. After deposition of Late Devonian marine strata in the Mixteco terrane, the amalgamated terranes were folded, weakly foliated, and metamorphosed at high temperatures during the (Early?) Carboniferous, probably during collision with a continental mass, and unconformably overlain by little-deformed Carboniferous to Permian(?) marine and continental strata. We interpret early(?)

Carboniferous deformation and metamorphism of the amalgamated Zapoteco and Mixteco terranes as a result of collision with and incorporation into the Gondwanan forearc. The Zapoteco-Mixteco composite terrane lacks deformation associated with late Paleozoic Ouachitan continental collision, so we infer that it was within the westernmost, west-facing part of the Gondwanan forearc and was swept westward or northwestward into the proto-Pacific ocean basin past the southern termination of the Tarahumara-Ouachita orogenic belt. By the Permian, the composite terrane lay west and perhaps north of modern México (Fig. 28D).

Late Proterozoic rocks in the northern Guachichil terrane, southern Guachichil terrane, and central Cuba may have been rifted from the southern end of the Grenville belt in the latest Proterozoic (p. 85). We propose that during Paleozoic time these basement blocks occupied the ocean basin south of southwestern North America, an indeterminate distance south of the continent, where Paleozoic shelfal strata in the northern Guachichil terrane may have been deposited on Grenville gneiss.

Available data permit at least three interpretations of the evolution of the Granjeno Schist of the northern Guachichil terrane. First, the schist may be a metamorphosed subduction complex that formed at a trench, either at the northern margin of Gondwana or within the ocean basin between Gondwana and North America. Serpentinite that structurally underlies the schist may have been derived from either upper plate asthenosphere or subducted oceanic lithosphere. Second, protoliths of the schist may have been deposited on the flank of a microcontinental block in the ocean basin between Gondwana and North America (label A in Fig. 30) and metamorphosed and deformed during accretion of the block to the Gondwanan forearc. The Granjeno Schist is now outboard (west) of the Permo-Triassic arc that developed at the western margin of Pangea; in either case, it was translated westward during the late Paleozoic. A third interpretation is that metamorphism and deformation of the protoliths, whether in a subduction complex or during collision, occurred in the ocean basin south of southwestern North America.

In the discussion of Premise 6 we proposed that basement rocks of probable or possible Late Proterozoic to early Paleozoic age in the Maya and Coahuiltecano terranes are remnants of Gondwana. We infer that middle to late Paleozoic metamorphism and plutonism that have been demonstrated or inferred in these terranes occurred in the arc or forearc of northern Gondwana (label B in Fig. 30). Mississippian plutonism and metamorphism and Pennsylvanian volcanism in Guatemala and Belize imply that much or most of the Maya terrane was part of the continental magmatic arc of northern Gondwanan. The Coahuiltecano terrane, which has yielded a few mid(?)-Paleozoic radiometric dates, bounds the Ouachita orogenic belt on the south and probably is a salient of the Gondwanan forearc.

On the strength of a single Rb-Sr isochron indicating Pennsylvanian plutonism, we speculate that the Chortis terrane also was part of the magmatic arc of northern Gondwana. However, given the incomplete understanding of the deformational, mag-

matic, and metamorphic history of the Chortis terrane, it certainly is possible that plutonism occurred in a tectonic setting unrelated to Ouachitan orogenesis.

Carboniferous and Early Permian

Formation of Pangea. Convergence between North America and Gondwana resulted in complete consumption of intervening oceanic lithosphere, diachronous Mississippian to mid-Permian collision between the two continents, and the formation of Pangea (Premise 5). During convergence but prior to continental collision, early Paleozoic basinal or "off-shelf" strata were stripped from the southern, oceanic, part of the North America plate and accreted to the forearc of Gondwana; such rocks form the cores of the Benton and Marathon uplifts in the Ouachitan orogen. Mississippian and Pennsylvanian flysch deposited on the outer shelf, slope, and proximal ocean floor of the North American passive margin also were accreted to the Gondwanan forearc and then driven northward onto the North American shelf; such rocks form the external zones of the Ouachitan orogen (King, 1975; Ross and Ross, 1985; Ross, 1986; Viele and Thomas, 1989). The North America–Gondwana collision is marked by the Alleghany orogen in eastern North America, the Ouachita-Marathon orogen in the southern United States, and the Tarahumara terrane in northern México. The Ouachitan orogeny in eastern México has been referred to as the Coahuilan orogeny (Guzmán and de Cserna, 1963).

Following Premises 2, 3, and 6, we postulate that the western margin of Pangea featured a nearly right-angle bend that framed an oceanic corner near what was to become México. This corner was bounded by the roughly north-south–trending (present coordinates) western margin of South America and by the east-southeast/west-northwest–trending passive margin of southwestern North America in Chihuahua and Sonora (Fig. 31). The Tarahumara-Ouachitan orogenic belt, which marks the collision between North America and Gondwana, terminated near this bend and never extended farther to the southwest because the north-south–trending western margin of South America prevented contact of the continental masses there. Because Ouachitan orogenesis probably did not affect rocks oceanward of this corner, we infer that terranes with Precambrian and Paleozoic basement that lack evidence of Pennsylvanian to Early Permian deformation, magmatism, and metamorphism ascribable to the Ouachitan orogeny were in this region.

Loading of the southern edge of North America by northward thrusting of the Ouachita orogenic belt produced a series of discontinuous foreland basins and forebulges on the buckling, foundering continental shelf from New England to Chihuahua, and perhaps as far west as central Sonora (Ross and Ross, 1985; Ross, 1986; Armin, 1987). The southern part of the Pedregosa basin in México (Fig. 24) displays an abrupt transition from cratonal strata to deeper water rocks at the beginning of the Permian, implying the onset of foreland loading (Armin, 1987; Handschy and others, 1987). Wolfcampian flysch was late tectonic to post-

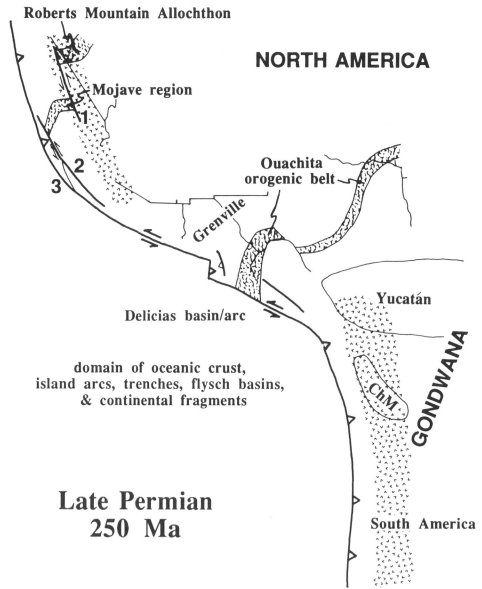

Figure 31. Late Permian (about 250 Ma) paleogeographic reconstruction, approximately synchronous with cessation of Ouachitan orogenesis. V pattern indicates active magmatic arc; irregular pattern, obducted deformed rocks of Roberts Mountain Allochthon and Ouachita orogenic belt; ChM, Chiapas Massif. Faults 1, 2, and 3 are discussed in text.

tectonic in the Marathon region and syntectonic in Chihuahua, suggesting diachroneity of the collision and termination of Ouachitan tectonism by late Early Permian (Leonardian) time (Ross and Ross, 1985; Ross, 1986; Thomas, 1985, 1989). Ouachitan orogenesis probably subjected Precambrian and Paleozoic rocks of North America to regional low-grade thermal metamorphism, producing late Paleozoic cooling ages (Denison and others, 1971; Mauger and others, 1983).

Marine siliciclastic strata of Carboniferous and Permian age that crop out in several Mexican terranes help determine whether the terranes were involved in the Ouachitan orogeny. In the Maya terrane, Late Mississippian(?) to Pennsylvanian flysch

and volcanogenic rocks and Pennsylvanian–Early Permian shale and limestone are cut by an Early Permian angular unconformity, indicating late Early Permian deformation, uplift, and erosion. These relations support our earlier inference that the Maya terrane was part of the Gondwanan forearc.

The northern and southern Guachichil terranes contain fault-bounded, thick, provisionally correlative Early Permian flysch that was derived from siliciclastic, volcanic, and carbonate sources. Unequivocal identification of a depositional environment will be difficult because the flysch units are rootless, having been faulted against Grenville basement and other pre-Mesozoic rocks in Permo-Triassic time. Possible depositional sites include the continental rise west of Pangea, continental foreland basins (e.g., Pedregosa basin) near the corner in the Pangean continental margin (Fig. 31), intracontinental successor basins above the cooled, subsided magmatic arc in Gondwana, and continental backarc basins in Gondwana.

In the southern Coahuiltecano terrane, coarse Late Pennsylvanian to Permian marine strata of the Las Delicias basin were derived from and deposited adjacent to and north of a continental calc-alkalic arc fringed by carbonate banks (McKee and others, 1988, 1990). These strata once may have been contiguous with the partly volcanogenic Permian flysch in the northern and southern Guachichil terranes (J. McKee, personal communication, 1990).

The Oaxacan Complex of the Zapoteco terrane and the Acatlán Complex of the Mixteco terrane are overlapped by marine, possibly shelfal, and continental Carboniferous-Permian siliciclastic rocks, but these strata do not constrain the late Paleozoic position of the Zapoteco-Mixteco block with respect to other blocks or to Pangea. The terranes are intruded by Permian granitoids, implying proximity to a magmatic arc.

Cordilleran tectonics and sedimentation. Deformation, magmatism, and metamorphism associated with the formation of Pangea had little effect on southwestern North America west and northwest of the Pedregosa basin. In the Great Basin and Mojave regions, late Paleozoic shallow- to deep-water sediments were deposited in shelf and off-shelf environments. In the Serí terrane, Late Mississippian to Permian shelfal rocks were deposited conformably on older Paleozoic shelfal rocks, and deformed Ordovician–Early Mississippian basinal rocks were overlain unconformably by Late Mississippian to Early Permian deep-water flysch (Ketner, 1986; Poole and Madrid, 1986, 1988).

In Pennsylvanian to Permian time, the western edge of the southern Cordillera was truncated at an inferred sinistral fault system of approximately northwest strike (Premise 10). The most inboard fault or faults in this system displaced part of the Roberts Mountain allochthon (RMA) roughly 400 km southeastward (present coordinates) from the northern Great Basin to what is now the western Mojave region (Fig. 31, fault 1), and may be the Mojave-Sonora Megashear (Walker, 1988; Stevens and others, 1992). Activity on this fault also displaced Paleozoic shelfal rocks south of the RMA that were to become the Serí terrane.

We speculate that concomitant sinistral slip occurred on at least two other, more outboard faults in the system (Fig. 31). We propose a hypothetical fault system, here named fault 3, at which basinal strata of the Serí terrane were displaced to a position adjacent to and outboard of shelfal strata of the Serí terrane in what is now central Sonora (Figs. 31, 32). Net displacement of the basinal strata was minimal if they were deposited near Sonora but correspondingly larger if the depocenters were more distant. The age of displacement on fault 3 probably was Permian to Early Triassic, based on the youngest basinal rocks and on the timing of thrusting of the basinal rocks onto shelfal rocks in central Sonora.

We propose another fault system, here named fault 2, that separates Paleozoic shelfal rocks of the Serí terrane in what is now Baja California, denoted as "Baja" in the figures, from apparently correlative rocks of the Serí terrane in what is now Sonora. To be consistent with our interpretation that the Paleozoic shelfal Baja rocks were displaced 1,500 to 2,500 km northwestward during the Late Cretaceous and Paleogene (p. 80), Paleozoic shelfal rocks in Baja must have been displaced southeastward 1,500 to 2,500 km on a sinistral fault system such as fault 2 prior to the mid-Mesozoic. Sinistral displacement on fault 2 may have begun in the Pennsylvanian, but we suspect that most displacement occurred in the Triassic and Jurassic, after major displacement on hypothetical fault 3 to the west.

Late Paleozoic continental truncation of the southern Cordillera indicates margin-parallel shearing between North America and an oceanic plate or plates to the west, possibly due to a change of relative plate motions accompanying collision of Gondwana with southern North America. Margin-parallel sinistral shearing apparently started prior to calc-alkalic magmatism along the western margin of Pangea, but may have been coeval with magmatism for much of the Permian. Relative motion with a large component of sinistral displacement continued into the early Mesozoic, implying thousands of kilometers of southward or southeastward displacement relative to North America of rocks in the ocean basin west of the continent. The long-term southward or southeastward sweep of the oceanic plates suggests that those Mexican terranes that arrived relatively late at the active margin of western México, such as the northern Tahué, eastern Nahuatl, and Pericú, lay far to the north and west of North America in the Paleozoic and subsequently were displaced southeastward relative to North America while embedded in oceanic lithosphere that converged with the continent, or on margin-parallel faults along the plate boundary such as fault 3 in Figure 31. The basinal protoliths of these terranes may have been similar to accretionary forearcs of the proto-Pacific basin such as the Sonomia and Golconda allochthons in the western United States.

Late Permian to Present: Overview

Oceanic lithosphere probably has been subducted with a large eastward component beneath the western margin of México since the Permian. There is no evidence of post-Ouachitan convergence in the Gulf of Mexico region, so the magmatic and tectonic effects of plate convergence and subduction must have been accommodated to the east, in what is now mainland México. In the reconstruction that follows, we infer that the continuous subduction of oceanic lithosphere beneath México resulted in southward and westward continental growth due to the accretion of continental fragments, island arcs, and intervening basins, and in southward and westward migration of the locus of subduction-related arc magmatism. Paleomagnetic and geologic data and plate motion models indicate that margin-parallel displacement of outboard terranes due to oblique convergence was sinistral (southward) during the Permian, Triassic, Jurassic, and earliest Cretaceous, and dextral (northward) during most of the Cretaceous and the Paleogene.

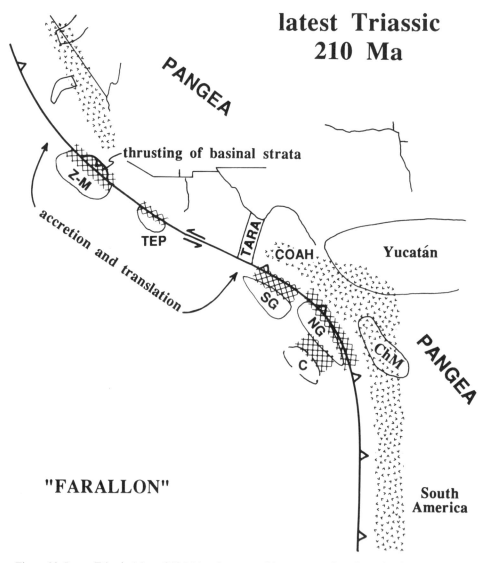

Figure 32. Latest Triassic (about 210 Ma) paleogeographic reconstruction. Cross-hatched pattern: zone of collision, obduction, or accretion. Emplacement of allochthonous basinal rocks in central Sonora is speculatively attributed to collision or transport of Zapoteco-Mixteco composite terrane. Abbreviations: C, central Cuba; COAH, Coahuiltecano terrane; NG, Northern Guachichil terrane; SG, Southern Guachichil terrane; TARA, Tarahumara terrane; TEP, Tepehuano terrane; Z-M, Zapoteco and Mixteco terranes. Other abbreviations and patterns as in Figure 31.

In most cases it is not possible to identify with confidence the exact position of inferred major tectonic features such as subduction zones and strike-slip faults at latitudes between Sonora and northern South America because of unresolved tectonic questions such as the organization of oceanic plates and relative motions among these plates and North America. For this reason, many structures, particularly those of Permian and early Mesozoic age, are depicted quite schematically.

Late Permian to Middle Triassic

Permo-Triassic arc. Calc-alkalic magmatism was initiated in a continental magmatic arc at the western margin of Pangea in the Early to Late Permian in northwestern South America and in the Late Permian to Early Triassic in the southwestern United States (Premise 8). The onset of magmatism records eastward subduction of oceanic lithosphere of one or more plates in the paleo-Pacific basin beneath western Pangea (Fig. 31). The onset of magmatism was roughly coeval with the mid-Permian cessation of Gondwana–North America convergence and probably reflects a major reorganization of relative plate motions. Permian convergence probably was left-oblique based on the sense of slip inferred in the southwestern United States (Avé Lallemant and Oldow, 1988). Sinistral displacement on the boundary probably contributed to the progressive southeastward displacement of the Serí terrane in the Baja sliver and more outboard Mexican terranes.

The Permo-Triassic arc is preserved intact in the southwestern United States and northern South America and is known from subcrops and rarer outcrops in the Coahuiltecano and Maya terranes of eastern México (see discussion of Premise 8). This arc was emplaced into continental crust on the western edge of Pangea starting in the Early Permian (Figs. 31, 32). The apparent continuity and linearity of the arc in eastern México implies eastward subduction of oceanic lithosphere beneath the western margin of Pangea during most of the Permian and Triassic. Subsequent displacement transverse to the Mexican reach of the arc may have occurred at the boundary between the Coahuiltecano and Maya terranes, where minor Late Jurassic sinistral slip probably occurred on the Mojave-Sonora Megashear, and between the Chiapas Massif and Veracruz region, which probably were separated during opening of the Gulf of Mexico (pp. 96–98). Because the arc is known chiefly in the subsurface, the magnitude and age of these and other transverse and longitudinal displacements of the arc cannot be ascertained.

We attribute the apparent discontinuity in the Permo-Triassic arc between the northern end of the Mexican reach in Coahuila and the southern end of the United States reach in the Mojave region to the original plate geometry, rather than to post-Triassic truncation and displacement of the arc. The boundary between North America and an oceanic plate or plates in the paleo-Pacific basin was a subduction zone in the western United States and eastern México, where the boundary trended roughly north-south, but was a transform fault in northern México, where the boundary trended east or east-southeast (present coordinates; Figs. 31, 32). A similar idea was discussed by J. H. Stewart in an oral presentation at the 1990 meeting of the Geological Society of America Cordilleran Section. We infer that relative plate motion was roughly parallel to this transform, but conceivably there may have been a margin-normal component of shortening or extension. The reconstruction depicts left-oblique convergence at the trench in the southwestern United States in order to account for inferred southeastward displacement of Mexican terranes such as the sliver of Serí terrane in Baja (Figs. 31, 32). We surmise that convergence south of the transform margin in eastern México was either roughly orthogonal or left-oblique.

Permo-Triassic arc rocks also crop out in the amalgamated Zapoteco and Mixteco terranes. They include Permian granitoids in both terranes, post–Early Permian, pre–Middle Jurassic ignimbrite in the Mixteco terrane, and Triassic(?) to Middle Jurassic dikes and sills in both terranes. Paleomagnetic data are interpreted to indicate that the terranes were at the latitude of Sonora at the end of the Triassic. Three interpretations of the origin of the Zapoteco-Mixteco arc rocks are permissible based on available data. First, the Zapoteco-Mixteco composite terrane underwent arc magmatism within the Mexican reach of arc in the Permo-Triassic, after which it was excised from the arc and translated to the latitude of Sonora in the early Mesozoic. Second, the Zapoteco-Mixteco composite terrane underwent arc magmatism in an arc unrelated to subduction beneath western Pangea, at an unspecified latitude some distance west of the Pangean margin,

perhaps requiring translation to the latitude of Sonora by the Late Triassic. Third, the Zapoteco-Mixteco composite terrane underwent arc magmatism within the southern end of the United States reach of the Permo-Triassic continental arc of western Pangea. The first option requires northwestward transport of the terranes relative to North America, which we consider unlikely because of evidence for southward margin-parallel displacements in the early Mesozoic (Avé Lallemant and Oldow, 1988). We cannot distinguish between the second and third options with available data, so we have omitted the terranes from the Late Permian reconstruction (Fig. 31) but shown them at the latitude of Sonora by the Late Triassic (Fig. 32).

Cordilleran deformation. The Permo-Triassic arc at the western margin of Pangea probably was paired with an east-dipping subduction zone that consumed one or more oceanic plates of the paleo-Pacific basin. Although the paleogeography of this basin is not known, its subsequent history suggests that it contained fragments of Grenville and Paleozoic continental basement, magmatic arcs, and flysch basins (Fig. 31). Some of these may have been separated from Pangea by substantial tracts of oceanic lithosphere, whereas others may have been nearby, perhaps enmeshed in a complex system of trenches similar to the modern southeastern Pacific.

Contractional deformation of Permo-Triassic age, reported from several parts of southwestern North America, probably resulted from convergence between North America and one or more oceanic plates to the west or south. In the Great Basin, Permo-Triassic collision of the Sonomia terrane with North America emplaced the forearc of Sonomia as the Golconda allochthon onto previously accreted Paleozoic rocks of the Roberts Mountain allochthon and its Paleozoic cover (Speed, 1979). The Mojave and Death Valley regions were affected by Late Permian deformation, including development of large thrust systems (Carr and others, 1984; Snow and others, 1991). Deformation in these regions probably was caused by margin-normal shortening caused by high-angle convergence to the west. In central Chihuahua, along the inferred transform fault boundary at the truncated southwestern margin of North America, east-vergent thrusting of Precambrian basement and Early Permian flysch occurred between the mid-Permian and Middle Jurassic (Handschy and Dyer, 1987). Thrusting may record either margin-parallel shortening at the transform boundary or margin-normal shortening east of a hypothetical north-south trench that offset the transform (Fig. 31). In the Serí terrane, Paleozoic basinal rocks may have been juxtaposed with Paleozoic shelfal rocks by late Paleozoic sinistral strike-slip faulting (Fig. 31); penetrative deformation of both units accompanied Late Permian to Middle Triassic overthrusting of the shelfal rocks by the basinal rocks (Poole and Madrid, 1986, 1988). The cause of this overthrusting is unknown. We have speculated that it is related to displacement of the Zapoteco-Mixteco composite terrane along the continental margin (Fig. 31), but the supposed northward vergence of thrusting is difficult to reconcile with generally eastward to southeastward displacement of oceanic plates relative to North America.

It is much more difficult to determine the tectonic setting and significance of Permo-Triassic deformation in other Mexican terranes. In the northern Tahué terrane, granitic gneiss apparently was metamorphosed and deformed during the Triassic (T. Anderson, personal communication, 1990). The plutonic protolith of the gneiss may have served as basement for weakly metamorphosed Carboniferous basinal strata, or may have been faulted against the Carboniferous rocks. In the northern Tepehuano terrane, protoliths of weakly metamorphosed, strongly deformed flysch and olistostromal mélange of pre–Middle Jurassic age (Taray Formation) may be coeval or correlative with Triassic marine rocks elsewhere in the terrane. We provisionally interpret the Taray Formation as a Triassic(?) subduction complex that formed in the forearc on the western margin of Pangea (p. 58). In the Juchatengo fault zone at the southern margin of the Zapoteco terrane, cataclastic granitoids and oceanic rocks of probable late Paleozoic age are provisionally interpreted as a late Paleozoic to Triassic(?) subduction complex and arc(?) that were accreted to the outboard margin of the amalgamated Zapoteco and Mixteco terranes.

Late Triassic to Late Jurassic

Breakup of Pangea

The breakup of Pangea began in the Late Triassic, but drifting of the North America and South America plates and formation of an intervening Caribbean basin did not begin until the earliest Cretaceous (Premises 15, 16). Pre-Cretaceous extension between the North America and South America cratons was accommodated heterogeneously, as indicated by the distribution of crustal thickness and depths to basement in eastern México and adjacent offshore regions (Buffler and Sawyer, 1985). Eastern México is underlain by thick continental crust that probably was thickened during Permo-Triassic arc magmatism. Offshore eastern México, a precipitous continental slope leads into the Gulf of México basin (Fig. 33). Oceanic lithosphere in the Gulf of México basin is circumscribed by a zone of thin transitional crust that is much wider to the northwest and southeast than to the northeast and southwest. The Gulf Coast region of the United States and the Yucatán platform are underlain by thick transitional crust. Marine seismic profiling indicates that the present pattern of crustal thicknesses of eastern and southeastern México and the Gulf of México was attained by the end of the Jurassic, although cooling and subsidence of deeper regions continued in the Cretaceous (R. Buffler, personal communication, 1988).

Onland stratigraphy indicates that pre-Cretaceous rifting and drifting of South America from North America occurred in two stages (Buffler and Sawyer, 1985; Dunbar and Sawyer, 1987; Winker and Buffler, 1988). In the first stage, Late Triassic to Early Jurassic rifting occurred in the Gulf Coast region of the United States and in the Guachichil and Maya terranes of eastern México, forming grabens and half-grabens that were filled with red beds and volcanic rocks. There are no indications of large

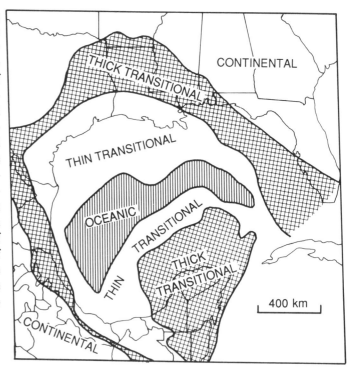

Figure 33. Crustal types and thicknesses in and adjacent to the Gulf of Mexico, based on Buffler and Sawyer (1985).

throw or great crustal attenuation. The first stage climaxed with development of a widespread Middle Jurassic unconformity that is interpreted to be due to a regional rise of asthenosphere below México and the Gulf of Mexico. In the second stage, Middle to Late Jurassic horizontal stretching produced the thin transitional crust around the periphery of the Gulf of Mexico basin. Closely spaced crustal thickness contours are roughly parallel to the strikes of the rift structures that formed during the first stage. The second stage climaxed with Late Jurassic oceanic spreading in the deep central basin.

Jurassic extension between the North America and South America cratons may not have been completely confined to the Gulf of Mexico region. Formation of a Pacific-Atlantic seaway between North America and South America by Middle Jurassic time, as indicated by biostratigraphic studies in southern México (Westermann and others, 1984), may signify crustal stretching southeast of the Maya terrane. The Cuicateco basin in southern México, which probably was floored by oceanic lithosphere only in its southern part, may have been an aborted northward-propagating continental rift associated with drifting of South America from the southern Maya terrane.

Pre-Cretaceous displacement due to oceanic spreading in the Gulf of Mexico in the direction of Cretaceous drift between North America and South America was about 450 km. Displacement in the same direction that was accommodated by stretching of continental crust in broad zones on the northwestern and southeastern sides of the basin exceeds that due to spreading, assuming validity of the stretching values calculated by Dunbar and Sawyer (1987). Thus, at least 1,000 km, or at least one-third of the ~3,000 km of northwest-southeast displacement between the two cratons, was accommodated prior to Cretaceous opening of the Caribbean basin. Displacements of individual blocks in the Gulf of Mexico region apparently were complex and did not follow small circle paths about the North America–South America Euler pole, given the evidence for local counterclockwise rotations of Jurassic age in northeastern México (Gose and others, 1982) and in the Yucatán block (Molina-Garza and others, 1992).

There is no consensus about the configuration of the spreading system in and adjacent to the Gulf of Mexico. In our recon-

struction, the Jurassic extensional displacement field in the Gulf of Mexico region was bounded to the west by a fault beneath the continental slope of eastern México (Fig. 34), similar to the Tamaulipas–Golden Lane fault of Pindell (1985). We view this fault as a roughly northwest-striking transform that connected a Jurassic spreading center in the Gulf of Mexico with a spreading center or trench south of México. The transform probably accommodated relative displacement between large, heterogeneous extension in the Gulf of Mexico basin and Yucatán platform to the east and little Jurassic extension in continental crust of eastern México to the west.

Displacement on the transform boundary probably was transtensional. Finite, but minor, horizontal extension must have occurred in order to produce the zone of thin transitional crust at

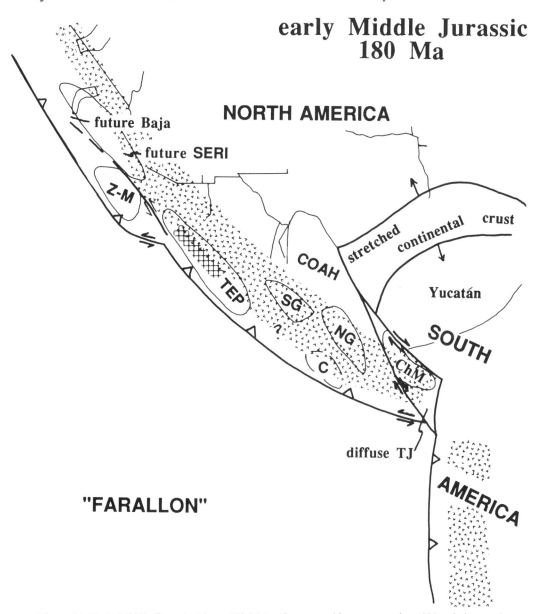

Figure 34. Early Middle Jurassic (about 180 Ma) paleogeographic reconstruction. Abbreviations and patterns as in Figure 32.

the western margin of the Gulf of Mexico basin. Transtension also may have been taken up by Late Triassic–Early Jurassic graben formation in the eastern Guachichil and northern Maya terranes, and by Middle to Late Jurassic formation of roughly margin-parallel basins and platforms throughout eastern México (Fig. 35). We place the transform fault beneath the continental slope off the coast of Veracruz, as did Pindell (1985), but we project the fault to the southeast along the western, rather than the eastern, side of the Chiapas Massif in view of the similarity of Paleozoic strata there and in northern Guatemala. This allows Jurassic southeastward displacement of both the Yucatán platform and the southern province of the Maya terrane relative to both the northern province of the Maya terrane and the amalgamated Zapoteco and Mixteco terranes. Displacement on the proposed north-striking Salina Cruz fault on the Isthmus of Tehuantepec, if any, is probably Tertiary and thus unrelated to the opening of the Gulf of Mexico (Salvador, 1987, 1988).

Conspicuous by its absence from our analysis of displacements during the opening of the Gulf of Mexico is the Mojave-Sonora Megashear (MSM). Many workers have inferred that the left-lateral MSM connected a spreading system in the Gulf of Mexico with the trench at the western margin of North America (Pindell and Dewey, 1982; Anderson and Schmidt, 1983; Klitgord and Schouten, 1987). We offer an alternate interpretation in which slip on the megashear is driven by transpression in the Cordillera, independent of extension in the Gulf of Mexico region (p. 103).

Activity in the Cordillera

Throughout the Mesozoic, oceanic lithosphere of one or more plates in the Pacific basin converged with and was subducted eastward beneath the western margin of Pangea. Protracted left-oblique convergence resulted in subduction, arc magmatism, accretion, east-west shortening, and southward translation of continental blocks, island arcs, and basinal strata during westward growth of continental México.

Arc magmatism. Late Triassic to Late Jurassic eastward subduction of oceanic lithosphere generated continental magmatic

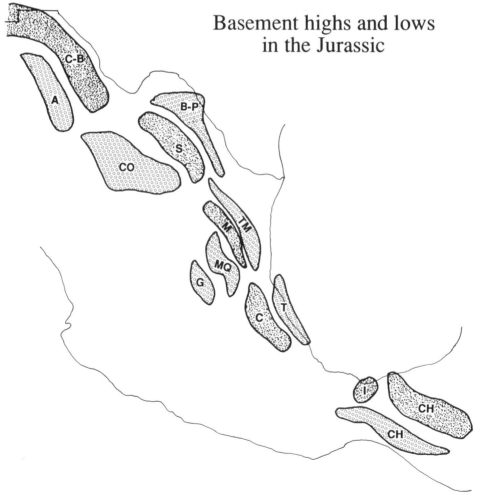

Figure 35. Jurassic basins (irregular stipple) and horsts (circle pattern) in northern and eastern México, modified after Winker and Buffler (1988). Abbreviations: A, Aldama; B-P, Burro-Picacho; C, Chicontepec; C-B, Chihuahua-Bisbee; CH, Chiapas; CO, Coahuila; G, Guaxcamá; I, Isthmian; M, Magicatzin; MQ, Miquihuana; S, Sabinas; T, Tuxpán; TM, Tamaulipas.

arcs in southwestern North America and northwestern South America. In the western Great Basin and Mojave region, Late Triassic to Late Jurassic magmatism continued near the older Permo-Triassic magmatic arc (Kistler, 1974; Dilles and Wright, 1988). In southern Arizona, continental arc magmatism began in the Early Jurassic or possibly the latest Triassic, indicating a change from a transform to a convergent plate margin (Figs. 32, 34) (Coney, 1978; Damon and others, 1981; Asmeron and others, 1990; Tosdal and others, 1990a). In northwestern South America, Late Triassic(?) and Jurassic magmatic rocks crop out in the Cordillera de Mérida and Sierra de Perijá in western Venezuela, the Guajira Peninsula and Sierra Nevada de Santa Marta in northern Colombia, the Cordillera Central and Cordillera Oriental in central Colombia (Goldsmith, 1971; Tschanz and others, 1974; Shagam and others, 1984; Case and others, 1990), in southern Ecuador (Feininger, 1987), in the southern part of the Eastern Cordillera of Peru (Cobbing and Pitcher, 1983), and in Chile north of about 28°S (Farrar and others, 1970; Aguirre, 1983).

Arc magmatism in the Mexican region occurred progressively farther westward during Permian to Jurassic time. A continental arc was constructed on continental crust in the Maya and eastern Coahuiltecano terranes in easternmost México in Permian and Triassic time (p. 95). We provisionally divide Late Triassic to Early Jurassic and more abundant Middle to Late Jurassic magmatic rocks of the Mexican region into two groups. Group 1 rocks formed in a Jurassic continental arc that was approximately coincident with the Permo-Triassic arc in the southern Maya terrane (e.g., Chiapas Massif) but was southwest of the Permo-Triassic arc in the northern Maya terrane and Coahuiltecano terrane (Fig. 34). Group 2 rocks formed in a continental magmatic arc west of the Group 1 arc or in one or more island arcs west of the continental margin. We interpret the Permian to Jurassic westward shift of the locus of magmatism to indicate that terranes comprising continental blocks and island arcs were accreted to the western margin of Pangea in the Mexican region after consumption of intervening ocean basins of unknown width and age at trenches along and west of the continental margin.

Group 1 rocks formed in a continental magmatic arc that connected the segments of the Late Triassic–Jurassic continental magmatic arc in the southwestern United States and northwestern South America (Figs. 34, 36) (Damon and others, 1984). The arc was cut by the Mojave-Sonora Megashear in the Late Jurassic, but there may have been little transverse offset because displacement probably was roughly parallel to the trend of the arc. We infer that, although margin-parallel displacement of the arc is not yet understood, the arc originally was continuous on continental crust between northern Sonora and Chiapas (Figs. 34, 36). In northern Sonora (North America), continental volcanic rocks (>180 to 170 Ma) intruded by 175- to 150-Ma plutons (Anderson and Silver, 1979) are correlated with coeval rocks in southern Arizona and were emplaced in the gap in the older Permo-Triassic arc, implying a change in relative plate motions. Trias-

sic(?) ignimbrite and Triassic to Middle Jurassic dikes in the Mixteco terrane probably were emplaced when the composite Zapoteco-Mixteco terrane was at the latitude of Sonora (Figs. 32, 34). Orthogneiss in the Chatino terrane has yielded Middle Jurassic to Early Cretaceous Rb-Sr isochron ages that have been interpreted as crystallization ages of synkinematic plutons intruded along the southern margin of the Zapoteco-Mixteco composite terrane (Morán-Zenteno and others, 1991). Middle and Late Jurassic S-type granites intruded the eastern subterrane of the Yuma terrane (Todd and others, 1991) prior to and perhaps during the early stages of its translation from northern to southern México (Fig. 37). Jurassic plutonic rocks in the northwestern corner of the Tepehuano terrane, Jurassic(?) arc volcanic rocks older than 160 Ma in the northeastern Tepehuano terrane, and rare Middle Jurassic granitoids in the Serí terrane probably are parts of the Jurassic arc that were displaced southeastward from the Mojave region or northwestern Sonora during Late Jurassic displacement on the Mojave-Sonora Megashear. In the southern Tepehuano terrane, volcanic rocks of the Upper Jurassic Chilitos Formation may have been derived from a continental arc. In the northern and southern parts of the Guachichil terrane, volcanic and plutonic rocks that yielded Jurassic K-Ar ages were penetrated by wells (López-Infanzón, 1986). In the southern Guachichil terrane and adjacent areas, tuff interbedded with Late Jurassic flysch and pelagic rocks (Longoria, 1984; Salvador, 1987; Longoria, 1988) was deposited as far southeast as Mexico City (Suter, 1987), but the source arc has not been identified. The Chiapas Massif in the southern Maya terrane contains Early to Middle Jurassic granitoids and Late Jurassic or older andesite.

Group 2 includes all other Late Triassic to Late Jurassic magmatic rocks in México and northern Central America. Several subterranes of the Cochimí terrane contain Middle Jurassic to early Early Cretaceous plutonic, volcanic, and volcaniclastic rocks that probably represent fragments of oceanic island arcs accreted to North America by the earliest Cretaceous. Granite pebbles from Tithonian-Aptian strata in one of the subterranes yielded K-Ar dates and a lower U-Pb discordia intercept of about 150 Ma, indicating derivation from Jurassic plutons. Both the eastern and western subterranes of the Yuma terrane contain minor volcanic rocks of inferred Late Triassic or Early Jurassic age interbedded with clastic rocks, and the western subterrane consists chiefly of Late Jurassic and more abundant Early Cretaceous volcanic arc rocks. In the Pericú terrane, protoliths of metasedimentary and minor metavolcanic rocks are probably Jurassic or older. Intermediate volcanic rocks of demonstrated and inferred Late Jurassic age are widespread throughout the Nahuatl terrane. The Chortis terrane is intruded by a Late Jurassic granodiorite and many other undated granitoids that also may be Jurassic. Precambrian rocks in central Cuba were intruded by potassic granite about 172 Ma (U-Pb, zircon) (Renne and others, 1989).

Group 2 magmatic rocks formed in a variety of tectonic environments that in many cases still are incompletely understood. The apparent absence of continental crust in the Cochimí,

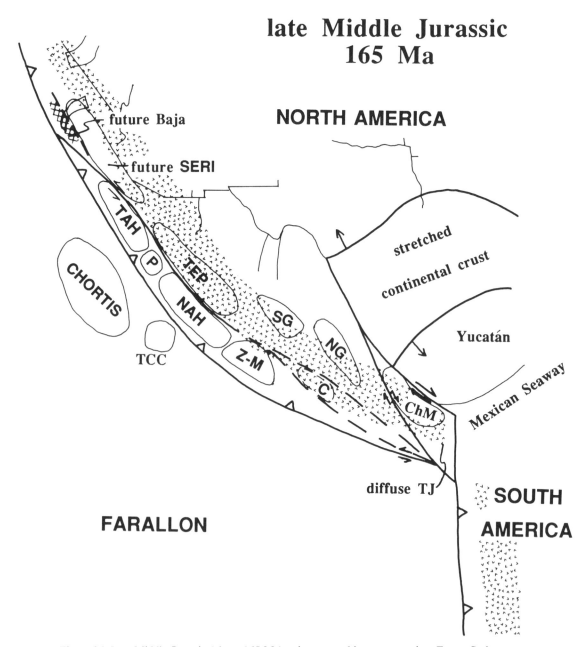

Figure 36. Late Middle Jurassic (about 165 Ma) paleogeographic reconstruction. Future Serí terrane and future Baja California block shown in northwestern corner. Abbreviations: NAH, Nahuatl terrane; P, Pericú terrane; TAH, Tahué terrane; TCC, Tierra Caliente Complex. Other abbreviations and patterns as in Figure 32.

Yuma, Pericú, and Nahuatl terranes suggests that Jurassic magmatic rocks in these terranes formed in one or more island arcs outboard of the continental arc at the western margin of North America. We have inferred that the latter three terranes originated in the northern proto-Pacific basin in the Paleozoic, but so little data pertain to the Jurassic positions of these arcs relative to one another and to the continent that we refrain from depicting them in the reconstruction (Figs. 34, 36). The Jurassic locations of the Chortis terrane and central Cuba are unknown, but because both have continental basement we speculate that they were parts of the North American continental arc that later were detached and displaced (Fig. 34).

Westward growth of Mexican continental crust. Eastward subduction of oceanic lithosphere during Permian to Jurassic time was accompanied by east-west shortening in a compressional or transpressional displacement field. Here we outline the tectonic history insofar as can be determined for each terrane that may have been accreted to western México during this time span.

Northern Guachichil terrane. The Precambrian Novillo

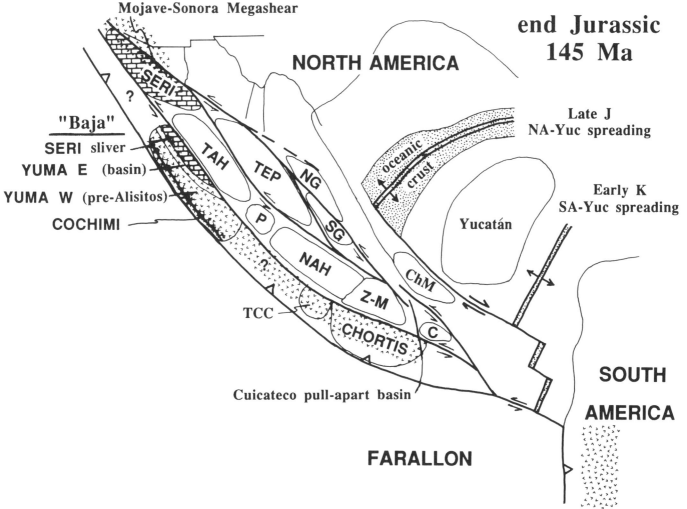

Figure 37. Latest Jurassic (about 145 Ma) paleogeographic reconstruction, shortly after displacement on Mojave-Sonora Megashear and shortly prior to jump of spreading center from North America–Yucatán to Yucatán–South America. Feature termed "Baja" composite terrane consists of Serí, Yuma, and Cochimí terranes. Block pattern, lower Paleozoic shelfal rocks in Serí terrane; stipple, oceanic crust in Gulf of Mexico region. Other patterns and abbreviations as in Figure 36.

Gneiss apparently was not affected by Ouachitan orogenesis, so we infer a late Paleozoic position in the ocean basin south of southwestern North America. We infer that Paleozoic miogeoclinal strata currently faulted against the gneiss are parautochthonous with respect to it. The Granjeno Schist was deformed and metamorphosed in an uncertain location in the Carboniferous and probably was attached to the gneiss and Paleozoic strata in Pennsylvania, Permian, or possibly earliest Triassic time. Early Permian flysch must have been attached during the Late Permian to Middle Triassic. Fault juxtaposition of all these rocks was completed prior to deposition of overlapping redbeds of the Upper Triassic–Lower Jurassic Huizachal Formation, which also may be present in the subsurface in the northern Maya terrane. Permian to Jurassic magmatic rocks are absent from the northern Guachichil terrane but are present in terranes to the east (Maya) and west (Tepehuano). Based on these relations, we infer (1) a late Paleozoic position in the ocean basin south of southwestern North America for some or all of the pre-Triassic rock units, (2) Permo-Triassic eastward subduction of the intervening ocean basin beneath western Pangea and arc magmatism in the Coahuiltecano and Maya terranes, (3) Middle to Late Triassic accretion of the pre-Triassic rock units to the western margin of Pangea, (4) a Late Triassic to Jurassic westward shift of magmatism from the western margin of Pangea to an unidentified arc west of the accreted northern Guachichil terrane, and (5) deposition of overlapping Late Triassic to Early Jurassic strata during the breakup of Pangea (Figs. 31, 32, 34). It is unclear whether the pre-Triassic rocks were amalgamated in the ocean basin west of central Pangea or were successively accreted directly to the western margin of Pangea.

Southern Guachichil terrane. The Precambrian Huiznopala Gneiss apparently was not affected by Ouachitan orogenesis, so we infer a late Paleozoic position in the ocean basin south of southwestern North America. Early Permian flysch was faulted against the gneiss prior to deposition of overlapping Late Triassic to Early Jurassic red beds. Permian-Jurassic magmatic rocks are absent. These relations imply a tectonic history very similar to that of the northern Guachichil terrane. We propose that the northern and southern Guachichil terranes were accreted to adjoining parts of the western Pangean margin, and we speculate that Late Jurassic sinistral displacement on a splay of the Mojave-Sonora Megashear displaced them to their present locations (Figs. 32, 37).

Las Delicias basin. The Las Delicias basin in the southern Coahuiltecano terrane contains Pennsylvanian and Permian volcanogenic strata derived from and deposited on the fringes of a continental magmatic arc (McKee and others, 1988, 1990). The identity and present location of the arc and the time of its separation from the basin are unknown. Strata in the Las Delicias basin are anomalous with respect to Paleozoic rocks in adjacent terranes, but Triassic granitoids intrude both the basin and the northern Coahuitelcano terrane, implying little relative displacement since the Triassic. We speculate that the Las Delicias basin and its source arc formed in northwestern Gondwana in the late Paleozoic, and that the basin was translated an unknown distance northward or northwestward along the western margin of Pangea in the Permo-Triassic (Fig. 31).

Zapoteco and Mixteco terranes. Basement rocks of these terranes were amalgamated in the Paleozoic, prior to their overlap by Carboniferous and Permian strata, and are inferred to have been displaced west from Pangea by the beginning of the Permian (p. 91). Other postamalgamation rocks in the terranes include a few Permian granitoids, Triassic(?) ignimbrite, Triassic to Middle Jurassic dikes, Early Jurassic nonmarine rocks and coal, and Early Jurassic–Cretaceous epicontinental strata. Remagnetization caused by Permian intrusions occurred while the terrane was at its present latitude with respect to North America, but anomalous directions from Permian to Oxfordian rocks in the Mixteco terrane imply about $15° ± 8°$ of southward translation in Oxfordian to Albian time, about 160 to 110 Ma. We infer that in the late Paleozoic the amalgamated terranes were on the periphery of a Permo-Triassic magmatic arc in the ocean basin west of Pangea, at or near their present latitude with respect to North America. Strongly deformed late Paleozoic oceanic rocks that crop out in the fault zone at the southern boundary of the terranes (Juchatengo subterrane of Mixteco terrane) may be a subduction complex that formed at the trench associated with this arc. We speculate that the amalgamated terranes were displaced northward during Permo-Triassic subduction of oceanic lithosphere beneath western Pangea and were attached to western Pangea at the latitude of northern México by the end of Triassic time, possibly causing emplacement of Paleozoic basinal rocks on Paleozoic shelfal rocks in central Sonora (Fig. 32). The locus of continental arc magmatism probably shifted west of the Mixteco terrane after collision, although sparse Triassic to Middle Jurassic

magmatic rocks imply that the terranes were on the periphery of an active arc (Figs. 32, 34, 36). Across most of the Zapoteco and Mixteco terranes, Jurassic strata were deposited in continental and epicontinental environments distant from magmatic arcs. The amalgamated Zapoteco and Mixteco terranes probably were translated at least 1,000 km southeastward during the Late Jurassic on sinistral faults that accommodated oblique convergence (Figs. 36, 37).

Tepehuano terrane. Protoliths of strongly deformed flysch and mélange of the Taray Formation in the northern part of the terrane may be coeval and correlative with Triassic flysch and metabasite in the southern part of the terrane. We provisionally interpret the Taray Formation as a subduction complex that formed during the Triassic to Early Jurassic in the forearc of western Pangea, which by this time included the accreted Guachichil terranes (Figs. 32, 34). The Taray Formation apparently was overthrust by Late Triassic(?) to Late Jurassic(?) calc-alkalic volcanic arc rocks prior to intrusion of a quartz porphyry pluton in the early Late Jurassic. The basement of the Jurassic arc is unknown. We speculate that the arc developed on the western edge of Pangean/North American crust of unknown thickness after the westward shift of magmatism from the Permo-Triassic arc in eastern México, and was thrust eastward over the Taray subduction complex in the Early to Middle Jurassic (Figs. 34, 36). Magmatism continued into the Late Jurassic.

Yuma terrane. The eastern subterrane of the Yuma terrane consists of Triassic to Jurassic flysch and sparse interbedded volcanic rocks that were strongly deformed and uplifted prior to mid-Cretaceous overthrusting by Late Jurassic to Early Cretaceous volcanic rocks of the western subterrane of the Yuma terrane. On its eastern margin, flysch of the eastern subterrane is juxtaposed with Paleozoic rocks of the Serí terrane at an enigmatic contact that we provisionally identify as a reverse or thrust fault or fault system. We infer that flysch of the eastern subterrane was deposited in a trench or forearc basin during the Triassic to Middle Jurassic and was deformed and uplifted during Jurassic accretion to allochthonous Paleozoic rocks of the Serí terrane in central México (Figs. 36, 37). The western subterrane developed as an outboard island arc in the Late Jurassic and was accreted to the Serí terrane and eastern Yuma subterrane in the late Early Cretaceous (p. 107).

Central Cuba. Rocks in central Cuba include Grenville(?) marble, schist, and quartzite metamorphosed about 950 to 900 Ma, potassic granite dated at 172 Ma, and unconformably overlying Late Jurassic conglomerate and limestone and Early Cretaceous chert and shale (Renne and others, 1989; Lewis and Draper, 1990). We speculate that the Proterozoic rocks were south of southwestern North America during the Paleozoic; were accreted to western Pangea in the Triassic; were underthrust by oceanic lithosphere and hosted arc magmatism during the Middle Jurassic; and were deformed, uplifted, and eroded by collision of outboard terranes or closure of a backarc basin in the Late Jurassic (Figs. 32, 34, 36). During the Late Jurassic and

Early Cretaceous, central Cuba was a stable shelf or platform that may have been contiguous with southern México (Fig. 37).

Tahué terrane. The late Paleozoic to Jurassic kinematic history of the Tahué terrane is enigmatic. The substrate of Mississippian to Permian clastic rocks, siliceous to intermediate volcanic rocks, chert, and thin carbonates was probably oceanic or transitional crust, based on initial $^{87}Sr/^{86}Sr$ ratios in Cretaceous and Tertiary plutons. Amphibolite-facies gneiss, which probably was metamorphosed and deformed in the Triassic, and undated quartz diorite gneiss have uncertain relation to one another and to the Paleozoic rocks. The gneiss may be transitional crustal basement of the Paleozoic strata, or may have evolved separately from the Paleozoic rocks and been juxtaposed with them after Triassic metamorphism and before Cretaceous magmatism. Volcanic and plutonic rocks were emplaced in the Tahué terrane from at least the Early Cretaceous until the Miocene. We infer that the Tahué terrane was accreted to mainland México in the Middle to Late Jurassic by closure of a now-obscure intervening basin (Fig. 36). We speculate that the paucity of Late Jurassic magmatic rocks in the Tahué terrane can be attributed to the passage of the Baja block as it migrated southeastward along the continental margin above a steeply dipping subduction zone (Fig. 37). Arc magmatism affected the Tahué terrane by the middle of the Early Cretaceous, indicating passage of the Baja block, westward migration of the arc due to shallowing of the subduction zone, or both.

Cochimí terrane. The structurally highest level of the northern part of this composite terrane consists of three distinct subterranes, each a fragment of an oceanic island arc, that were accreted to the western margin of North America in the latest Jurassic or earliest Cretaceous (Fig. 37). The Choyal subterrane is a Middle Jurassic island arc that was accreted to North America in the Late Jurassic; its short lifespan may reflect relative proximity to the continental margin as a fringing arc. The Vizcaíno Norte and Vizcaíno Sur subterranes are Late Triassic to Late Jurassic island arcs that were accreted to North America in the latest Jurassic or earliest Cretaceous; the long duration of arc magmatism and the lack of terrigenous detritus indicate formation in a large ocean basin and protracted subduction of oceanic lithosphere of that basin prior to accretion. Paleomagnetic studies infer that the terranes were accreted to México about 1,500-2,000 km south of their present position (p. 22).

Sinistral displacement. Relative motions between North America and oceanic plates to the west are poorly constrained for the Jurassic and Early Cretaceous, but it is likely that the Late Jurassic increase in the rate of northward absolute motion of North America (May and Butler, 1986) initiated or accelerated sinistral displacement at the western margin of the continent. Geologic and paleomagnetic studies suggest that convergence included a significant left-oblique component that was resolved on sinistral strike-slip faults in the arc and forearc (Avé Lallemant and Oldow, 1988; Beck, 1989; Wolf and Saleeby, 1991). Most Mesozoic paleomagnetic poles from Mexican terranes are dis-

placed to the left of the reference apparent polar wander path for North America, implying sinistral displacements and counterclockwise rotations relative to the craton (Urrutia-Fucugauchi and others, 1987). We discuss evidence for sinistral displacement on the Mojave-Sonora Megashear, the most inboard of the sinistral fault systems, and for southeastward displacement of the amalgamated Mixteco and Zapoteco terranes and the Serí terrane in Baja California.

Mojave-Sonora Megashear. The existence and position of, and evidence of displacement on, the Mojave-Sonora Megashear are best documented in northern Sonora (Premise 11). To the northwest, in California, the record of the megashear has been erased nearly completely by superposed episodes of thrusting, extension, and strike-slip displacement. Southeastward from Sonora, we speculate that displacement on the megashear was partitioned among several southeast-striking splays (Figs. 22, 37) that we informally term the megashear fault zone. Constituent faults of the megashear fault zone in northeastern México include but are not limited to the inferred Coahuiltecano-Maya terrane boundary, the Guachichil-Tepehuano terrane boundary, a buried fault separating the northern Guachilchil terrane from the southern Guachichil terrane, and perhaps the San Marcos fault in the southern Coahuiltecano terrane.

Late Jurassic contraction and extension near the trace of the megashear and megashear fault zone may record displacement at right (restraining) and left (releasing) steps or bends, respectively. Late Jurassic transpression has been inferred in Zacatecas (Anderson and others, 1991) and northern Sonora (Connors and Anderson, 1989; Tosdal and others, 1990b). Late Jurassic transtension near the trace of the megashear in southeastern California has been inferred to explain the formation of ensimatic pull-apart or rift basins in which clastic and basaltic protoliths of the Pelona-Orocopia schist were deposited (Tosdal and others, 1990a). Late Jurassic ophiolites of the Klamath Mountains and Sierran foothills may have formed in similar rift basins along the northwestward projection of the megashear in central and northern California (Harper and others, 1985). Local transtension along the northwestward projection of the megashear may also be indicated by intrusion ages of 165 to 147 Ma of dikes from southeastern California to the Sierran foothills (James, 1989; Wolf and Saleeby, 1991). Transtension in the southern part of the megashear fault zone may have created the Cuicateco basin in southern México (Figs. 36, 37), which probably formed in the Middle to Late Jurassic as a rift basin within continental crust (Ortega-Gutiérrez and others, 1990). Late Jurassic to Early Cretaceous rocks of the Cuicateco terrane may be analogous to rocks of the Josephine or Smartville ophiolites of California, or to the protoliths of the Pelona-Orocopia schist.

Although displacement on the megashear is widely cited as Late Jurassic, available data are consistent with several hundred kilometers of late Paleozoic sinistral slip (Stevens and others, 1992), with reactivation and several hundred kilometers of displacement in the Late Jurassic. Our reconstruction portrays all left-lateral displacement on the megashear as Late Jurassic, but it can be easily modified to accommodate partitioning of displacement into late Paleozoic and Late Jurassic episodes.

Displacement of amalgamated Mixteco and Zapoteco terranes. Paleomagnetic data from Permian, Bathonian to Oxfordian, and Albian rocks in the Mixteco terrane are interpreted to indicate that the amalgamated Mixteco and Zapoteco terranes were translated 15° ± 8° southward to their present latitude during Oxfordian to Albian time, about 160 to 110 Ma (Urrutia-Fucugauchi and others, 1987; Ortega-Guerrero and Urrutia-Fucugauchi, 1989). This interpretation supersedes a previous estimate of 20° to 30° of southward translation. Based on the revised estimate, we have inferred accretion of the terranes to western Pangea at the latitude of northern México. We speculate that southeastward translation of the amalgamated Zapoteco and Mixteco terranes was accommodated by Late Jurassic slip on an outboard strand of the megashear fault zone and earliest Cretaceous slip on a fault outboard of the megashear (Fig. 36,

37). Temporal partitioning and rates of displacement are unknown. By Albian time, and perhaps earlier in the Early Cretaceous, the Zapoteco-Mixteco composite terrane was in its current position with respect to the Cuicateco terrane and southern Maya terrane (Fig. 38).

Displacement of Serí terrane in Baja California. Paleozoic miogeoclinal rocks of the Serí terrane in northeastern Baja California probably were deposited in contiguity with correlative rocks of the Serí terrane in northern Sonora (Gastil and others, 1978; Gastil and Miller, 1984). This relation apparently conflicts with numerous paleomagnetic studies that have inferred about 1,500 km of Late Cretaceous and Cenozoic northward displacement of Baja California and a mid-Cretaceous position in central or southern México (p. 80). The geologic correlations can be reconciled with the paleomagnetic data only if the sliver of Serí terrane in northeastern Baja California was translated southward from an initial position adjacent to Sonora to central México on sinistral strike-slip faults prior to the mid-Cretaceous.

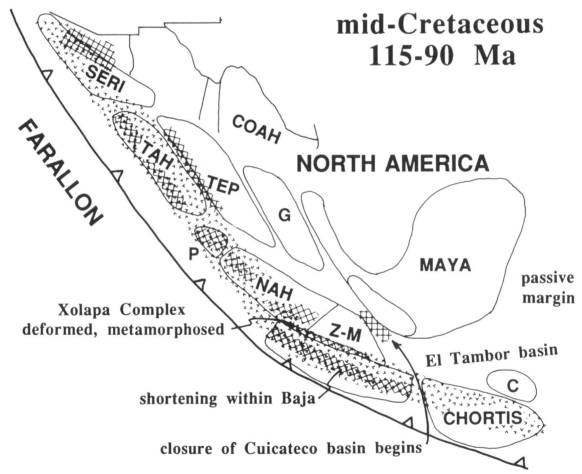

Figure 38. Mid-Cretaceous (about 115-90 Ma) paleogeographic reconstruction. Southeastward displacement of Baja composite terrane and Chortis ceases by about 100 Ma. Xolapa Complex (part of Chatino terrane) is deformed and metamorphosed at southern margin of Zapoteco-Mixteco composite terrane. Shortening is coeval with arc magmatism in Pericú terrane, Xolapa Complex, and Baja composite terrane; shortening also occurs in Cuicateco basin and in Serí, North America, Tahué, western Tepehuano, and Nahuatl terranes. G, Guachichil terrane. Other abbreviations and patterns as in Figure 37.

The Serí terrane in Baja California probably was trimmed from the continent by a sinistral strike-slip fault and translated southward to central México beginning in late Paleozoic time, coeval with similar displacement in the Mojave region. On the far western edge of North America, the western part of the Serí terrane was especially susceptible to translations accompanying oblique convergence, which probably was sinistral in the late Paleozoic to Jurassic. Earlier, we proposed a few hundred kilometers of southward displacement of the Serí terrane in Baja during the late Paleozoic (Fig. 31). We infer an additional 1,000 to 1,500 km southward displacement during the Middle Jurassic to Early Cretaceous (Figs. 36–38), at least partly coeval with the southward displacement of the amalgamated Zapoteco and Mixteco terranes and displacement on the megashear fault zone.

Jurassic deposition in and near the Gulf of Mexico

Prior to Late Jurassic oceanic spreading in the Gulf of Mexico, the Yucatán platform probably was about 500 km north-northwest of its current position, adjacent to stretched crust of the U.S. Gulf Coast (Fig. 36). In the Callovian (late Middle Jurassic), these areas formed a flat-lying, gently subsiding region that periodically was flushed by marine water, leading to the accumulation of the thick Louann-Isthmian salts. The presence of Callovian shallow-marine strata in northern Veracruz supports a Pacific (western) origin for these waters; Tethyan (eastern) waters may not have reached the Gulf of Mexico until late Kimmeridgian to Tithonian time (Salvador, 1987).

In the Oxfordian (early Late Jurassic), concomitant with oceanic spreading in the Gulf of Mexico and southeastward displacement of the Yucatán platform and southern province of the Maya terrane, transgression and subsidence resulted in the deposition of shallow-marine strata in the Maya terrane on the margins of the Gulf of Mexico and in the Coahuiltecano, Guachichil, and Tepehuano terranes in eastern continental México (López-Ramos, 1981; Enos, 1983; Young, 1983; Salvador, 1987; de Cserna, 1989). Platform carbonates accumulated on a series of topographic highs at the western margin of the growing Gulf of Mexico (Fig. 35), which probably was the site of a transform boundary (Figs. 34, 36). In Oxfordian time, fine-grained pelagic carbonates, shale, and local evaporites known as the Santiago Formation and the Zuloaga Formation (or Zuloaga Group; Götte and Michalzik, 1991) in central and northeastern México, respectively, were deposited in intraplatform shelf basins. These strata are approximately correlative with the Norphlet, Smackover, and Haynesville Formations in the U.S. Gulf Coast (Imlay, 1943). Continued transgression caused marine deposition in the Chihuahua Trough starting in the late Kimmeridgian to Tithonian (Tovar-R., 1981; Dickinson and others, 1986; Araujo-Mendieta and Casar-González, 1987; Salvador, 1987) and greater marine influence around emergent islands and peninsulas in eastern México (Fig. 35). Clastic sediments accumulated near the margins of emergent areas such as the Coahuila platform and Tamaulipas platform, and red beds were deposited across the Yucatán Penin-

sula (Viniegra-Osorio, 1981). Late Jurassic nonmarine strata were deposited south of the marine environments in the Chiapas-Guatemala region.

Cretaceous to Paleogene

With the cessation of extension in the Gulf of Mexico region and sinistral displacement on the Mojave-Sonora Megashear by the beginning of the Early Cretaceous, the Serí (in Sonora), Tarahumara, Coahuiltecano, Maya, Guachichil, Tepehuano, and Tahué terranes had attained their approximate current positions with respect to North America and thus were part of the North America plate (Fig. 38). Sinistral displacement of more outboard terranes such as the amalgamated Zapoteco and Mixteco terranes may have continued until the Albian. All terranes were inboard of a major trench at which oceanic lithosphere of the Farallon and perhaps other plates was subducted beneath North America during Cretaceous and Paleogene time (Fig. 38). Subduction was accompanied by arc magmatism, late Early to early Late Cretaceous collision and accretion, Early Cretaceous sinistral and Late Cretaceous dextral translation of outboard terranes, and Late Cretaceous–Paleogene Laramide orogenesis.

Cretaceous magmatic arc

During the Early Cretaceous, a continental magmatic arc was active near the southwestern margin of México (Fig. 38). In the Serí terrane in Sonora, Early Cretaceous silicic and intermediate volcanic rocks are intruded by Late Cretaceous granitoids that are younger to the east (Anderson and Silver, 1969). In northern Sonora, volcanogenic Early Cretaceous strata were deposited on North American basement in a backarc basin (Almazán-Vázquez and others, 1987). In the Tahué terrane, basalts and andesites interbedded with Aptian-Cenomanian carbonates are intruded by syntectonic early Late Cretaceous granitoids and post-tectonic late Late Cretaceous and Paleogene granitoids (Henry and Fredrikson, 1987). In the Pericú terrane, metamorphic country rocks were intruded by late Early Cretaceous mafic to intermediate plutonic rocks during strong regional compression; undeformed high-K granitoids intruded these rocks in the Late Cretaceous (Ortega-Gutiérrez, 1982; Aranda-Gómez and Pérez-Venzor, 1989). In the Nahuatl terrane, intermediate to siliceous volcanic and volcaniclastic rocks and interbedded Early Cretaceous siliciclastic rocks and carbonates are intruded by mid-Cretaceous to Paleogene granitoids and overlain by Late Cretaceous rhyolites and red beds. Geophysical data imply that the western part of the Trans-Mexican Volcanic Belt between the Tahué and Nahuatl terranes is underlain by Cretaceous-Tertiary(?) granitic rocks (Campos-Enríquez and others, 1990). In the Yuma terrane, for which we infer a position adjacent to southern México by the end of the Early Cretaceous (Fig. 38), intermediate to siliceous volcanic rocks of the Alisitos and Santiago Peak Formations were erupted in the western subterrane during the Late Jurassic and Early Cretaceous (Gastil and

others, 1975). The Yuma terrane is intruded by syntectonic plutons of the western, Early Cretaceous part of the Peninsular
Ranges batholith, whereas Paleozoic miogeoclinal rocks of the
Serí terrane are intruded by late to posttectonic plutons of the
eastern, Late Cretaceous part of batholith (Silver and others,
1975; Silver, 1979; Todd and others, 1988). In the Chatino terrane, orthogneisses have yielded U-Pb and Rb-Sr dates of 160 to
128 Ma that may be interpreted to indicate Late Jurassic and
Early Cretaceous intrusion of granitoids prior to deformation and
metamorphism. Magmatic rocks were not emplaced in the Chatino and adjacent Zapoteco-Mixteco composite terrane from the
late Early Cretaceous until the earliest Tertiary, about 110 to 60
Ma. The Chortis terrane, for which we infer a position south of
Guerrero in the Late Cretaceous (p. 109) (Fig. 39), is intruded by
abundant mid-Cretaceous to Paleogene granitoids (Horne and
others, 1976; Gose, 1985, and references therein).

During the Late Cretaceous and Paleogene the continental arc in northern México progressively broadened from a
narrow zone in western Sonora and Sinaloa to a wider zone
extending eastward into Chihuahua, Durango, and Zacatecas
(Anderson and Silver, 1969; Clark and others, 1980, 1982;

Damon and others, 1981). This broadening or migration of arc
magmatism probably was caused by progressive shallowing of
the Benioff zone beneath western North America (Coney and
Reynolds, 1977).

The magmatic history of southwestern México is consistent
with a more southerly Early Cretaceous position of terranes of the
Baja California peninsula (terranes in Baja hereafter are referred
to as Baja composite terrane or simply Baja; analogous to Peninsular Ranges terrane of Bottjer and Link, 1984; Lund and Bottjer,
1991). Cretaceous intrusive rocks older than about 115 Ma are
absent from the southern Tahué, Pericú, and Nahuatl terranes. We
suggest that these three terranes were inboard (northeast) of the
magmatic arc that intruded the Yuma and Serí terranes in the
Baja composite terrane (Fig. 37). Mid-Cretaceous intrusion of the
Tahué, Pericú, and Nahuatl terranes may have been delayed until
the Baja California block was transported sufficiently far to the
southeast (Fig. 38), and also may indicate shallowing of the subduction zone and eastward migration of arc magmatism. A central Mexican position of the Baja California block until the Late
Cretaceous obviates the need for parallel volcanic arcs in
northwestern México, which would be required if Baja were in its

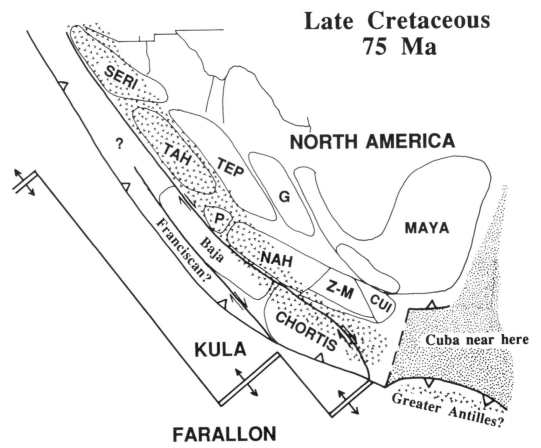

Figure 39. Late Cretaceous (about 75 Ma) paleogeographic reconstruction. Speculative plate boundaries
southwest and southeast of continent. Stipple pattern indicates oceanic lithosphere created by rifting of
South America from Yucatán platform; CUI, Cuicateco terrane. Other abbreviations and patterns as in
Figure 38.

current position in the Early Cretaceous because a Jurassic–Early Cretaceous magmatic arc crops out in Sonora (cf. Rangin, 1978; Gastil, 1983).

Collision and accretion

The continental magmatic arc in western México indicates that oceanic lithosphere of the Farallon and perhaps other plates was subducted beneath North America throughout Cretaceous time. The orthogonal component of convergence caused shortening and accretion of several terranes at the western margin of México.

Cochimí composite terrane. Triassic to Jurassic island arc subterranes of the Cochimí composite terrane probably were accreted to the western margin of the Yuma terrane in the Late Jurassic and Early Cretaceous (Fig. 37). Accretion terminated arc magmatism in each terrane. The accreted arc terranes and the Yuma terrane were overlapped by forearc-basin flysch in the Albian. Paleomagnetic studies infer as much as 1,500 km of Late Cretaceous and early Cenozoic northward translation with respect to North America.

The Cochimí composite terrane also contains Late Triassic to mid-Cretaceous ocean-floor basalt and overlying pelagic and clastic sedimentary rocks that were subducted, metamorphosed to blueschist facies, and underplated to North America near the equator during mid-Cretaceous time (Sedlock, 1988a, c; Baldwin and Harrison, 1989; Hagstrum and Sedlock, 1990, 1991, 1992). After about 1,000 km of late Early Cretaceous and Late Cretaceous northward transport, the blueschists were juxtaposed with the previously accreted arc subterranes of the Cochimí terrane in the Late Cretaceous. During later Cretaceous and Paleogene time, the entire Cochimí composite terrane and the Yuma terrane were translated about 1,500 km northward with the rest of Baja California (Figs. 39, 40) (Hagstrum and others, 1985; Sedlock, 1988b, c; D. Smith and Busby-Spera, 1989, and unpublished manuscript; Hagstrum and Sedlock, 1990, 1991, 1992).

Yuma terrane. We have speculated that Mesozoic basinal strata of the eastern Yuma subterrane were accreted to allochthonous Paleozoic rocks of the Serí terrane in the central Mexican forearc in the Middle or Late Jurassic. In the Late Jurassic, arc magmatism at this latitude jumped westward from the Mexican continental arc, east of the amalgamated eastern Yuma subterrane and the allochthonous part of the Serí terrane, to an island arc in the western subterrane of the Yuma terrane (Figs. 36, 37). In other words, the arc jump transferred the eastern Yuma subterrane and the allochthonous part of the Serí terrane from the forearc to the backarc.

In Early to early Late Cretaceous time, the Late Jurassic to Early Cretaceous island arc in the western Yuma subterrane closed with and then was thrust eastward beneath the amalgamated eastern Yuma subterrane and allochthonous Serí terrane, which were now part of North America (Fig. 38) (Griffith and Goetz, 1987; Goetz and others, 1988; Todd and others, 1988). In southern California and Baja California north of the Agua Blanca

fault, the island arc in the western Yuma subterrane was thrust eastward beneath the deformed, uplifted, and eroded eastern subterrane of the Yuma terrane in the Early Cretaceous. South of the Agua Blanca fault, the island arc in the western Yuma subterrane underthrust Paleozoic miogeoclinal rocks of the Serí terrane and the eastern subterrane of the Yuma terrane about 105 to 100 Ma, based on U-Pb ages of deformed and undeformed plutons that stitch the suture. Vergence and sense of displacement of the suture are deduced from steeply plunging lineations, asymmetric microstructural kinematic indicators, and the inferred level of crustal exposure. Plutonism in the Peninsular Ranges batholith migrated eastward across the suture during the mid-Cretaceous.

Pericú terrane. In the Pericú terrane, prebatholithic metasedimentary rocks were intruded by mafic plutons and low-K granitoids beginning about 115 Ma. Penetrative, roughly east-west shortening of the metasedimentary and plutonic rocks apparently began shortly after intrusion of the plutonic rocks and ceased by about 95 Ma (Fig. 38). Syntectonic emplacement is supported by concordant foliation in the metasedimentary and deformed plutonic rocks. S-C fabrics in thick mylonites indicate left-reverse displacement on north- to northeast-striking, east-dipping shear zones (Aranda-Gómez and Pérez-Venzor, 1989). Compression ceased by about 95 Ma, as indicated by intrusion of undeformed high-K granitoids that crosscut the older rocks and structures.

Cuicateco terrane. The Cuicateco terrane contains earliest Cretaceous or older sedimentary rocks that were deposited in a basin of enigmatic origin, here named the Cuicateco basin. In the southern part of the terrane, the sedimentary rocks are faulted against the Early Cretaceous Chontal island(?) arc, a subterrane of the Cuicateco terrane (Carfantan, 1981). Along most of its western margin, the Cuicateco terrane contains deformed and metamorphosed granitoids and volcanic rocks of uncertain origin.

Opening of the Cuicateco basin began prior to earliest Cretaceous deposition of sediments, probably in the Late Jurassic and perhaps as early as the Middle Jurassic. Carfantan (1983) proposed that the basin formed by aborted intracontinental rifting during drifting of North America from South America, but this seems unlikely because the northwest-southeast trend of the basin was parallel to the probable drifting direction. We propose two other, possibly interdependent, origins of the basin. First, the basin may have formed by Middle to Late Jurassic transtension on the southern reach of the transform fault at the western margin of the extending Gulf of Mexico region (Figs. 36, 37). We have inferred that this fault is buried near the western margin of the Chiapas Massif, i.e., the eastern margin of the Cuicateco terrane. Second, the basin may have formed by Late Jurassic wrenching at a releasing step in the megashear fault zone.

There are at least two interpretations of the original tectonic setting of the strongly deformed metaigneous rocks at the western margin of the Cuicateco terrane; both are consistent with the two proposed interpretations of the origin of the Cuicateco basin. The rocks may be products of anorogenic magmatism associated with continental rifting. Alternatively, the rocks may have formed in

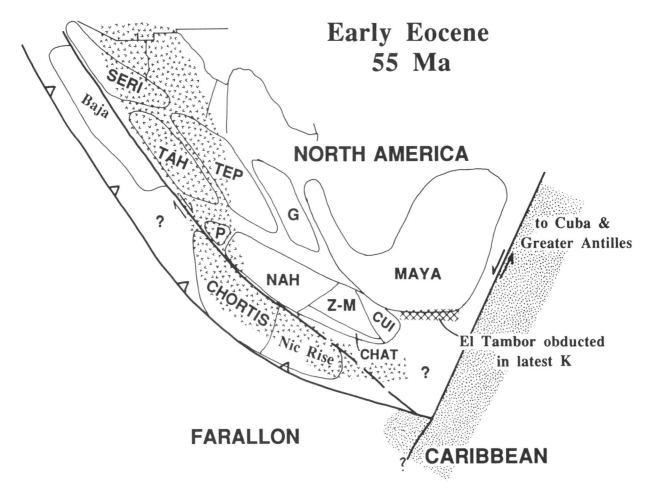

Figure 40. Early Eocene (about 55 Ma) paleogeographic reconstruction. CHAT indicates Chatino terrane; Nic Rise, Nicarague Rise. Other abbreviations and patterns as in Figure 39.

the continental arc on the western margin of México. The latter option implies that a strand or strands of the megashear coincided with the Jurassic arc in southern México, as has been postulated for the megashear in northern México.

Extension in the Gulf of Mexico region and southward displacement of the Chiapas Massif in the southern Maya terrane with respect to the Cuicateco basin terminated by the end of the Jurassic, when the locus of Pangean rifting shifted from the northwestern to the southeastern margin of the Yucatán block (Figs. 36, 37). Southeastward displacement of the amalgamated Zapoteco and Mixteco terranes with respect to the Cuicateco basin probably ceased by Albian or earlier Cretaceous time. In the Albian, these two continental blocks began to converge, closing the intervening Cuicateco basin (Delgado-Argote, 1989). Pervasive northeast-southwest shortening, intrusion by synkinematic plutons, and weak metamorphism of the basin peaked in the Turonian (Carfantan, 1983). Deformation of synorogenic Campanian to Maastrichtian flysch indicates that closure continued throughout the Cretaceous. By earliest Paleogene time, the Zapoteco terrane was thrust eastward over the Cuicateco terrane

on the Juárez suture, and the Cuicateco terrane was thrust eastward over the adjacent Maya terrane, causing internal deformation of the Maya terrane. An alternate model calls for earlier (pre-Cretaceous) thrusting of the Zapoteco terrane eastward over basment rocks of the Maya terrane, prior to opening of the Cuicateco basin.

Other Cordilleran terranes. Mid-Cretaceous (pre-Laramide) thrusting, folding, and metamorphism have been noted in other terranes near the southwestern margin of México. Cretaceous shortening in the Serí and North America terranes includes not only a latest Cretaceous-Paleogene Laramide event but also unrelated(?), early Late Cretaceous, east- to northeast-vergent thrusting (Roldán-Quintana, 1982; Pubellier and Rangin, 1987; Rodríguez-Castañeda, 1988; Nourse, 1990; Siem and Gastil, 1990). Large elongational strains in northern Sonora and northeastern Baja California may indicate a genetic link with coeval, generally south-vergent shortening in the Maria fold and thrust belt of southeastern California and southwestern Arizona (Reynolds and others, 1986).

In the Tahué terrane, the Jurassic–Early Cretaceous magmatic arc was deformed and metamorphosed during east-vergent overthrusting by a partly conformable ophiolite sequence. The ophiolite probably formed in a forearc basin (Servais and others, 1982, 1986) or possibly a backarc basin (Ortega-Gutiérrez and others, 1979). Mylonites developed along some thrusts. Deformation was inferred to be of Laramide age by Servais and others (1982, 1986), but we infer a mid-Cretaceous age because Late Cretaceous plutons older than 85 Ma are syntectonic (Fig. 38).

In the Tepehuano terrane, disrupted mafic and ultramafic magmatic rocks of probable Late Jurassic to Early Cretaceous age were shortened and thrusted to the north-northeast over volcaniclastic rocks of the backarc basin. The lithology, structure, and geochemistry of the magmatic rocks are more consistent with an ophiolitic origin (Servais and others, 1982) than an oceanic island arc origin (Monod and others, 1990). Thrusting probably occurred during Early Cretaceous, pre-Albian closure of a backarc basin east of the Early Cretaceous arc in the Tahué terrane (Fig. 38).

In the Nahuatl terrane, Paleozoic(?) to early Mesozoic metamorphic rocks of the Tierra Caliente Complex (TCC) exhibit subhorizontal foliation, axial surfaces, and thrust faults that indicate eastward to northeastward tectonic transport of uncertain age. The TCC is overlain by low-grade metamorphic rocks in the lower part of the Upper Mesozoic Assemblage (UMA) that were strongly deformed during east-vergent thrusting of mid-Cretaceous age. We speculate that the TCC was accreted to the Mexican continental margin by the end of the Jurassic (Fig. 37), and that the TCC and the lower part of the UMA together underwent east-vergent to northeast-vergent folding and thrusting in the late Early Cretaceous (Fig. 39). After deposition of Albian-Coniacian platform carbonates of the upper part of the UMA, the entire UMA underwent Laramide thrusting.

In the Chatino terrane, metasedimentary rocks of the Xolapa Complex have a complex history of deformation, intrusion, and metamorphism. The pre–Late Jurassic history of the terrane is completely unknown. The metasedimentary rocks may have been intruded in the Late Jurassic and Early Cretaceous by tonalitic plutons derived at least partly from Proterozoic sources such as those in the Mixteco and Zapoteco terranes. These plutons may have been strongly deformed and metamorphosed during thrusting of the Chatino terrane over the Mixteco and Zapoteco terranes during the late Early to Late Cretaceous (Fig. 38).

In our reconstruction, rocks deformed during the mid-Cretaceous as described above form a quasi-continuous belt that roughly coincided with the arc and forearc at the southwestern margin of México (Fig. 38). The cause or causes of this deformation have not been demonstrated but may include collision and accretion of outboard terranes and changes in relative plate motions. The collision hypothesis is awkward because the continuity of the deformed belt requires later excision of unidentified and unaccounted for terranes from the entire length of the western margin of México. The plate motion hypothesis may be supported by the apparent increase in Farallon–North America con-

vergence throughout the Early Cretaceous at the latitude of central México (Table 18), which may have disrupted the western margin of the overriding North America plate.

Southern Maya terrane. Late Cretaceous collision at the southern edge of the Maya terrane resulted in the Campanian-Maastrichtian northward obduction of the El Tambor Group, a Cretaceous ophiolite and forearc assemblage (Donnelly and others, 1990a). The tectonic setting and kinematics of the collision are controversial. One popular model infers Cretaceous subduction of oceanic lithosphere southwest of the passive southern margin of the Maya terrane beneath a northeastward-migrating arc terrane that probably included much of modern Cuba; latest Cretaceous collision of this arc with the Maya terrane; and latest Cretaceous and Cenozoic northeastward to eastward translation of this arc into the Caribbean region (Pindell and Dewey, 1982; Burke and others, 1984; Pindell and others, 1988; Pindell and Barrett, 1990). A necessary element of this model is that the Chortis terrane was adjacent to the southwestern margin of México west of the collision zone and has undergone 1,000 to 2,000 km of postcollision sinistral slip on the southern boundary of the Maya terrane (Figs. 39–42). This model is supported by interpretations of 1,100 to 1,400 km of post–late Eocene left slip on the Cayman Trough (Macdonald and Holcombe, 1978; Burke and others, 1984; Rosencrantz and Sclater, 1986); by the presence of Grenville basement in central Cuba, which has no correlative within or on the periphery of the Caribbean; and by the absence of late Early Cretaceous to earliest Tertiary magmatic rocks in the Chatino and Zapoteco-Mixteco terranes. The latter terranes may have been shielded from arc magmatism by an outboard terrane such as Chortis.

An alternate interpretation of the collision is that the Chortis terrane collided with the southern margin of the Maya terrane, presumably after closure of a small ocean basin (Donnelly, 1989). According to this model, postcollision sinistral displacement of the Chortis terrane would be no more than a few hundred kilometers, in agreement with onland investigations of the major faults in central Guatemala (Deaton and Burkart, 1984a; Burkart and others, 1987). A thickened crustal root beneath the Motagua fault (T. Donnelly, unpublished gravity study) seems to be more consistent with the Chortis collision model than with a model involving collision and subsequent removal of an island arc. The gravity data also indicate that slabs of the El Tambor Group dip to the north, implying northward subduction of oceanic lithosphere beneath the Maya terrane prior to collision and southward as well as northward vergence during collision. This interpretation counters arguments for the westward transfer of ≥1,100 km of sinistral displacement in the Cayman Trough by arguing for less cumulative displacement in the trough or large-scale east-west crustal extension of the Chortis terrane, particularly of the submerged Nicaragua Rise (Donnelly, 1989). We provisionally accept the island arc alternative and the implied mobility of the Chortis terrane, but we do not discount the possibility or likelihood of the alternate interpretation and discuss it further on pages 116–117.

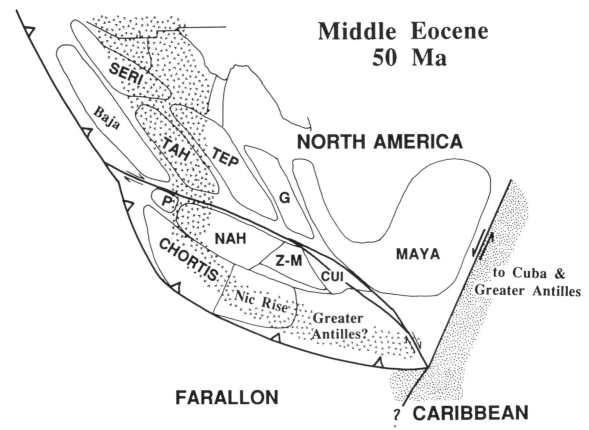

Figure 41. Middle Eocene (about 50 Ma) paleogeographic reconstruction. Abbreviations and patterns as in Figure 39.

According to either interpretation, north-south shortening of the El Tambor Group may have accommodated only the orthogonal component of possibly oblique collision. During collision, Paleozoic basement rocks in the southern Maya terrane were pervasively deformed and metamorphosed to greenschist facies, and some fragments may have been metasomatized and metamorphosed under blueschist- or eclogite-facies conditions prior to incorporation in serpentinite (Harlow, 1990). Detritus derived from the El Tambor Group has been found in Campanian (Sepur Group), Maastrichtian, and Paleocene flysch that overlie Paleozoic basement rocks and older Mesozoic strata in the southern Maya terrane. These strata also contain volcanic clasts that must have been derived from an island arc or continental arc (Chortis?) to the south.

Dextral displacement

Many terranes in the western United States and Canada may have been derived from sources at the latitude of México and translated hundreds to thousands of kilometers northward along the western continental margin during the Late Cretaceous and early Cenozoic, probably in response to right-oblique convergence between North America and subducting oceanic plates in the Pacific (McWilliams and Howell, 1982; Champion and

others, 1984; Page and Engebretson, 1984; Engebretson and others, 1985; Tarduno and others, 1985, 1986; Stock and Hodges, 1989; Umhoefer and others, 1989). In this section, we summarize evidence for Late Cretaceous and Paleogene dextral displacement along the western margin of México.

Baja California. Paleomagnetic studies of rocks of diverse age and lithology in the Yuma and Cochimí terranes have concluded that Baja California was translated 10° to 25° northward between mid-Cretaceous and Miocene time (Patterson, 1984; Hagstrum and others, 1985, 1987; Filmer and Kirschvink, 1989; Flynn and others, 1989; D. Smith and Busby-Spera, 1989, and unpublished manuscript). In Premise 12, we postulate that about 15° of northward translation of Baja (including displacement due to the opening of the Gulf of California) has occurred since about 90 Ma. Although paleomagnetic data have not yet been collected from the Serí terrane in northeastern Baja California or the southern Yuma terrane, we infer that these areas have undergone similar displacement because plutons of the Peninsular Ranges batholith stitch the Serí sliver to the Yuma terrane by about 95 Ma and probably intrude the entire Yuma terrane. Paleomagnetic studies of the Magdalena subterrane of the Cochimí terrane are in progress. The blueschist-facies subduction complex of the Cochimí terrane appears to have undergone more dextral displace-

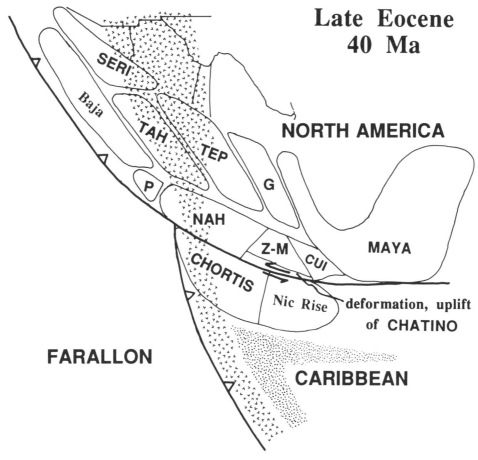

Figure 42. Late Eocene (about 40 Ma) paleogeographic reconstruction. Abbreviations and patterns as in Figure 39.

ment than the rest of the Baja terranes (Hagstrum and Sedlock, 1990, 1992). We have not inferred large translation of the Pericú terrane because it is separated from the Yuma terrane by the La Paz fault, a major structure with a multi-phase displacement history. Paleomagnetic studies of rocks in the Pericú terrane may help determine the relation of the Pericú and Yuma terranes.

Plate motion models predict that most margin-parallel dextral displacement at the latitude of central México occurred between 90 and 60 Ma, though additional dextral slip may have occurred during the Paleogene (p. 84). Baja California probably was near its pre–Gulf of California position adjacent to Sonora and Sinaloa by the early Eocene, based on a provenance link of distinctive conglomerate clasts in the Yuma terrane with source rocks in Sonora, and on paleomagnetic results from Eocene strata in the Yuma terrane (Bartling and Abbott, 1983; Flynn and others, 1989). Other paleomagnetic data imply that the final arrival of Baja at its pre–Gulf of California position may not have occurred until about 40 Ma (Lund and others, 1991b). In our reconstruction (Figs. 38, 39), the Baja California composite terrane (Yuma, Cochimí, and western Serí terranes) is translated northward about 1,200 km from 85 to 55 Ma. The minimum

average rate of tangential plate motion of 40 mm/yr is lower than the maximum average rate suggested by Engebretson and others (1985). An additional 2° to 3° of northward translation of Baja occurred during late Cenozoic opening of the Gulf of California.

Cretaceous and Paleogene northward displacement of the Baja block has important implications for Cretaceous arc magmatism on the western margin of North America (Fig. 38). If the Baja block was in southern México during the Cretaceous, the Alisitos–Santiago Peak–Peninsular Range batholith rocks would have been part of the North American magmatic arc in a region otherwise lacking such an arc. If Baja instead was in its present position during the Cretaceous, the arc rocks of Baja would have been positioned outboard of and parallel to coeval arc and forearc rocks in Sonora and Sinaloa. The hypothesis of northward transport of Baja in the Late Cretaceous and Paleogene thus neatly fills a void in the continental arc in southern México and avoids a potential pairing of arcs in northwestern México.

Chortis. Our reconstruction implies that dextral faults that accommodated Late Cretaceous displacement of Baja California also accommodated displacement of similar age and magnitude of the Chortis terrane (Figs. 38–40). Our reconstruction predicts that translation of Chortis at the southern margin of México was to the west or west-northwest and thus unlikely to be resolved by paleolatitude calculations. Paleomagnetic data indicate a complex history of rotation but little latitudinal displacement of Chortis during the Cretaceous (Gose, 1985). The Cretaceous dextral slip and Cenozoic sinistral slip of Chortis shown in our reconstruction roughly (but not exactly) correspond to times of clockwise and counter-clockwise rotations of Chortis as determined by Gose (1985).

Trans-Mexican Volcanic Belt. In order to remove a kink in various mineralization belts and to better align the Cretaceous granitoids of the Tahué, Pericú, and Nahuatl terranes, we postulate early Cenozoic dextral slip on a fault system that is now concealed by the late Cenozoic Trans-Mexican Volcanic Belt (TMVB). In our reconstruction, about 400 km of dextral displacement occurred on the TMVB fault system from 55 to 40 Ma, implying an average rate of 25 mm/yr (Figs. 40, 41). Gastil and Jensky (1973) inferred 175 km of latest Cretaceous and early Cenozoic dextral slip and 260 km of Neogene dextral slip on the TMVB, but Neogene dextral displacement seems unlikely in view of reconstructed plate motions (Stock and Hodges, 1989) and evidence for Neogene sinistral slip along the southern margin of México (Burkart and others, 1987). Our reconstruction assumes that dextral slip on the TMVB occurred during and after northward translation of Baja California.

Mexican origin of terranes in the United States and Canada. In this section, we compare displaced terranes north of México with coastal terranes of México to which they may have been adjacent in mid-Cretaceous time. Displacement of some terranes is disputed on the basis of geologic arguments that are beyond the scope of this volume.

The following tectonic history accounts for paleomagnetic, biostratigraphic, and geologic data from terranes of southwestern Canada and southeastern Alaska. (Monger and others, 1982; Hillhouse and Gromme, 1984; Irving and others, 1985; Gehrels and Saleeby, 1987; Umhoefer and others, 1989). The Stikine, Cache Creek, and Quesnel terranes were amalgamated to form Superterrane I in the eastern Pacific at the latitude of the northwestern United States in the Early or Middle Jurassic time. Superterrane I was translated southward to the latitude of northwestern México by the mid-Cretaceous. The Wrangellia and Alexander terranes were in the eastern Pacific basin near but not necessarily adjacent to southern México and northwestern South America by the Late Triassic, and were amalgamated to form Superterrane II at the latitude of México in the Late Jurassic to Early Cretaceous. After mid-Cretaceous accretion of Superterrane II to Superterrane I at the latitude of northwestern México, both superterranes were translated 2,500 to 4,000 km northward along western North America in the Late Cretaceous and Paleogene. The Wrangellia terrane contains late Paleozoic to Jurassic volcanic rocks that may have formed at an arc in or near the oceanic corner west of central Pangea (Fig. 31). The Alexander terrane probably was exotic with respect to North America until the Triassic (Gehrels and Saleeby, 1987), but Middle(?) Jurassic

granitoids formed at the latitude of southern México in either the continental Jurassic arc or an outboard arc (Group 1 and Group 2 Jurassic arc rocks, respectively). The amalgamated superterranes were intruded by Cretaceous granitoids of the Coast Plutonic Complex while at the latitude of northwestern México (references in Irving and others, 1985), perhaps adjacent to the Serí and Tahué terranes (Figs. 37, 38).

The Salinia block in the central California Coast Ranges consists of Cretaceous granitic basement and overlying Late Cretaceous sedimentary rocks that have yielded anomalous magnetic inclinations suggesting about 2,500 km of latest Cretaceous to Paleogene northward translation (Champion and others, 1984). Restoration of this displacement places Salinia adjacent to the mid-Cretaceous continental arc in the Nahuatl and Mixteco terranes. However, petrologic, isotopic, and paleontologic data (e.g., James and Mattinson, 1988) seem to indicate that the granitic basement of Salinia formed south of the Sierra Nevada and north of the Peninsular Ranges batholith.

In the Coast Ranges of California, paleomagnetic data have been interpreted to indicate more than 2,000 km of Late Cretaceous to Eocene northward displacement of oceanic rocks including Jurassic ophiolite, Late Cretaceous sedimentary rocks, and mid-Cretaceous limestone blocks of the Franciscan Complex (Alvarez and others, 1980; McWilliams and Howell, 1982; Courtillot and others, 1985; Tarduno and others, 1985, 1986; M. Fones and others, unpublished manuscript). We speculate that in the mid-Cretaceous the ophiolitic rocks and flysch, and perhaps other parts of the Franciscan Complex, were part of the forearc of the continental arc at the latitude of central México (e.g., Fig. 39). Limestone blocks probably were derived from seamounts in the Pacific basin that accreted to North America in the latest Cretaceous or Paleogene.

Late Cretaceous–Paleogene shortening: Laramide (Hidalgoan) orogeny

During latest Cretaceous to about middle Eocene time, east-northeast–west-southwest to northeast-southwest shortening produced a roughly north-northwest–trending foreland fold and thrust belt in most of eastern and central México (Fig. 43) during the Hidalgoan orogeny (Guzmán and de Cserna, 1963). Deformation was synchronous and kinematically and spatially akin to Laramide foreland deformation in the Cordillera north of the border (e.g., Davis, 1979), so we substitute the term Laramide for Hidalgoan in this work. Laramide deformation in México and coeval eastward migration of the locus of arc magmatism may have been caused by progressive shallowing of subducted oceanic lithosphere, as has been postulated for the southwestern United States (Coney and Reynolds, 1977).

The age and kinematics of Laramide deformation in México are well understood in and near the Sierra Madre Oriental province, where post-Laramide volcanism and tectonism were minimal. Parallel fold ridges and valleys trend roughly north-south in east-central México and beneath the Gulf coastal plain

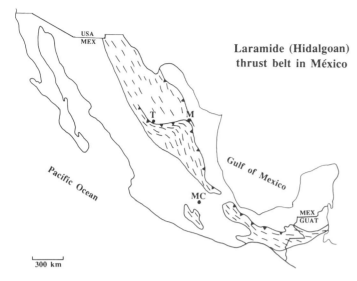

Laramide (Hidalgoan) thrust belt in México

Figure 43. Map of Laramide (Hidalgoan) thrust belt, after Campa-Uranga (1985b). Barbed lines indicate major deformation fronts; short line segments, structural trends. Western margin of continent has not been paleogeographically restored.

(Mossman and Viniegra-Osorio, 1976; Suter, 1984, 1987) and in the Chihuahua tectonic belt in northern Chihuahua (Lovejoy, 1980; Corbitt, 1984; Dyer and others, 1988); they trend east-west in the intervening Monterrey-Torreón transverse system (de Cserna, 1956; Tardy, 1975; Padilla y Sánchez, 1985, 1986; Quintero-Legorreta and Aranda-García, 1985). Tectonic transport normal to these trends—i.e., to the north, northeast, and east—locally was controlled by preexisting basement highs such as the Coahuila platform. Deformation was thin-skinned, with only local involvement of crystalline basement rocks. Detached sheets locally rotated about vertical axes during foreland thrusting (Kleist and others, 1984). Cumulative transverse displacement was 40 to 200 km, with up to 30% shortening, but spatial and temporal partitioning of displacement are poorly understood. In our pre-Cenozoic reconstruction, we arbitrarily restored about 150 km of margin-normal Laramide shortening, all at the eastern edge of the fold and thrust belt.

Late Cretaceous to Paleogene deformation is less well understood in southern and western México, in large part due to the overprint of subsequent tectonism and magmatism. Northeast- to east-vergent folding and thrusting of probable latest Cretaceous to Paleogene age in the Cuicateco, Mixteco, and Nahuatl terranes were synchronous with and may have been continuous with Laramide deformation in the Sierra Madre Oriental (Ferrusquía-Villafranca, 1976; Carfantan, 1981, 1983; Johnson and others, 1990). In the Tahué terrane, roughly east-west compression caused eastward thrusting and mylonitization of ophiolitic and forearc basin strata and greenschist metamorphism of underlying Jurassic-Cretaceous arc rocks (Servais and others, 1982, 1986). In the Serí and North America terranes, Late Cretaceous to Paleogene Laramide shortening was superposed on early Late Cre-

taceous contractional structures (p. 108); tectonic transport directions are reported to both the north and south (Roldán-Quintana, 1982; Pubellier and Rangin, 1987; Rodríguez-Castañeda, 1988; Nourse, 1990; Siem and Gastil, 1990).

Although the age of Laramide deformation still is not well constrained in many parts of México, available data are consistent with a progressive decrease in age to the northeast, as in the western United States (de Cserna, 1989). For example, shortening probably occurred in the Late Cretaceous in Sonora and Sinaloa but in the late Paleocene to middle Eocene near Monterrey and in the Veracruz subsurface (Mossman and Viniegra-Osorio, 1976; Padilla y Sánchez, 1985, 1986).

Laramide deformation corresponded in time and space with Late Cretaceous–Paleogene magmatism, which occurred west of the thrust front (López-Infanzón, 1986). Paleogene nonmarine conglomerate and finer-grained clastic rocks, often referred to as molasse, that locally crop out in the Tepehuano and Nahuatl terranes probably were deposited atop unconformity surfaces during or immediately after Laramide orogenesis (Edwards, 1955; Córdoba, 1988).

Depositional patterns

The following summary of the Cretaceous depositional history of eastern México is distilled from Viniegra-Osorio (1971, 1981), Enos (1983), Young (1983), Cantú-Chapa and others (1987), Winker and Buffler (1988), and de Cserna (1989), to whom the reader is referred for more detailed accounts. The Late Jurassic arrangement of isolated ephemeral islands, shallow-water platforms, and intraplatform deeper water basins at the western and southern margins of the Gulf of Mexico persisted into the Early Cretaceous. Neocomian transgression transformed Jurassic peninsulas (Coahuila, Tamaulipas) to islands and reduced the area of smaller islands. The Tethyan affinity of reptilian fossils in Neocomian strata in the northern Mixteco terrane indicates exchange of oceanic waters between the Gulf of Mexico basin and the Tethyan realm to the south and east (Ferrusquía-Villafranca and Comas-Rodríguez, 1988). A likely paleogeography includes a nearly closed basin, bounded on the north by the U.S. Gulf Coast, on the west by the Cretaceous arc in western México, and on the south by the southern Maya terrane and Zapoteco-Mixteco terranes, with a single oceanic outlet in the eastern Gulf of Mexico (see Winker and Buffler, 1988, Fig. 17).

After an influx of terrigenous material during late Aptian transgression, mid-Cretaceous (Albian-Cenomanian) carbonates were deposited over a larger part of México than in Neocomian time. Steep-sided, high-relief carbonate-evaporite platforms bounded by reef complexes shed very coarse carbonate detritus, notable also as prolific hydrocarbon reservoirs, into adjacent deep-water basins. The decline in the number and size of carbonate banks around the Gulf of Mexico in Comanchean time (90 to 85 Ma) may have been caused by thermal contraction of stretched crust (Winker and Buffler, 1988). Late Cretaceous (post-Turonian) deposition in eastern México was characterized by an increasing terrigenous component, shallowing of basins, and eastward migration of the shoreline. Throughout the Cretaceous, evaporites and carbonates accumulated on the stable Yucatán platform. We interpret the similarity of Cretaceous carbonates and clastic rocks in the Chortis terrane and coeval deposits of the open marine shelf in eastern México to indicate that the regions were contiguous (Figs. 38, 39).

In western México, Cretaceous sediments probably were deposited in a backarc basin or basins on the eastern flank of the continental arc in the Serí and Tahué terranes. Servais and others (1982) inferred an active backarc basin environment for Cretaceous terrigenous marine strata in Durango, Zacatecas, and Guanajuato (Tepehuano terrane) that apparently interfingered eastward with carbonate and shale of the open marine shelf in eastern México. A backarc setting has been proposed for Neocomian to Turonian terrigenous and carbonate strata in northeastern Durango and southeastern Chihuahua (Araujo-Mendieta and Arenas-Partida, 1986). In Chihuahua and northern Sonora (North America terrane), Kimmeridgian (Late Jurassic) to Cretaceous strata of the Chihuahua Group and Bisbee Group accumulated in the Chihuahua Trough, a north- to northwest-trending basin on the eastern flank of the continental arc (Fig. 35) (Cantú-Chapa and others, 1985; Dickinson and others, 1986; Araujo-Mendieta and Estavillo-González, 1987; Salvador, 1987; Scott and González-León, 1991). The oldest volcanogenic debris observed in these strata is Early Cretaceous in the Serí and North America terranes in Sonora (Araujo-Mendieta and González, 1987; Jacques-Ayala, 1989), Turonian in the Tepehuano terrane (Tardy and Maury, 1973), and Campanian-Maastrichtian in the Parras basin of the Coahuiltecano terrane (McBride and others, 1974). Late Cretaceous flysch in the Yuma terrane and Chatino terrane probably accumulated in intraarc depositional settings.

Cenozoic

The distribution of Cenozoic magmatic rocks indicates that oceanic lithosphere was subducted beneath southern and western México during much or all of Cenozoic time. Dextral slip on the TMVB and on the fault system east of Baja California and Chortis ceased during the Eocene or perhaps Oligocene. Major Cenozoic tectonism included shortening during the Laramide orogeny, sinistral displacement of the Chortis block along the southern margin of México, Basin and Range extension, and extensional deformation and dextral faulting associated with the opening of the Gulf of California.

Magmatism

Northern México. Magmatism affected most of northern México during the Paleogene, but the composition and style of emplacement of magmatic rocks are quite variable. The areal distribution of Cenozoic magmatic rocks generally is interpreted to reflect the slow eastward migration of magmatism and the progressive widening of the continental arc during the early Ce-

nozoic, followed by the comparatively rapid return of the arc to the western margin of the continent by the end of Oligocene time. Below, we briefly summarize several temporally and petrologically distinct suites of magmatic rocks in the region.

Continental arc magmatism associated with eastward subduction of oceanic lithosphere beneath North America was well established in northwestern México by the Late Cretaceous. Latest Cretaceous to mid-Tertiary continental magmatic rocks in México are traditionally divided into an andesitic lower volcanic sequence and a rhyolitic upper volcanic sequence with abundant ignimbrites (McDowell and Keizer, 1977). However, this division appears to be meaningful only in northwestern México, where a major unconformity has been mapped between the two units. A major unconformity has not been recognized in northeastern México and adjacent Trans-Pecos Texas, where magmatism began in the earliest Tertiary and continued into the Oligocene and locally the early Miocene.

The composition and distribution of magmatism in northern México changed after the Oligocene ignimbrite flare-up in the Sierra Madre Occidental. Continental arc magmatism was confined to a narrow arc at the western margin of the continent that became inactive by the end of the middle Miocene. Late Oligocene to Miocene basaltic andesites in the Sierra Madre Occidental may have been emplaced in an extensional backarc basin. Miocene and younger alkalic basalts in and east of the Sierra Madre Occidental probably were emplaced during Basin and Range extension.

The "lower volcanic complex" of the Sierra Madre Occidental (McDowell and Keizer, 1977) consists of Late Cretaceous and Paleogene calc-alkalic volcanic rocks and associated granitoids in the North America, Serí, Tahué, and Tepehuano terranes (Tables 10, 12, 13, 15). These rocks are widely presumed to have formed in a continental magmatic arc above the subducting Farallon plate, and the general eastward decrease in crystallization age may indicate progressive shallowing of the Benioff zone (Coney and Reynolds, 1977; Clark and others, 1980, 1982; Damon and others, 1981). A paucity of crystallization ages between 50 and 40 Ma led to the postulation of a "magma gap" in western México at this time, but studies in the last decade have identified widespread Eocene magmatic rocks in northwestern México (Aguirre-Díaz and McDowell, 1991) and in Trans-Pecos Texas (Henry and McDowell, 1986). Eocene volcanic rocks exhibit a broad spectrum of compositions and volcanic styles typical of arc magmatism, including intermediate flows and felsic ash-flow tuffs. Eocene rocks probably are more abundant in the Sierra Madre Occidental than currently realized; Eocene silicic tuffs, flows, and calderas may have been mistakenly grouped with the overlying Oligocene ignimbrite unit (McDowell and others, 1990; Aguirre-Díaz and McDowell, 1991). By 50 Ma, a broad, roughly north-south calc-alkalic magmatic arc stretched from the western Sierra Madre Occidental east to Coahuila, San Luis Potosí, and Guanajato and south to Chortis, which we infer was adjacent to southwestern México (Fig. 41). Magmatism in this region abated by about 40 Ma but continued locally until 36 to 34 Ma (Wark and others, 1990, and references therein). In some parts of western México (e.g., Tepehuano terrane), the lower volcanic complex was uplifted, tilted, and eroded prior to late Eocene(?) deposition of nonmarine sandstone and conglomerate.

In Trans-Pecos Texas, subduction-related magmatism began about 48 Ma and continued until about 31 Ma. Magmatic rocks were emplaced a great distance from the trench and are more alkalic than rocks from arcs near trenches, but trace element analyses and age relations strongly suggest that the rocks are subduction-related (James and Henry, 1991). An arc environment is also supported by paleostress measurements that indicate emplacement in a compressional regime (see below). Rocks dated at 48 to 31 Ma are divided into western alkalic-calcic and eastern alkalic belts; several calderas have been mapped in each belt (Henry and Price, 1984). Magma chemistry is consistent with a history of derivation from a deep, partly dehydrated subducted slab, a small degree of partial melting relative to typical arcs, and interaction of rising melts with a thick mantle wedge and thick continental crust (James and Henry, 1991).

Cretaceous to Eocene magmatic rocks in western México are overlain unconformably by a kilometer-thick sequence of Oligocene calc-alkalic rhyolite ignimbrite and subordinate andesite, dacite, and basalt of the "upper volcanic supergroup" (McDowell and Keizer, 1977; McDowell and Clabaugh, 1979; Cameron and others, 1980). These rocks were erupted throughout the Sierra Madre Occidental and perhaps as far west as east-central Baja California (Hausback, 1984). At least a dozen calderas have been mapped, but several hundred may be present based on comparisons with other ignimbrite provinces (Swanson and McDowell, 1984). Rhyolitic and alkalic basaltic rocks were emplaced in Trans-Pecos Texas starting about 31 Ma, apparently in response to a change from compression to east-northeast–west-southwest tension (James and Henry, 1991). Most petrologic and isotopic studies imply that mid-Tertiary rhyolitic magmas are differentiates of mantle-derived basalts mixed with a small crustal component (Cameron and Hanson, 1982; Gunderson and others, 1986). The transition from andesitic to rhyolitic volcanism may have been a consequence of the slowing or cessation of Farallon–North America plate convergence: the resulting decrease in the flux of mantle melts into the lower crust may have caused ascending intermediate magmas to stall, accumulate, and differentiate to produce rhyolite (Wark and others, 1990; Wark, 1991). Nd and Sr isotope ratios in silicic tuffs and lower crustal xenoliths can be interpreted to indicate that the rhyolitic magmas were derived from crustal anatexis (Ruiz and others, 1988a), but others have argued for small or negligible crustal components (Cameron and Cameron, 1985; Cameron and others, 1991; Wark, 1991).

During and after the final stages of explosive rhyolitic volcanism, basaltic andesites referred to as SCORBA (Southern Cordilleran Basaltic Andesites) were erupted throughout the northern Sierra Madre Occidental and southwestern United States (Cameron and others, 1989). SCORBA lavas are included in the upper volcanic supergroup of McDowell and Keizer (1977). K-Ar ages from SCORBA samples range from 32 to 17 Ma, with a westward decrease from >24 Ma in the Basin and Range province to 24 to 17 Ma in the Sierra Madre Occidental. Most of the SCORBA suite is older than a Miocene continental arc at the western margin of northern México and thus, probably was erupted in an arc rather than backarc setting (Cameron and others, 1989). The mid-Tertiary transition from silicic volcanism (andesite, rhyolite) to mafic volcanism (basaltic andesite) coincided with the onset of regional east-northeast–west-southwest tension across northern México (p. 117). The sheetlike form of 36- to 30-Ma ignimbrites in northern México and southwestern United States apparently attests to weak regional stresses and the absence of fault-generated topography, but younger SCORBA lavas clearly were erupted in an extensional regime, probably during Basin and Range extension (Cameron and others, 1989, and references therein).

By about 24 Ma, arc magmatism in northern México had shifted westward to a narrow chain of coalescing stratovolcanoes composed chiefly of calc-alkalic andesite with minor basalt and dacite (Damon and others, 1981; Clark and others, 1982; Hausback, 1984; Sawlan and Smith, 1984; Sawlan, 1991). The axial core of the arc crops out in coastal Baja California between latitude 29° and 25°N, as indicated by vent-facies rocks as much as 2 km thick. North and south of this strip, the axial core may be submerged within the Gulf of California, or a broader, more diffuse arc complex may have developed instead of an axial core (Hausback, 1984; Sawlan, 1991). Arc magmatism persisted until about 16 Ma north of about 28°N and until about 11 Ma in Baja California Sur. Waning orogenic magmatism persisted until about 10 Ma, overlapping in time and space with the onset of alkalic volcanism related to initial rifting in the Gulf of California starting about 13 Ma (Sawlan, 1991). The alkalic magmatic rocks ("bajaites") have unusual trace element characteristics that may be attributed to the presence of recently-subducted very young oceanic lithosphere, and that appear to have derived at least partly from MORB (Saunders and others, 1987; Sawlan, 1991). Late Cenozoic tholeiitic rocks in central Baja and within the Gulf of California record a transition from older intraplate tholeiite to current MORB that mirrors the transition from ensialic to oceanic rifting (Moore and Carmichael, 1991; Sawlan, 1991). Scattered late Miocene to Quaternary calc-alkalic volcanic rocks may have been produced by remelting of the subduction-modified source of the Miocene calc-alkalic magmas.

The salient aspects of the Late Cretaceous and Cenozoic magmatic evolution of northern México appear to be related to changes in plate boundaries and relative plate motions. The Late Cretaceous to Eocene lower volcanic sequence and a probably coeval intrusive suite record arc magmatism related to eastward subduction of oceanic lithosphere of the Farallon and Kula(?) plates beneath the continent. The eastward decrease in the age of the onset of arc magmatism and the widening of the region affected by arc magmatism in northern México probably were caused by progressive shallowing of the Benioff zone beneath western North America (Damon and others, 1981; Clark and others, 1982). The progressive decrease in age and increase in buoyancy of subducted Farallon lithosphere probably resulted in regional uplift and the development of one or more unconformities during the Eocene in the Sierra Madre Occidental. Some or most of this uplift was concomitant with part of Laramide orogenesis, which also is attributed to the progressively shallower angle of subduction of Farallon lithosphere. The decline at about 40 Ma in the emplacement of andesitic and rhyolitic magmatic arc rocks in the Sierra Madre Occidental may reflect a decrease in or possibly the cessation of Farallon–North America convergence (Ward, 1991). Such a change may have caused transtensional stretching of the hanging wall of this subduction zone, providing space in the upper and middle crust for the accumulation of magmas that ultimately were erupted as rhyolite ignimbrite in the Oligocene (Ward, 1991). Oligocene to early Miocene alkalic magmatic rocks in eastern México and Trans-Pecos Texas and SCORBA magmas in the Mexican Basin and Range province also were emplaced into stretched crust. By the early Miocene, Cocos (ex-Farallon) lithosphere was being subducted at a relatively steep angle beneath western México (Severinghaus and Atwater, 1990), producing a narrow magmatic arc in eastern Baja California. This arc was extinguished in the late Miocene by the growth of the Pacific-North America transform plate boundary (p. 118).

Southern Mexico. Crystallization ages of Cenozoic magmatic arc rocks in southern México are progressively younger to the east. The Nahuatl terrane contains Late Cretaceous, Paleocene, Eocene, and Oligocene plutonic and rarer volcanic rocks. The age of Cenozoic plutons in the Chatino terrane ranges from about 45 Ma (Eocene) in the western part of the terrane to about 12 Ma (Miocene) in the east. Calc-alkalic volcanic rocks in the Mixteco, Zapoteco, and Cuicateco terranes were erupted in the Oligocene and Miocene (Ortega-Gutiérrez and others, 1990). In Chiapas (southern Maya terrane), plutons in the northern part of the magmatic arc of Central America are as old as about 6 Ma (Damon and Montesinos, 1978). These data have been interpreted to indicate progressive eastward lengthening of the subduction zone along the southern México continental margin, probably due to eastward displacement of the Chortis terrane in Eocene and later time (p. 117). Late Cretaceous and Paleogene granitoids and volcanic rocks in the Chortis terrane probably were contiguous with coeval arc rocks in west-central and northwestern México.

Eastern México. Isolated Tertiary alkalic magmatic rocks crop out in eastern México (Coahuiltecano and Maya terranes), with crystallization ages decreasing southward from Oligocene in Tamaulipas to Quaternary at San Andres Tuxtla (Bloomfield and Cepeda-Dávila, 1973; Barker, 1977, 1979; Cantagrel and Robin, 1979). These rocks and post–31-Ma alkalic rocks in Trans-Pecos Texas have been ascribed to a roughly north-south belt related to continental rifting, but recent geochemical and geochronologic results support a different interpretation. Late Oligocene and early Miocene alkalic intrusive rocks in Tamaulipas and post–31-Ma alkalic magmatic rocks in Trans-Pecos Texas probably formed during intraplate extension (James and Henry, 1991). Miocene and younger magmatic rocks in Veracruz, Hidalgo, and Puebla probably were produced in arc and backarc settings (Cantagrel and Robin, 1979; López-Infanzón and Nelson, 1990; Nelson and others, 1991). Mid-Miocene rocks are calc-alkalic and may be related to subduction of the Cocos plate, whereas late Miocene to Recent rocks are alkalic and calc-alkalic and probably were erupted in an extensional backarc setting.

Trans-Mexican Volcanic Belt. The late Cenozoic Trans-Mexican Volcanic Belt (TMVB) cuts across the northwest-southeast structural grain of México (Fig. 3). Intermediate volcanic rocks range in age from late Miocene to Quaternary and are younger southward across the axis of the belt (Cantagrel and Robin, 1979; Nixon and others, 1987). Despite anomalous arc-trench distance and the discontinuous distribution of volcanic rocks in the TMVB, the calc-alkalic composition of most volcanic rocks probably indicates a genetic link with subduction of the Cocos and Rivera plates (Nixon, 1982; Nixon and others, 1987). The modern TMVB may be the site of sinistral or transtensional displacement (Shurbet and Cebull, 1984; Cebull and Shurbet, 1987; Urrutia-Fucugauchi and Böhnel, 1987; Johnson and Harrison, 1989; DeMets and Stein, 1990).

Alkalic volcanic rocks have been erupted in the Colima, Chapala, and Zacoalco grabens in the western TMVB (Fig. 1) since the mid-Pliocene. Suggested causes of continental extension and alkalic volcanism in this region include (1) interaction of the transform boundary between the subducting Cocos and Rivera plates with overriding continental lithosphere (Nixon, 1982; Nixon and others, 1987), (2) continental rifting during the protracted eastward shifting of the Pacific–Cocos–North America (Rivera) triple junction (Gastil and others, 1979; Luhr and others, 1985; Allan, 1986; Allan and others, 1991), (3) and passive extension caused by southeastward displacement of coastal southern México during oblique convergence (DeMets and Stein, 1990).

Displacement of the Chortis block

As part of the Caribbean plate, the Chortis terrane has moved eastward with respect to México along a major sinistral plate boundary since 10 Ma or earlier (Burkart and others, 1987). According to our reconstruction, the Chortis terrane was translated westward to northwestward during the Cretaceous to Paleocene, so it was transferred from the North America plate to the Caribbean plate after ~60 and prior to 10 Ma.

An Eocene age of the transfer of the Chortis terrane is supported by the age of truncation of southern México, by the age of sinistral transtension inferred to accompany the uplift of the Chatino terrane, and by reconstructions of Cenozoic displacements on the northern margin of the Caribbean plate east of Central America. The progressive eastward decrease in the age of Tertiary magmatism in southern México probably resulted from northward subduction beneath progressively eastward sectors of southern México and progressive sinistral displacement of a buffering land mass on the southern side of the Maya terrane (Fig. 42) (Malfait and Dinkelman, 1972; Damon and others, 1983; Wadge and Burke, 1983; Pindell and others, 1988; Pindell and Barrett, 1990). Structural and geochronologic data from the Xolapa Complex in the Chatino terrane at the southern edge of México have been interpreted to indicate uplift of the complex on shallowly north-dipping normal faults in response to sinistral transtension on a major boundary to the south (Robinson and others, 1989; Ratschbacher and others, 1991; Robinson, 1991). Geophysical and geologic studies of the Cayman Trough and the northern margin of the Caribbean plate in the eastern Caribbean have concluded that at least 1,100 km of post-Eocene sinistral slip was transferred westward on the southern margin of México (Macdonald and Holcombe, 1978; Rosencrantz and Sclater, 1986; Pindell and others, 1988; Rosencrantz and others, 1988; Pindell and Barrett, 1990). In central Guatemala, this sinistral displacement presumably was accommodated within the Motagua fault zone, where highly deformed, locally mylonitic rocks in a shear zone 10 to 25 km wide are inferred to record post-Cretaceous sinistral displacement of large but undetermined magnitude (Erikson, 1990). Minor left-lateral strike-slip faulting of late Eocene to Miocene age has been documented in southern Guerrero and southern Oaxaca (Johnson and others, 1988; Robinson, 1991), but almost all eastward displacement of the Chortis terrane south of México probably was accommodated on offshore faults.

In an alternate model of the evolution of the Caribbean region, Chortis was accreted to the southern Maya terrane in the latest Cretaceous and subsequently underwent negligible pre–late Miocene displacement with respect to México (Donnelly, 1989). The strongest support for this model is the similarity of Cretaceous and Paleogene rock units, including undeformed Eocene clastic rocks, across the Motagua fault zone. According to this model, mylonitic fabrics in the Motagua fault zone were formed during latest Cretaceous oblique collision of the Chortis terrane with the southern Maya terrane. Mid-Tertiary sinistral displacement in Guerrero and Oaxaca may be reconciled with this model by invoking eastward displacement of coastal slivers during highly oblique subduction, but the model does not satisfactorily explain the age pattern of Tertiary magmatism in southern México.

Extension

Cenozoic extension in the Basin and Range province of the United States has been divided into (1) east-northeast–west-southwest extension from 30 to 10 Ma, the so-called pre–Basin and Range, and (2) northwest-southeast extension since 10 Ma, the so-called true Basin and Range (Zoback and others, 1981). Basin and Range extension in the United States is physically continuous with extension that affected most of northern and central México, and the timing and geometry of Mexican extension appear to be similar to, although not as well understood as, the history of extension in the U.S.

The belt of mid-Tertiary metamorphic core complexes in the southwestern United States probably continues south into México, as indicated by late Cenozoic detachment faulting, subhorizontal penetrative ductile deformation, synkinematic magmatism, and brittle extension of upper plate rocks in metamorphic core complexes in the northern Serí terrane and adjacent North America in Sonora (Anderson and others, 1980; Davis and others, 1981; Nourse, 1990; Siem and Gastil, 1990, 1991). Possible causes of mid-Tertiary extension include a reduction in compressional stress on the Farallon–North America plate boundary to the west due to changes in relative plate motions, collapse of a crustal welt thickened during Late Cretaceous–Paleogene Laramide orogenesis, and lowering of the viscosity of the crust by the strong mid-Tertiary magmatic pulse (Coney and Reynolds, 1977; Spencer and Reynolds, 1986; Coney, 1987).

Much of northern and central México was affected by high-angle normal faulting typical of the Miocene-Recent Basin and Range extension in the Great Basin (Fig. 1), but extension appears to have started 10 to 15 m.y. earlier in México. Late Cenozoic net extension is minimal in the Sierra Madre Occidental but significant in extended provinces to its east and west. In the following discussion we refer to these areas as the eastern and western Mexican Basin and Range provinces; the western province also has been called the Gulf of California extensional province (Stock and Hodges, 1989, 1990).

The timing and kinematics of extension have been documented in only a few parts of the eastern Mexican Basin and Range. In north-central México and Trans-Pecos Texas, roughly east-northeast–west-southwest compression was supplanted by east-northeast–west-southwest Basin and Range extension by 28 and probably by 31 Ma, based on the geometry of dikes and veins; major normal faulting did not begin until about 24 Ma (Seager and Morgan, 1979; Dreier, 1984; Henry and Price, 1986; Aguirre-Díaz and McDowell, 1988; Henry and others, 1990, 1991). This mid-Oligocene stress reorientation was coeval with and probably caused the change to alkalic magmatism (p. 115). North-northwest–trending horsts and grabens formed in Sonora and Chihuahua in the Miocene (Roldán-Quintana and González-León, 1979). North-northwest–trending horsts and grabens in central México may indicate that late Cenozoic Basin and Range extension affected crust as far south as and possibly south of the TMVB (Henry and Aranda-Gómez, 1992). Cenozoic normal displacement on faults north of the TMVB has been episodic, with Quaternary reactivation of many older faults; faulting caused moderate tilt of fault blocks and in many areas was associated with eruption of alkalic basalts (Henry and Aranda-Gómez, 1992).

In the western Mexican Basin and Range province, Miocene extension has been documented on the periphery of and on islands within the Gulf of California. In western Sinaloa between the Gulf of California and the Sierra Madre Occidental, east-northeast–west-southwest extension began as early as 32 Ma, major tilting and faulting began after about 17 Ma, and net extension is estimated to be 20 to 50%, depending on the deep geometry of normal faults (Henry, 1989). Geometrically similar tilting and faulting indicating east-northeast–west-southwest extension began by 17 Ma and terminated by 9 Ma in parts of coastal Sonora (Gastil and Krummenacher, 1977), began by 12 Ma in coastal northeastern Baja California (Dokka and Merriam, 1982; Stock and Hodges, 1989, 1990), began by 15 to 13 Ma on Isla Tiburón (Neuhaus and others, 1988), and probably began after about 12 Ma in the La Paz region (Hausback, 1984).

Miocene faulting around the periphery of the Gulf of California was physically continuous with Basin and Range extension in the southwestern United States and probably occurred in a similar tectonic setting, namely, adjacent to the evolving Pacific–North America transform margin (Henry, 1989; Stock and Hodges, 1989). An uncertain, but probably small, component of slightly transtensional Pacific–North America relative motion (Table 18) was accommodated in the gulf region during the Miocene, resulting in continental stretching, rifting, and thinning, subsidence, and the development by a marine proto-gulf embayment, the so-called "proto–Gulf of California." The distribution and fauna of Miocene marine strata indicate that a narrow seaway had formed by about 14 Ma over the eastern half of the modern gulf and parts of coastal Sonora and Sinaloa, and that this seaway opened to the south near the mouth of the modern Gulf of California where it may have fed the 14.5- to 13-Ma Magdalena fan at the base of the slope (Moore, 1973; Lozáno-Romen, 1975; Yeats and Haq, 1981; Aguayo-C., 1984; Smith and others, 1985; Gastil and Fenby, 1991; Smith, 1991). It is unlikely that crust in the western Mexican Basin and Range province was thickened by Laramide orogenesis or arc magmatism, implying that extension was not caused by collapse of a thick crustal welt (Henry, 1989). Miocene dextral slip has not been documented in or adjacent to the proto-gulf, but a reconstruction of kinematics of the evolving Pacific–North America plate boundary zone predicts about 60 km of dextral displacement on faults that were linked to the San Andreas fault to the north (Sedlock and Hamilton, 1991).

The least principal stress and extension directions changed from east-northeast–west-southwest to roughly northwest-southeast in the latest Miocene, coeval with the initiation of rifting along the axis of the modern Gulf of California (Gastil and Krummenacher, 1977; Angelier and others, 1981; Henry, 1989). Subsidence of the gulf has spurred gulfward displacement of continental blocks from the topographic rims of the gulf along late Cenozoic and possibly active detachment faults (Fenby and Gastil, 1991; Gastil and Fenby, 1991). Post-Miocene faulting in the Gulf of California region includes dextral displacement on northwest-striking strike-slip faults and normal displacement on down-to-the-gulf normal faults (Gastil and Krummenacher, 1977; Angelier and others, 1981). Locally, the geometry of faulting is complex, probably reflecting the reactivation of older faults and the interaction of translational and extensional components of the plate boundary zone (e.g., Umhoefer and others, 1991).

Recent geochronologic and geomorphologic studies contradict the long-held notions that Baja has been uplifted rapidly and tilted westward in the late Cenozoic (e.g., Beal, 1948). Apatite fission-track studies of batholithic rocks in northern Baja indicate less than 2 km of uplift and erosion since the mid-Tertiary (Cerveny and others, 1991). Studies of emerged Pleistocene marine terraces indicate uniform uplift of Baja, with no westward tilting, at about 100 ± 50 mm/1,000 yr since 1 Ma and perhaps since about 5 Ma (Ortlieb, 1991). This rate is about an order of magnitude lower than in parts of coastal California in the U.S., suggesting that Baja and Alta California may be independent structural blocks, perhaps decoupled along the Agua Blanca and other faults in northern Baja California.

Late Cenozoic tectonic history of northwestern México

Latest Oligocene to middle Miocene calc-alkalic magmatism in the continental arc in eastern Baja was related to subduction of the Farallon (Cocos) plate beneath Baja California. As the Pacific–Cocos–North America triple junction migrated southward along the western edge of the continent, convergence was supplanted by dextral transform motion by 16 Ma west of northern Baja California and by 12 Ma west of southern Baja California (Lonsdale, 1989, 1991). From 12 to about 5 Ma, the triple junction was south of Cabo San Lucas and was connected to the Mendocino triple junction via a transform boundary in which most slip occurred on offshore faults such as the Tosco-Abreojos fault zone (Spencer and Normark, 1979; Lonsdale, 1989, 1991; Sedlock and Hamilton, 1991). Late Miocene dextral displacement also may have occurred on strike-slip faults in the proto–Gulf of California (see above) and on proposed but unidentified faults that cut the Baja peninsula (see below). Late Miocene (8 to 5.5 Ma) extension near the tip of Baja produced an embayment in the continental margin, accounting for the greater width of the gulf near its mouth (about 450 km) than at its north end (about 300 km). About 5.5 Ma, the Pacific–Cocos–North America triple junction jumped eastward to this embayment, initiating the opening of the modern Gulf of California (Curray and Moore, 1984). As rifting and spreading in the gulf progressed, most dextral transform displacement was transferred from the Tosco-Abreojos fault zone to faults within the modern gulf that connected northward with the San Andreas fault.

Alternate models for the opening of the southern Gulf of California have been summarized by Lyle and Ness (1991). Each of these models proposes that creation of oceanic crust and drifting within the gulf began by 9 to 8.3 Ma, thereby accounting for the different northern and southern widths of the gulf but also requiring the accommodation of about 150 km of late Miocene dextral slip on a fault or faults that cut the Baja California peninsula along or north of the northwestward projection of the Pescadero transform fault (cf. Humphreys and Weldon, 1991). The simplest test of these models is whether 150 km of Miocene dextral slip can be demonstrated on transpeninsular faults. Although the entire peninsula has not yet been mapped in detail, the age and distribution of units certainly are known well enough to dismiss the possibility that a single fault accommodated all 150 km of hypothesized displacement. Displacement may have been distributed on several or dozens of faults, but as the number of faults increases so does the likelihood that one or more faults would have been recognized. Not only are these models incompatible with the geology of Baja as currently understood, but also they imply that the change to northwest-southeast extension in the gulf region occurred 3 to 4 m.y. earlier than indicated by field studies (p. 118). The models of Lyle and Ness (1991) are inconsistent with available observations, but more stringent testing requires further mapping and structural analysis.

Late Cenozoic shortening in southern México

Offshore geophysical studies have documented late Cenozoic folding and thrusting of Cretaceous and Cenozoic strata around the southern margin of the Gulf of Mexico. East-west shortening in the 600-km-long, north-south–trending Mexican Ridges foldbelt off the coast of Tamaulipas and northern Veracruz has been ascribed to either gravity sliding and growth faulting or east-west crustal shortening (Buffler and others, 1979). The southern end of this offshore foldbelt, which also is known as Cordillera Ordóñez, currently may be overthrust to the northeast by the continental shelf of southern Veracruz (De Cserna, 1981). Other evidence for late Cenozoic northeast-southwest shortening in southern México includes a southwest-verging fold and thrust belt on the northeast side of the Chiapas Massif that was active from middle Miocene to Pliocene time, and active northwest-trending folds off the north coast of Tabasco and Campeche (de Cserna, 1989). We speculate that the locus of shortening is above a proto-subduction zone where much of southern México overrides the lithosphere of the Gulf of Mexico basin.

SUMMARY

In Part 2 of this volume, we have presented a speculative model of the Late Precambrian to Cenozoic tectonic evolution of México that is based on the geologic and tectonic history of constituent tectonostratigraphic terranes described in Part 1, geophysical and plate motion constraints, and formal premises. Some of the salient features of the model are summarized below.

1. Grenville basement in eastern and southern México is considered to be far-traveled with respect to the southern termination of the Grenville belt in North America.

2. The late Paleozoic Ouachitan suture that marks the collision of North America and Gondwana does not and did not extend into central México.

3. The Permo-Triassic continental arc on the western margin of Pangea affected only the far eastern edge and far northwestern corner of México; most of what is now México was a complex assemblage of arcs, continental blocks, and basins in the oceanic region west and south of the Pangean continental arc.

4. Continental México grew most markedly toward its present form during the Late Triassic and Jurassic as terranes were episodically accreted to its southern and western flanks.

5. Mesozoic southward and westward continental growth was accompanied by a southward and westward shift of the locus of arc magmatism.

6. The tectonically active southern and western margins of México were sites of large margin-parallel translations of terranes resulting from oblique convergence of México with oceanic lithosphere to the west. Convergence and terrane translation were sinistral from the Triassic(?) until the Early Cretaceous, and dextral from the mid-Cretaceous to the Paleogene.

7. Jurassic stretching and rifting in the Gulf of Mexico was not kinematically related to sinistral faulting on the Mojave-Sonora Megashear; instead, slip on the megashear and on other, more outboard, fault systems was controlled by left-oblique convergence of México with plates in the Pacific basin.

8. Paleomagnetic data that indicate about 15° of northward latitudinal displacement of Baja in the Late Cretaceous and Paleogene can be reconciled with geologic correlations only by postulating an earlier episode of southward displacement during left-oblique convergence.

9. The Cretaceous reconstruction is consistent with postulated origins at Mexican latitudes of terranes in the western United States and Canada.

10. The Caribbean plate, including the Chortis block, has been translated 1,000 to 2,000 km eastward on strike-slip faults along the southern margin of México since about 45 Ma.

11. Basin and Range extension has affected most of México north of about 20°N.

Our reconstruction is not a unique explanation of the geologic and tectonic history of México, but it does provide an internally consistent framework for interpreting geoscientific data from the region. Many aspects of the model are testable; we look forward to the acquisition of new data and the formulation of new interpretations and hypotheses that will almost certainly result in modifications to the model.

REFERENCES CITED

Abbott, P. L., and Smith, T. E., 1989, Sonora, México, source for the Eocene Poway Conglomerate of southern California: Geology, v. 17, p. 329–332.

Abbott, P. L., Kies, R. P., Krummenacher, D., and Martin, D., 1983, Potassium-argon ages of rhyolite bedrock and conglomerate clasts in Eocene strata, northwestern México and southern California, *in* Anderson, D. W., and Rymer, M. J., eds., Tectonics and sedimentation along faults of the San Andreas system: Pacific Section, Society of Economic Mineralogists and Paleontologists, p. 59–66.

Adams, M. A., 1979, Stratigraphy and petrography of the Santiago Peak Volcanics east of Rancho Santa Fe, California [M.S. thesis]: San Diego, California, San Diego State University, 123 p.

Aguayo-C., J. E., 1984, Estudio de los sedimentos terrígenos de la Cuenca de Guaymas, Golfo de California, noroeste de México: Revista del Instituto Mexicano del Petróleo, v. 16(4), p. 5–19.

Ague, J. J., and Brandon, M. T., 1992, Tectonic tilting and large-scale northwards translation of the Peninsular Ranges batholith: Geological Society of America Abstracts with Programs, v. 24, p. 2.

Aguirre, L., 1983, Granitoids in Chile: Geological Society of America Memoir 159, p. 293–316.

Aguirre-Díaz, G., and McDowell, F. W., 1988, Nature and timing of faulting in the southern Basin and Range, central-eastern Durango, México: EOS Transactions of the American Geophysical Union, v. 69, p. 1412–1413.

Aiken, C.L.V., Schellhorn, R. W., and de la Fuente, M. F., 1988, Gravity of northern México, *in* Stratigraphy, tectonics, and resources of parts of Sierra Madre Occidental province, México, Clark, K. F., Goodell, P. C., and Hoffer, J. M., eds., El Paso Geological Society, Field Conference

Guidebook, p. 119–133.

Alaniz-Alvarez, S., and Ortega-Gutiérrez, F., 1988, Constituye el Complejo Xolapa realmente las raices de un arco?: Unión Geofísica Mexicana, Colima, GEOS, Número extraordinario, Epoca 2, FQIT 15/57.

Alatorre, A. E., 1988, Stratigraphy and depositional environments of the phosphorite-bearing Monterrey Formation in Baja California Sur, México: Economic Geology, v. 83, p. 1918–1930.

Alba, C. M., and Chávez, R., 1974, K-Ar ages of volcanic rocks from central Sierra Peña Blanca, Chihuahua, México: Isochron/West, v. 10, p. 2.

Alencaster, G., 1961, Paleontología del Triásico Superior de Sonora: Instituto de Geología, Universidad Nacional Autónoma de México, Paleontología Mexicana número 11, 38 p.

Allan, J. F., 1986, Geology of the Colima and Zacoalco grabens, SW México: Late Cenozoic rifting in the Mexican volcanic belt: Geological Society of America Bulletin, v. 97, p. 473–485.

Allan, J. F., Nelson, S. A., Luhr, J. F., Carmichael, I.S.E., Wopat, M., and Wallace, P. J., 1991, Pliocene-Holocene rifting and associated volcanism in southwest México: An exotic terrane in the making, in Dauphin, J. P., and Simoneit, B.R.T., eds., The Gulf and Peninsular Province of the Californias: American Association of Petroleum Geologists, Memoir 47, p. 425–445.

Allison, E. C., 1955, Middle Cretaceous gastropoda from Punta China, Baja California, México: Journal of Paleontology, v. 29, p. 400–432.

Almazán-Vázquez, E., 1988a, Geoquímica de las rocas volcánicas de la Formación Alisitos del Arroyo La Bocana en el Estado de Baja California Norte: Revista del Instituto de Geología, Universidad Nacional Autónoma de México, v. 7, p. 78–88.

—— , 1988b, Marco paleosedimentario y geodinámico de la Formación Alisitos en la Península de Baja California: Revista del Instituto de Geología, Universidad Nacional Autónoma de México, v. 7, p. 41–51.

Almazan-Vázquez, E., and 6 others, 1987, Stratigraphic framework of Mesozoic strata in northern Sonora, México: Geological Society of America Abstracts with Programs, v. 19, p. 570.

Altamirano-R., F. J., 1972, Tectónica de la porción meridional de Baja California Sur: Sociedad Geológica Mexicana, II Convención Nacional, Libro de Resúmenes, p. 113–114.

Alvarez, W., Kent, D. V., Premoli-Silva, I., Schweickert, R. A., and Larson, R., 1980, Franciscan Complex limestone deposited at 17° south paleolatitude: Geological Society of America Bulletin, v. 91, p. 476–484.

Alzaga-Ruiz, H., and Pano-Arciniega, A., 1989, Origen de la Formación Chivillas y presencia del jurásico tardío en la región de Tehuacán, Puebla, México: Revista del Instituto Mexicano del Petróleo, v. 21(1), p. 5–15.

Anderson, J. G., and 6 others, 1986, Strong ground motion from the Michoacán, México, earthquake: Science, v. 233, p. 1043–1049.

Anderson, J. L., 1983, Proterozoic anorogenic granite plutonism of North America, in Medaris, L. G., Jr., Byers, C. W., Mickelson, D. M., and Shanks, W. C., eds., Proterozoic geology; Selected Papers from an International Proterozoic Symposium: Geological Society of America Memoir 161, p. 133–154.

Anderson, P. V., 1984, Prebatholithic strata of San Felipe: The recognition of Cordilleran miogeosynclinal deposits in northeastern Baja California, México: Pacific Section, Society of Economic Mineralogists and Paleontologists, San Diego meeting, p. 29–30.

Anderson, T. H., and Schmidt, V. A., 1983, The evolution of Middle America and the Gulf of Mexico–Caribbean Sea region during Mesozoic time: Geological Society of America Bulletin, v. 94, p. 941–966.

Anderson, T. H., and Silver, L. T., 1969, Mesozoic magmatic events of the northern Sonora coastal region, México: Geological Society of America Abstracts with Programs, v. 1, p. 3–4.

—— , 1971, Age of granulite metamorphism during the Oaxacan orogeny, México: Geological Society of America Abstracts with Programs, v. 3, p. 492.

—— , 1977a, Geochronometric and stratigraphic outlines of the Precambrian rocks of northwestern México: Geological Society of America Abstracts with Programs, v. 9, p. 880.

—— , 1977b, U-Pb isotope ages of granitic plutons near Cananea, Sonora: Economic Geology, v. 72, p. 827–836.

—— , 1979, The role of the Mojave-Sonora Megashear in the tectonic evolution of northern Sonora, in Anderson, T. H., and Roldán-Quintana, J., eds., Geology of Northern Sonora: Geological Society of America, Field Trip Guidebook, p. 59–68.

—— , 1981, An overview of Precambrian rocks in Sonora, México: Revista del Instituto de Geología, Universidad Nacional Autónoma de México, v. 5, p. 131–139.

Anderson, T. H., Burkart, B., Clemons, R. E., Bohnenburger, O. H., and Blount, D. N., 1973, Geology of the western Altos Cuchumantes, northwestern Guatemala: Geological Society of America Bulletin, v. 84, p. 805–826.

Anderson, T. H., Eells, J. L., and Silver, L. T., 1979, Precambrian and Paleozoic rocks of the Caborca region, Sonora, México, in Anderson, T. H., and Roldán-Quintana, J., eds., Geology of Northern Sonora: Geological Society of America, Field Trip Guidebook, p. 1–22.

Anderson, T. H., Silver, L. T., and Salas, G. A., 1980, Distribution and U-Pb isotope ages of some lineated plutons, northwestern México: Geological Society of America Memoir 153, p. 269–283.

Anderson, T. H., Erdlac, R. J., Jr., and Sandstrom, M. A., 1985, Late-Cretaceous allochthons and post-Cretaceous strike-slip displacement along the Cuilco-Polochic fault, Guatemala: Tectonics, v. 4, p. 453–475.

—— , 1986, Reply to comment on "Late Cretaceous allochthons and post-Cretaceous strike-slip displacement along the Cuilco-Chixoy-Polochic fault, Guatemala" by Anderson, T. H., Erdlac, R. J., Jr., and Sandstrom, M. A.: Tectonics, v. 5, p. 473–475.

Anderson, T. H., McKee, J. W., and Jones, N. W., 1990, Jurassic(?) mélange in north-central Mexico: Geological Society of America Abstracts with Programs, v. 22, p. 3.

—— , 1991, A northwest trending, Jurassic fold nappe, northernmost Zacatecas, México: Tectonics, v. 10, p. 383–401.

Angelier, J., Colletta, B., Chorowicz, J., Ortlieb, L., and Rangin, C., 1981, Fault tectonics of the Baja California peninsula and the opening of the Sea of Cortez, México: Journal of Structural Geology, v. 3, p. 347–357.

Aranda-García, M., Gómez-Luna, M. E., and Contreras y Montero, B., 1987, El Jurásico Superior (Kimeridgiano-Titoniano) en al área Santa María del Oro, Durango, México: Revista del Sociedad Mexicana Paleontológica, v. 1, p. 75–87.

Aranda-García, M., Quintero-Legorreta, O., and Martínez-Hernández, E., 1988, Palinomorfos del Jurásico de la Formación Gran Tesoro, Santa María del Oro, Durango: Revista del Instituto de Geología, Universidad Nacional Autónoma de México, v. 7, p. 112–115.

Aranda-Gómez, J. J., and Pérez-Venzor, J. A., 1988, Estudio geoológico de Punta Coyotes, Baja California Sur: Revista del Instituto de Geología, Universidad Nacional Autónoma de México, v. 7, p. 1–21.

—— , 1989, Estratigrafía del complejo cristalino de la región de Todos Santos, Estado de Baja California Sur: Revista del Instituto de Geología, Universidad Nacional Autónoma de México, v. 8, p. 149–170.

Aranda-Gómez, J. J., Aranda-Gómez, J. M., and Nieto-Samaniego, A. F., 1989, Consideraciones acerca de la evolución tectónica durante el Cenozoico de la Sierra de Guanajuato y la parte meridional de la Mesa Central: Revista del Instituto de Geología, Universidad Nacional Autónoma de México, v. 8, p. 33–46.

Araujo-Mendieta, J., and Arenas-Partida, R., 1986, Estudio tectónico-sedimentario en el Mar Mexicano, estados de Chihuahua y Durango: Sociedad Geológica Mexicana Boletín, v. 47(2), p. 43–87.

Araujo-Mendieta, J., and Casar-González, R., 1987, Estratigrafía y sedimentología del jurásico superior en la Cuenca de Chihuahua, norte de México: Revista del Instituto Mexicano del Petróleo, v. 19(1), p. 6–29.

Araujo-Mendieta, J., and Estavillo-González, C. F., 1987, Evolución tectónica sedimentaria del jurásico superior y cretácico inferior en el NE de Sonora, México: Revista del Instituto Mexicano del Petróleo, v. 19(3), p. 4–67.

Arden, D. D., Jr., 1975, Geology of Jamaica and Nicaragua Rise, in Nairn, A.E.M., and Stehli, F. G., eds., The ocean basins and margins: Volume 3: The Gulf of Mexico and the Caribbean: New York, Plenum Press,

p. 617–661.

Arellano, A.R.V., 1956, Relaciones del Cámbrico de Caborca, especialmente con la base del Paleozoico, *in* Rodgers, J., ed., El Sistema Cámbrico, su paleogeografía y el problema de su base; pt II: Australia, América: México, D. F., XX Congreso Geológico Internacional, p. 509–527.

Armin, R. A., 1987, Sedimentology and tectonic significance of Wolfcampian (Lower Permian) conglomerates in the Pedregosa basin: Southeastern Arizona, southwestern New Mexico, and northern México: Geological Society of America Bulletin, v. 99, p. 42–65.

Ashby, J. R., 1989, A resume of the Miocene stratigraphic history of the Rosarito Beach basin, northwestern Baja California, México, *in* Abbott, P. L., ed., Geologic Studies in Baja California, Volume 63: Pacific Section, Society of Economic Paleontologists and Mineralogists, p. 27–36.

Asmerom, Y., Zartman, R. E., Damon, P. E., and Shafiqullah, M., 1990, Zircon U-Th-Pb and whole-rock Rb-Sr age patterns of lower Mesozoic igneous rocks in the Santa Rita Mountains, southeast Arizona: Implications for Mesozoic magmatism and tectonics in the southern Cordillera: Geological Society of America Bulletin, v. 102, p. 961–968.

Atwater, T., 1970, Implications of plate tectonics for the tectonic evolution of western North America: Geological Society of America Bulletin, v. 81, p. 3513–3536.

—— , 1989, Plate tectonic history of the northeast Pacific and western North America, *in* Winterer, E. L., Hussong, D. M., and Decker, R. W., eds., Decade of North American Geology, Volume N: The Eastern Pacific region: Geological Society of America, p. 21–72.

Aubouin, J., and 6 others, 1982, The Middle America trench in the geological framework of Central America, *in* Initial Reports of the Deep Sea Drilling Project, Volume 67: Washington, D.C., U.S. Government Printing Office, p. 747–755.

Avé Lallemant, H. G., and Oldow, J. S., 1988, Early Mesozoic southward migration of Cordilleran transpressional terranes: Tectonics, v. 7, p. 1057–1075.

Avila-Angulo, R., 1990, Lower Jurassic volcanic and sedimentary rocks of the Sierra López, west-central Sonora, México: Geological Society of America Abstracts with Programs, v. 22, p. 4.

Bagby, W. C., 1979, Geology, geochemistry and geochronology of the Batopilas quadrangle, Sierra Madre Occidental, Chihuahua, México [Ph.D. dissertation]: Santa Cruz, University of California, 271 p.

Balch, D. C., Bartling, S. H., and Abbott, P. L., 1984, Volcaniclastic strata of the Upper Jurassic Santiago Peak Volcanics, San Diego, California, *in* Crouch, J. K., and Bachman, S. B., eds., Tectonics and sedimentation along the California margin, Volume 38: Pacific Section, Society of Economic Paleontologists and Mineralogists, p. 157–170.

Baldwin, S. L., and Harrison, T. M., 1989, Geochronology of blueschists from west-central Baja California and the timing of uplift of subduction complexes: Journal of Geology, v. 97, p. 149–163.

—— , 1992, The P-T-t history of blocks in serpentinite-matrix mélange, west-central Baja California: Geological Society of America Bulletin, v. 104, p. 18–31.

Baldwin, S. L., Harrison, T. M., and Fitz Gerald, J. D., 1990, Diffusion of ^{40}Ar in metamorphic hornblende, Contributions to Mineralogy and Petrology, v. 105, p. 691–703.

Baldwin, S. L., Harrison, T. M., and Sedlock, R. L., 1987, ^{40}Ar/^{39}Ar geochronology of high-pressure blocks in mélange, Cedros Island, Baja California: Geological Society of America Abstracts with Programs, v. 19, p. 356.

Ballard, M. M., van der Voo, R., and Urrutia-Fucugauchi, J., 1989, Paleomagnetic results from the Grenvillian-aged rocks from Oaxaca, México: Evidence for a displaced terrane: Precambrian Research, v. 42, p. 343–352.

Bally, A. W., Scotese, C. R., and Ross, M. I., 1989, North America: Plate tectonic setting and tectonic elements, *in* Bally, A. W., and Palmer, A. R., eds., Decade of North American Geology, Volume A: The geology of North America; An overview: Geological Society of America, p. 1–15.

Barker, D. S., 1977, Northern Trans-Pecos magmatic province: Introduction and comparison with the Kenya rift: Geological Society of America Bulletin, v. 88, p. 1421–1427.

—— , 1979, Genozoic magmatism in the Trans-Pecos province: Relation to Rio Grande rift, *in* Riecker, R. E., ed., Rio Grande rift: Tectonics and magmatism: Washington, D.C., American Geophysical Union, p. 382–392.

Barker, R. W., and Blow, W. H., 1976, Biostratigraphy of some Tertiary formations in the Tampico-Misantla embayment, México: Journal of Foraminiferal Research, v. 6, p. 39–58.

Barnes, D. A., 1982, Basin analysis of volcanic arc derived, Jura-Cretaceous sedimentary rocks, Vizcaíno Peninsula, Baja California Sur, México [Ph.D. dissertation]: Santa Barbara, University of California, 240 p.

Barnes, D. A., and Mattinson, J. M., 1981, Late Triassic–Early Cretaceous age of eugeoclinal terranes, western Vizcaíno Peninsula, Baja California Sur, México: Geological Society of America Abstracts with Programs, v. 13, p. 43.

Barnes, L. G., 1992, The fossil marine vertebrate fauna of the latest Miocene Almejas Formation, Isla Cedros, Baja California, México: First International Meeting on Geology of the Baja California Peninsula, Memoir Universidad Autonoma de Baja California Sur, La Paz, B.C.S., p. 147–166.

Barrier, E., Bourgois, J., and Michaud, F., 1990, Le système de rifts actifs du point triple de Jalisco: vers un proto-golfe de Jalisco: Comptes Rendus l'Académie de Sciences, Paris, v. 310, p. 1513–1520.

Barth, A. P., Tosdal, R. M., and Wooden, J. L., 1992, Initiation of the Mesozoic Cordilleran arc in southern California: Geological Society of America Abstracts with Programs, v. 24, p. 6.

Barthelmy, D. A., 1979, Regional geology of the El Arco porphyry copper deposit, Baja California, *in* Baja California geology: Abbott, P. L., and Gastil, R. G., eds., San Diego, Department of Geological Sciences, San Diego State University, p. 127–138.

Bartling, W. A., and Abbott, P. L., 1983, Upper Cretaceous sedimentation and tectonics with reference to the Eocene, San Miguel Island and San Diego area, California, *in* Larue, D. K., and Steel, R. J., eds., Cenozoic marine sedimentation, Pacific margin, USA: Pacific Section, Society of Economic Mineralogists and Paleontologists, p. 133–150.

Bartolini, C., and Stewart, J. H., 1990, Stratigraphy and structure of Paleozoic oceanic strata in Sierra El Aliso, central Sonora, México: Geological Society of America Abstracts with Programs, v. 22, p. 6.

Bartolini, C., Morales, M., Damon, P., and Shafiqullah, M., 1992, K-Ar ages of tilted Tertiary volcanic rocks associated with continental conglomerates, Sonoran Basin and Range province, México: Geological Society of America Abstracts with Programs, v. 24, p. 6.

Bateson, J. H., 1972, New interpretation of the geology of the Maya Mountains, British Honduras: American Association of Petroleum Geologists Bulletin, v. 56, p. 956–963.

Bateson, J. H., and Hall, I.H.S., 1971, Revised geologic nomenclature for pre-Cretaceous rocks of British Honduras: American Association of Petroleum Geologists Bulletin, v. 55, p. 529–530.

—— , 1977, Geology of the Maya Mountains, Belize: Great Britain, Institute of Geological Sciences, Overseas Memoir 3, 42 p.

Bazan-B., S., 1987, Genesis de las pegmatitas del arco insular de Telixtlahuaca: Geomimet, v. 149, p. 54–67.

Beal, C. H., 1948, Reconnaissance of the geology and oil possibilities of Baja California, Mexico: Geological Society of America Memoir 31, 138 p.

Beauvais, L., and Stump, T. E., 1976, Corals, molluscs, and paleogeography of Late Jurassic strata of the Cerro Pozo Serna, Sonora, México: Palaeogeography, Palaeoclimatology, Palaeoecology, v. 19, p. 275–301.

Beck, M. E., Jr., 1989, Paleomagnetism of continental North America: Implications for displacement of crustal blocks with the western Cordillera, Baja California to British Columbia, *in* Pakiser, L. C., and Mooney, W. D., eds., Geophysical framework of the continental United States: Geological Society of America, Memoir 172, p. 471–492.

—— , 1991, Case for northward transport of Baja and coastal southern California: Paleomagnetic data, analysis, and alternatives: Geology, v. 19, p. 506–509.

Beck, R. J., and Thrash, L. T., 1991, State companies dominate non-U.S. OGJ100: Oil and Gas Journal, v. 89, n. 39, p. 72–77.

Becker, K., and Fisher, A. T., 1991, A brief review of heat-flow studies in the

Guaymas basin, Gulf of California, *in* Dauphin, J. P., and Simoneit, B.R.T., eds., The Gulf and Peninsular Province of the Californias: American Association of Petroleum Geologists Memoir 47, p. 709–720.

Bellon, H., Maury, R. C., and Stephan, J.-F., 1981, Dioritic basement, site 493: Petrology, geochemistry, and geodynamics, *in* Initial Reports of the Deep Sea Drilling Project, Volume 66: Washington, D.C., U.S. Government Printing Office, p. 723–730.

Bertrand, J., Delaloye, M., Fontignie, D., and Vuagnat, M., 1978, Ages (K-Ar) sur diverse ophiolitiques et roches associees de la Cordillere Centrale du Guatemala: Schweizerische Mineralogische und Petrographische Mitteilungen, v. 58, p. 405–413.

Bevis, M., and Isacks, B. L., 1984, Hypocentral trend surface analysis: Probing the geometry of the Benioff zones: Journal of Geophysical Research, v. 89, p. 6153–6170.

Bickford, M. E., 1988, The formation of continental crust: Part 1. A review of some principles; Part 2. An application to the Proterozoic evolution of southern North America: Geological Society of America Bulletin, v. 100, p. 1375–1391.

Bickford, M. E., Van Schmus, W. R., and Zietz, I., 1986, Proterozoic history of the midcontinent region of North America: Geology, v. 14, p. 492–496.

Blair, T. C., 1987, Tectonic and hydrologic controls on cyclic alluvial fan, fluvial, and lacustrine rift-basin sedimentation, Jurassic–lowermost Cretaceous Todos Santos Formation, Chiapas, México: Journal of Sedimentary Petrology, v. 57, p. 845–862.

—— , 1988, Mixed siliciclastic-carbonate marine and continental syn-rift sedimentation, Upper Jurassic–lowermost Cretaceous Todos Santos and San Ricardo Formations, western Chiapas, México: Journal of Sedimentary Petrology, v. 58, p. 623–636.

Blake, M. C., Jr., Jayko, A. S., Moore, T. E., Chavez, V., Saleeby, J. B., and Seel, K., 1984, Tectonostratigraphic terranes of Magdalena Island, Baja California Sur, *in* Frizzell, V. A., Jr., ed., Geology of the Baja California Peninsula, Volume 39: Pacific Section, Society of Economic Paleontologists and Mineralogists, p. 183–191.

Blickwede, J. F., 1981, Petrology and stratigraphy of Triassic(?) "Nazas Formation," Sierra de San Julian, Zacatecas, México: American Association of Petroleum Geologists Bulletin, v. 65, p. 1012.

Bloomfield, K., and Cepeda-Dávila, L., 1973, Oligocene alkaline igneous activity in NE México: Geological Magazine, v. 110, p. 551–555.

Blount, J. G., Walker, N. W., and Carlson, W. D., 1988, Geochemistry and U-Pb zircon ages of mid-Proterozoic metaigneous rocks from Chihuahua, México: Geological Society of America Abstracts with Programs, v. 20, p. A205.

Bobier, C., and Robin, C., 1983, Palemagnetisme de la Sierra Madre Occidentale dans les etats de Durango et Sinaloa (Mexique): Variations du champ ou rotations de blocs au Paleocene et au Neogene?: Geofísica Internacional, v. 22, p. 57–86.

Boehlke, J. E., and Abbott, P. L., 1986, Punta Baja Formation, a Campanian submarine canyon fill, Baja California, México, *in* Abbott, P. L., ed., Cretaceous stratigraphy western North America: Pacific Section, Society of Economic Paleontologists and Mineralogists, p. 91–101.

Böhnel, H., Alva-Valdivia, L., González-Huesca, S., Urrutia-Fucugauchi, J., Morán-Zenteno, D., and Schaaf, P., 1989, Paleomagnetic data and the accretion of the Guerrero terrane, southern México continental margin, in Hillhouse, J. W., ed., Deep structure and past kinematics of accreted terranes: American Geophysical Union Geophysical Monograph 50, p. 73–92.

Boles, J. R., 1986, Mesozoic sedimentary rocks in the Vizcaíno Peninsula–Isla de Cedros area, Baja California, México, *in* Abbott, P. L., ed., Cretaceous stratigraphy western North America: Pacific Section, Society of Economic Paleontologists and Mineralogists, p. 63–77.

Boles, J. R., and Landis, C. A., 1984, Jurassic sedimentary mélange and associated facies, Baja California, México: Geological Society of America Bulletin, v. 95, p. 513–521.

Bond, G. C., and Kominz, M. A., 1984, Construction of tectonic subsidence curves for the early Paleozoic miogeocline, southern Canadian Rocky Mountains: Implications for subsidence mechanisms, age of breakup, and

crustal thinning: Geological Society of America Bulletin, v. 95, p. 155–173.

Bonneau, M., 1969 (1970), Una nueva área cretácica fosilífera en el Estado de Sinaloa: Sociedad Geológica Mexicana Boletín, v. 32, p. 159–167.

Bottjer, D. J., and Link, M. H., 1984, A synthesis of Late Cretaceous southern California and northern Baja California paleogeography, *in* Crouch, J. K., and Bachman, S. B., eds., Tectonics and sedimentation along the California margin, Volume 38: Pacific Section, Society of Economic Paleontologists and Mineralogists, p. 171–188.

Bourgois, J., Azéma, J., Baumgartner, P. O., Tournon, J., Desmet, A., and Aubouin, J., 1984, The geologic history of the Caribbean–Cocos Plate boundary with special reference to the Nicoya ophiolite complex (Costa Rica) and DSDP results (Legs 67 and 84 off Guatemala): Tectonophysics, v. 108, p. 1–32.

Bourgois, J., and others, 1988, Fragmentation en cours du bord Ouest du Continent Nord Américain: Les frontières sous-marines du Bloc Jalisco (Mexique) de fracture de Rivera au large du Mexique: Comptes Rendus, l'Académie des Sciences, Paris, v. 307, p. 1121–1130.

Bowin, C. O., 1976, Caribbean gravity field and plate tectonics: Geological Society of America Special Paper 169, 79 p.

Bridges, L. W., 1964a, Stratigraphy of Mina Plomosas–Placer de Guadalupe area, Geology of Mina Plomosas–Placer de Guadalupe area, Chihuahua, México: West Texas Geological Society, Field Trip Guidebook, Publication 64-50, p. 50–59.

—— , 1964b, Structure of Mina Plomosas–Placer de Guadalupe area, Geology of Mina Plomosas–Placer de Guadalupe area, Chihuahua, México: West Texas Geological Society, Field Trip Guidebook, Publication 64-50, p. 60–61.

Brown, M. L., and Dyer, R., 1987, Mesozoic geology of northwestern Chihuahua, México, *in* Dickinson, W. R., and Klute, M. A., eds., Mesozoic rocks of southern Arizona and adjacent areas, Volume 18: Arizona Geological Society Digest, p. 381–394.

Brown, M. L., and Handschy, J. W., 1984, The tectonic framework of Chihuahua, *in* Geology and petroleum potential of Chihuahua, México, Volume 84-80: West Texas Geological Society, Field Trip Guidebook, p. 161–173.

Brune, J. N., Simons, R. S., Rebollar, C., and Reyes, A., 1979, Seismicity and faulting in northern Baja California, *in* Abbott, P. L., and Elliott, J. W., eds., Earthquakes and other perils of the San Diego region: Geological Society of America, Field Trip Guide, San Diego State University, San Diego, California, p. 83–100.

Brunner, P., 1979, Microfacies y microfósiles permo-triásicos en el área El Antimonio, Sonora, México: Revista del Instituto Mexicano del Petróleo, v. 11(1), p. 6–41.

Bryant, B., Gastil, G., and Roldán-Quintana, J., 1985, Detachment faulting on opposite sides of the Gulf of California: Geological Society of America Abstracts with Programs, v. 17, p. 345.

Buch, I. P., 1984, Upper Permian(?) and Lower Triassic metasedimentary rocks, northeastern Baja California, México, *in* Frizzell, V. A., Jr., ed., Geology of the Baja California Peninsula, Volume 39: Pacific Section, Society of Economic Paleontologists and Mineralogists, p. 31–36.

Buesch, D. C., 1984, The depositional environment and subsequent metamorphism of the Santiago Peak volcanic rocks, Camp Pendleton, California [M.S. thesis]: Los Angeles, California State University, 113 p.

Buffler, R. T., and Sawyer, D. S., 1985, Distribution of crust and early history, Gulf of Mexico basin: Transactions of the Gulf Coast Geological Society, v. 35, p. 333–344.

Buffler, R. T., Schaub, F. J., Watkins, J. S., and Worzel, J. L., 1979, Anatomy of the Mexican Ridges, southwestern Gulf of Mexico, *in* Watkins, J. S., Montadert, L., and Dickerson, P. W., eds., Geological and geophysical investigations of continental margins: American Association of Petroleum Geologists Memoir 29, p. 319–327.

Buitron, B. E., 1984, Late Jurassic bivalves and gastropods from northern Zacatecas, México, and their biogeographic significance, *in* Westermann, G.E.G., ed., Jurassic-Cretaceous biochronology and paleogeography of North America: Geological Association of Canada Special Paper 27, p. 89–98.

Burbach, G. V., Frohlich, C., Pennington, W. D., and Matumoto, T., 1984,

Seismicity and tectonics of the subducted Cocos plate: Journal of Geophysical Research, v. 89, p. 7719–7735.

Burchfiel, B. C., and Davis, G. A., 1981, Mojave Desert and environs, *in* Ernst, W. G., ed., The geotectonic development of California, Rubey Volume I: Englewood Cliffs, New Jersey, Prentice-Hall, p. 217–252.

Burkart, B., 1978, Offset across the Polochic fault of Guatemala and Chiapas, México: Geology, v. 6, p. 328–332.

—— , 1983, Neogene North America–Caribbean plate boundary across northern Central America: Offset along the Polochic fault: Tectonophysics, v. 99, p. 251–270.

—— , 1990, Contrast in effects of Late Cretaceous convergence between Chiapas, México, and Guatemala: Geological Society of America Abstracts with Programs, v. 22, A338.

Burkart, B., and Self, S., 1985, Extension and rotation of crustal blocks in northern Central America and effect on the volcanic arc: Geology, v. 13, p. 22–26.

Burkart, B., Clemons, R. E., and Crane, D. C., 1973, Mesozoic and Cenozoic stratigraphy of southeastern Guatemala: American Association of Petroleum Geologists Bulletin, v. 57, p. 63–73.

Burkart, B., Deaton, B. C., Dengo, C., and Moreno, G., 1987, Tectonic wedges and offset Laramide structures along the Polochic fault of Guatemala and Chiapas, México: Reaffirmation of large Neogene displacement: Tectonics, v. 6, p. 411–422.

Burke, K., Cooper, C., Dewey, J. F., Mann, P., and Pindell, J. L., 1984, Caribbean tectonics and relative plate motion, *in* Bonini, W. E., Hargraves, R. B., and Shagam, R., eds., The Caribbean–South American plate boundary and regional tectonics: Geological Society of America Memoir 162, p. 31–63.

Busby-Spera, C. J., 1988, Evolution of a Middle Jurassic back-arc basin, Cedros Island, Baja California: Evidence from a marine volcaniclastic apron: Geological Society of America Bulletin, v. 100, p. 218–233.

Busby-Spera, C. J., and Boles, J. R., 1986, Sedimentation and subsidence styles in a Cretaceous forearc basin, southern Vizcaíno Peninsula, Baja California (México), *in* Abbott, P. L., ed., Cretaceous stratigraphy western North America: Pacific Section, Society of Economic Paleontologists and Mineralogists, p. 71–90.

Busby-Spera, C. J., Morris, W. R., and Smith, D., 1988, Syndepositional normal faults across the width of the Cretaceous forearc region of Baja California (México): Geological Society of America Abstracts with Programs, v. 20, p. 147.

Butler, R. F., Dickinson, W. R., and Gehrels, G. E., 1991, Paleomagnetism of coastal California and Baja California: Alternatives to large-scale northward transport: Tectonics, v. 10, p. 561–576.

Callaway, J. M., and Massare, J. A., 1989, *Shastasauras altispinus* (Ichthyosauria, Shastasauridae) from the Upper Triassic of the El Antimonio district, northwestern Sonora, México: Journal of Paleontology, v. 63, p. 930–939.

Cameron, K. L., and Cameron, M., 1985, Rare earth element, [87]Sr/[86]Sr, and [143]Nd/[144]Nd compositions of Cenozoic orogenic dacites from Baja California, northwestern México, and adjacent west Texas: Evidence for the predominance of a subcrustal component: Contributions to Mineralogy and Petrology, v. 91, p. 1–11.

Cameron, K. L., and Hanson, G. N., 1982, Rare earth element evidence concerning the origin of voluminous mid-Tertiary rhyolitic ignimbrites and related volcanic rocks, SMO, Chihuahua, México: Geochimica et Cosmochimica Acta, v. 46, p. 1489–1503.

Cameron, K. L., Cameron, M., Bagby, W. C., Moll, E. J., and Drake, R. E., 1980, Petrologic characteristics of mid-Tertiary volcanic suites, Chihuahua, México: Geology, v. 8, p. 87–91.

Cameron, K. L., Cameron, M., Harmon, R. S., and Barreiro, B., 1991, Role of crustal interaction in the genesis of ignimbrites from Batopilas, northcentral Sierra Madre Occidental (SMO), México: Geological Society of America Abstracts with Programs, v. 23, A332.

Cameron, K. L., Nimz, G. J., Kuentz, D., Niemeyer, S., and Gunn, S., 1989, Southern Cordilleran Basaltic Andesite suite, southern Chihuahua, México: A link between Tertiary continental arc and flood basalt magmatism in North America: Journal of Geophysical Research, v. 94, p. 7817–7840.

Campa-Uranga, M. F., 1978, La evolución tectónica de Tierra Caliente, Guerrero: Sociedad Geológica Mexicana Boletín, v. 39, p. 52–64.

—— , 1985a, Metalogenesis y tectónica de placas: Ciencias, v. 1, p. 22–29.

—— , 1985b, The Mexican thrust belt, *in* Howell, D. G., ed., Tectonostratigraphic terranes of the Circum-Pacific region: Circum-Pacific Council for Energy and Mineral Resources, Earth Science Series, no. 1, p. 299–313.

Campa-Uranga, M. F., and Coney, P. J., 1983, Tectono-stratigraphic terranes and mineral resource distributions of México: Canadian Journal of Earth Sciences, v. 20, p. 1040–1051.

Campa-Uranga, M. F., Campos, M., Flores, R., and Oviedo, R., 1974, La secuencia mesozoica volcánica sedimentaria metamorfizada de Ixtapán de la Sal, México-Teloloapán: Sociedad Geológica Mexicana Boletín, v. 35, p. 7–28.

Campos-Enríques, J. O., Arroyo-Esquivel, M. A., and Urrutia-Fucugauchi, J., 1990, Basement Curie isotherm and shallow-crustal structure of the Trans-Mexican Volcanic Belt from aeromagnetic data: Tectonophysics, v. 172, p. 77–90.

Cantagrel, J. M., and Robin, C., 1979, K-Ar dating on eastern Mexican volcanic rocks—Relations between the andesitic and alkaline provinces: Journal of Volcanology and Geothermal Research, v. 5, p. 99–114.

Cantú-Chapa, A., 1979, Biostratigrafía de la Serie Huasteca (Jurásico Medio) en el subsuelo de Poza Rica, Veracruz: Revista del Instituto Mexicano del Petróleo, v. 11, p. 14–24.

Cantú-Chapa, C. M., Sandoval-Silva, R., and Arenas-Partida, R., 1985, Evolución sedimentario del cretácico inferior en el norte de México: Revista del Instituto Mexicano del Petróleo, v. 17(2), p. 14–37.

Carballido-Sánchez, E. A., and Delgado-Argote, L. A., 1989, Geología del cuerpo serpentinítico de Tehuitzingo, Estado de Puebla—Interpretación preliminar de su emplazamiento: Revista del Instituto de Geología, Universidad Nacional Autónoma de México, v. 8, p. 134–148.

Carfantan, J.-C., 1981 (1984), Evolución estructural del sureste de México: Paleogeografía e historia tectónica de las zonas internas mesozoicas: Revista del Instituto de Geología, Universidad Nacional Autónoma de México, v. 5, p. 207–216.

—— , 1983, Les ensembles géologiques du Mexique meridional. Evolution géodynamique durante le Mésozoique et le Cénozoique: Geofísica Internacional, v. 22, p. 9–37.

—— , 1986, Du systeme Cordillerain nord-americain au domaine Caraibe—Etude geologique du Mexique meridional [Ph.D.thesis]: Chambery, France, Universite de Savoie, 556 p.

Carpenter, R. H., 1954, Geology and ore deposits of the Rosario mining district and the San Juancito mountains, Honduras, Central America: Geological Society of America Bulletin, v. 65, p. 23–38.

Carr, M. D., Christiansen, R. L., and Poole, F. G., 1984, Pre-Cenozoic geology of the El Paso Mountains, southwestern Great Basin, California: A summary, *in* Lintz, J., Jr., ed., Western geological excursions, Volume 4: Reno, Nevada, Department of Geological Sciences, Mackey School of Mines, p. 84–93.

Carr, M. J., Stoiber, R. E., and Drake, C. L., 1974, The segmented nature of some continental margins, *in* Burk, C. A., and Drake, C. L., eds., The geology of continental margins: New York, Springer-Verlag, p. 105–114.

Carrasco, V. B., 1978, Estratigrafía de unas lavas almohadilladas y rocas sedimentarias del Cretácico Inferior en Tehuacán, Puebla: Sociedad Geológica Mexicana Boletín, v. 39, p. 13.

Carrillo, M., and Martínez, E., 1981 (1983), Evidencias de facies continentales en la Formación Matzitzi, Estado de Puebla: Revista del Instituto de Geología, Universidad Nacional Autónoma de México, v. 5, p. 117–118.

Carrillo-Bravo, J., 1961, Geología del Anticlinorio Huizachal-Peregrina al N-W de Ciudad Victoria, Tamaulipas: Asociación Mexicana de Geologos Petroleros Boletín, v. 13, p. 1–98.

—— , 1965, Estudio geológico de una parte del Anticlinorio de Huayacocotla: Asociación Mexicana de Geologos Petroleros Boletín, v. 17, p. 73–96.

Carrillo-Chavez, A., 1991, Las alteraciones como guías mineralógicas en yacimientos de oro tipo dike falla: Los Uvares, B.C.S., ejemplo característico: First International Meeting on Geology of the Baja California Peninsula, Universidad Autónoma de Baja California Sur, La Paz, B.C.S., p. 14.

Carrillo-Martínez, M., 1971, Geología de la Hoja San José de Gracía Sinaloa (Tésis Profesional): Mexico City, Universidad Nacional Autónoma de México Faculty Ingeniería, 154 p.
—— , 1990, Geometría estructural de la Sierra Madre Oriental, entre Peñamiller y Jalpan, Estado de Querétaro: Revista del Instituto de Geología, Universidad Nacional Autónoma de México, v. 9, p. 62–70.
Case, J. E., and Holcombe, T. L., 1980, Geologic-tectonic map of the Caribbean region: U.S. Geological Survey Miscellaneous Investigations Series, Map I-1100, scale 1:2,500,000.
Case, J. E., Shagam, R., and Giegegack, R. F., 1990, Geology of the northern Andes; An overview, in Dengo, G., and Case, J. E., eds., Decade of North American Geology, Volume H: The Caribbean region: Geological Society of America, p. 177–200.
Castillo-Rodriguez, H., 1988, Zur Geologie des kristallinen Grundgebirges der Sierra Madre Oriental–insbesondere des Granjeno-Schiefer-Komplexes–im Sudteil des Huizachal-Peregrina-Antiklinoriums (Raum Ciudad Victoria, Bundesstaat Tamaulipas, Mexiko) [Ph.D. thesis]: Münster, Germany, Universität Münster, 138 p.
Castrejón, F., Porres, A., and Nava, F., 1988, A velocity model perpendicular to the Acapulco trench region based on RESMAC data: Geofísica Internacional, v. 27(1), p. 131–147.
Castro-Mora, J., Schlaepfer, C. J., and Rodriquez, E. M., 1975, Estratigrafía y microfacies del mesozoico de la Sierra Madre del sur, Chiapas: Asociación Mexicano Geológicos Petroleros Boletín, v. 27, p. 1–95.
Cebull, S. E., and Shurbet, D. H., 1987, Mexican volcanic belt: An intraplate transform?: Geofísica Internacional, v. 26, p. 1–13.
Cebull, S. E., Shurbet, D. H., Keller, G. R., and Russell, L. R., 1976, Possible role of transform faults in the development of apparent offsets in the Ouachita-southern Appalachian tectonic belt: Journal of Geology, v. 84, p. 107–114.
Centeno-García, E., Ortega-Gutiérrez, F., and Corona-Esquivel, R., 1990, Oaxaca fault: Cenozoic reactivation of the suture between the Zapoteco and Cuicateco terranes, southern México: Geological Society of America Abstracts with Programs, v. 22, p. 13.
Cerveny, P. F., Dorsey, R. J., and Burns, B. A., 1991, Apatite and zircon fission-track ages from the Sierra San Pedro Mártir, eastern Peninsular Ranges, Baja California, México: Geological Society of America Abstracts with Programs, v. 23, p. 12.
Champion, D. E., Howell, D. G., and Gromme, C. S., 1984, Paleomagnetic and geologic data indicating 2500 km of northward displacement for the Salinian and related terranes, California: Journal of Geophysical Research, v. 89, p. 7736–7752.
Champion, D. E., Howell, D. G., and Marshall, M., 1986, Paleomagnetism of Cretaceous and Eocene strata, San Miguel Island, California, borderland and the northward translation of Baja California: Journal of Geophysical Research, v. 91, p. 11557–11570.
Chávez-Quirarte, R., 1982, El Cretácico superior en el área del proyecto hidroeléctrico de San Juan Tetelcingo, Guerrero: Libro guía de la excursión geológica a la cunca del alto Río Balsas: Sociedad Geológica Mexicana, p. 55–58.
Chiodi, M., Monod, O., Busnardo, R., Gaspard, D., Sánchez, A., and Yta, M., 1988, Une discordance ante albienne datee par une faune d'Ammonites et de Brachiopades de type tethysien au Mexique central: Geobios, v. 21, p. 125–135.
Clark, K. F., 1976, Geologic section across Sierra Madre Occidental, Chihuahua to Topolobampo, México: New Mexico Geological Society Special Publication 6, p. 26–38.
Clark, K. F., Damon, P. E., Schutter, S. R., and Shafiqullah, M., 1980, Magmatismo en el norte de México en relación a los yacimientos metalíferos: Geomimet, v. 106, p. 51–71.
Clark, K. F., Foster, C. T., and Damon, P. E., 1982, Cenozoic mineral deposits and subduction-related magmatic arcs in México: Geological Society of America Bulletin, v. 93, p. 533–544.
Cobbing, E. J., and Pitcher, W. S., 1983, Andean plutonism in Peru and its relationship to volcanism and metallogenesis at a segmented plate edge: Geological Society of America Memoir 159, p. 277–291.

Cocheme, J.-J., and Demant, A., 1991, Geology of the Yécora area, northern Sierra Madre Occidental, México, in Pérez-Segura, E., and Jacques-Ayala, C., eds., Studies of Sonoran geology: Geological Society of America Special Paper 254, p. 81–94.
Cohen, K. K., Anderson, T. H., and Schmidt, V. A., 1986, A paleomagnetic test of the proposed Mojave-Sonora megashear in northwestern México: Tectonophysics, v. 131, p. 23–51.
Cojan, I., and Potter, P., 1991, Depositional environment, petrology, and provenance of the Santa Clara Formation, Upper Triassic Barranca Group, eastern Sonora, México, in Pérez-Segura, E., and Jacques-Ayala, C., eds., Studies of Sonoran geology: Geological Society of America Special Paper 254, p. 37–50.
Coney, P. J., 1978, The plate tectonic setting of southeastern Arizona, Land of Cochise: New Mexico Geological Society, 29th Field Conference, p. 285–290.
—— , 1983, Un modelo tectónico de México y sus relaciones con América del Norte, América del Sur y el Caribe: Revista del Instituto Mexicano del Petróleo, v. 15(1), p. 6–15.
—— , 1987, The regional tectonic setting and possible causes of Cenozoic extension in the North American Cordillera, in Coward, M. P., Dewey, J. F., and Hancock, P. L., eds., Continental extensional tectonics: Geological Society of London Special Publication 28, p. 177–186.
Coney, P. J., and Campa-Uranga, M. F., 1987, Lithotectonic terrane map of Mexico (west of the 91st meridian): U.S. Geological Survey, Miscellaneous Field Studies Map MF-1874-D, scale 1:2,500,000.
Coney, P. J., and Reynolds, S. J., 1977, Cordilleran Benioff zones: Nature, v. 270, p. 403–406.
Coney, P. J., Jones, D. L., and Monger, J.W.H., 1980, Cordilleran suspect terranes: Nature, v. 299, p. 329–333.
Connors, C. D., and Anderson, T. H., 1989, Expression and structural analysis of the Mojave-Sonora Megashear in northwestern Sonora, México: Geological Society of America Abstracts with Programs, v. 21, A91.
Contreras y Montero, B., Martínez, C. A., and Gómez-Luna, M. E., 1988, Bioestratigrafía y sedimentología del Jurásico Superior en San Pedro del Gallo, Durango, México: Revista del Instituto Mexicano del Petróleo, v. 20(3), p. 5–49.
Cooper, G. A., and Arellano, A.R.V., 1946, Stratigraphy near Caborca, northwest Sonora, México: American Association of Petroleum Geologists Bulletin, v. 30, p. 606–611.
Copeland, P., and Bowring, S. A., 1988, U-Pb zircon and $^{40}Ar/^{39}Ar$ ages for Proterozoic rocks, west Texas: Geological Society of America Abstracts with Programs, v. 20, p. 95–96.
Corbitt, L. L., 1984, Tectonics of fold and thrust belt of northwestern Chihuahua, in Geology and petroleum potential of Chihuahua, México, Volume 84-80: West Texas Geological Society, Field Trip Guidebook, p. 174–180.
Córdoba, D. A., 1965, Hoja Viesca 13R-1(9), con resumen de la geología de la Hoja Apizolaya, Estados de Zacatecas y Durango: Carta Geológica de México, serie de 1:100,000.
—— , 1988, Estratigrafía de las rocas volcánicas de la región entre Sierra de Gamón y Laguna de Santaiguillo, Estado de Durango: Revista del Instituto de Geología, Universidad Nacional Autónoma de México, v. 7, p. 136–147.
Córdoba, D. A., and Silva-Mora, L., 1989, Marco geológico del área de Revolución-Puerta de Cabrera, Estado de Durango: Revista del Instituto de Geología, Universidad Nacional Autónoma de México, v. 8, p. 111–122.
Corona-Esquivel, R. J., 1981 (1984), Estratigrafía de la región de Olinalá-Tecocoyunca, noreste del Estado de Guerrero: Revista del Instituto de Geología, Universidad Nacional Autónoma de México, v. 5, p. 1–16.
Corona-Esquivel, R., Ortega-Gutiérrez, F., Martínez-Reyes, J., and Centeno-García, E., 1988, Evidencias de levantamiento tectónico asociado con el sismo del 19 de septiembre de 1985, en la regió de Caleta de Campos, Estado de Michoacán: Revista del Instituto de Geología, Universidad Nacional Autónoma de México, v. 7, p. 106–111.
Cossio-Torres, T., 1988, Zur Geologie des kristallinen Grundgebirges der Sierra Madre Oriental–insbesondere des Novillo-Gneis-Komplexes–im Sudteil des

Huizachal-Peregrina-Antiklinoriums (Raum Ciudad Victoria, Bundesstaat Tamaulipas, Mexiko) [Ph.D. thesis]: Münster, Universität Münster, 99 p.

Couch, R., and Woodcock, S., 1981, Gravity and structure of continental margins of southwestern México and northwestern Guatemala: Journal of Geophysical Research, v. 86, p. 1829–1840.

Couch, R. W., Ness, G. E., Victor, L., Shanahan, S., and Troseth, S. C., compilers, 1985, Free-air gravity anomalies, southern México to Costa Rica margin, *in* Ladd, J. W., and Buffler, R. T., eds., Middle America Trench off western Central America: Woods Hole, Massachusetts, Marine Science International, Ocean Basins and Margins Drilling Program Atlas Series, Atlas 7, sheet 2.

Couch, R. W., and others, 1991, Gravity anomalies and crustal structure of the Gulf and Peninsular Province of the Californias, *in* Dauphin, J. P., and Simoneit, B.R.T., eds., The Gulf and Peninsular Province of the Californias: American Association of Petroleum Geologists Memoir 47, p. 25–45.

Courtillot, V., Feinberg, H., Ragaru, J. P., Kerguelen, R., McWilliams, M., and Cox, A., 1985, Franciscan Complex limestone deposited at 24°N: Geology, v. 13, p. 107–110.

Criscione, J. J., Davis, T. E., and Ehlig, P., 1978, The age of sedimentation/diagenesis for the Bedford Canyon Formation and the Santa Monica Formation in southern California: A Rb/Sr evaluation, *in* Howell, D. G., and McDougall, K., eds., Mesozoic paleogeography of the western United States, Pacific Coast Paleogeography Symposium 2: Pacific Section, Society of Economic Mineralogists and Paleontologists, p. 385–396.

Crouch, J. K., 1979, Neogene tectonic evolution of the western Transverse Ranges and the California Continental Borderland: Geological Society of America Bulletin, v. 90, p. 338–345.

Cuevas-Pérez, E., 1983, Evolución geológica mesozoica del Estado de Zacatecas, México: Zentralblatt für Geologie und Paleontologie Teil I (3/4), p. 190–201.

——, 1985, Geologie des Alteran mesozoikums in Zacatecas und San Lusi Potosi, Mexiko [Ph.D. dissertation]: Marburg, Germany, Universität Marburg, 189 p.

Cunningham, A. B., and Abbott, P. L., 1986, Sedimentology and provenance of the Upper Cretaceous Rosario Formation south of Ensenada, Baja California, México, *in* Abbott, P. L., ed., Cretaceous stratigraphy western North America: Pacific Section, Society of Economic Paleontologists and Mineralogists, p. 103–118.

Curray, J. R., and Moore, D. G., 1984, Geologic history of the Gulf of California, *in* Crouch, J. K., and Bachman, S. B., eds., Tectonics and sedimentation along the California margin, Volume 38: Pacific Section, Society of Economic Paleontologists and Mineralogists, p. 17–35.

Curray, J. R., Moore, D. G., and others, 1982, Initial Reports of the Deep Sea Drilling Project: Washington, D.C., U.S. Government Printing Office, v. 64, 507 p. (Part 1) and 1313 p. (Part 2).

Dallmeyer, R. D., 1982, Pre-Mesozoic basement of the southeastern Gulf of Mexico: Geological Society of America Abstracts with Programs, v. 14, p. 471.

Dalziel, I.W.D., 1991, Pacific margins of Laurentia and East Antarctica–Australia as a conjugate rift pair: Evidence and implications for an Eocambrian supercontinent: Geology, v. 19, p. 598–601.

Damon, P. E., and Coney, P. J., 1983, Rate of movement of nuclear Central America along the coast of México during the last 90 Ma: Geological Society of America Abstracts with Programs, v. 15, p. 553.

Damon, P. E., and Montesinos, E., 1978, Late Cenozoic volcanism and metallogenesis over an active Benioff zone in Chiapas, Mexico: Arizona Geological Society Digest, v. 11, p. 155–168.

Damon, P. E., Livingston, D. E., Mauger, R. L., Giletti, B. J., and Pantoja-Alor, J., 1962, Edad del Precámbrico "Anterior" y de otras rocas del Zocalo de la región de Caborca–Altar de la parte noroccidental de Estado de Sonora: Boletín del Instituto de Geología, Universidad Nacional Autónoma de México, v. 64, p. 11–44.

Damon, P. E., Shariqullah, M., and Clark, K. F., 1981, Age trends of igneous activity in relation to metallogenesis in the southern Cordillera, *in* Dickinson,

W. R., and Payne, W. D., eds., Relations of tectonics to ore deposits in the southern Cordillera, Volume 14: Arizona Geological Society Digest, p. 137–154.

——, 1983, Geochronology of the porphyry copper deposits and related mineralization of México: Canadian Journal of Earth Sciences, v. 20, p. 1052–1071.

Damon, P. E., Shafiqullah, M., and Roldán-Quintana, J., 1984, The Cordilleran Jurassic arc from Chiapas (southern Mexico) to Arizona: Geological Society of America Abstracts with Programs, v. 16, p. 482.

Dauphin, J. P., and Ness, G. E., 1991, Bathymetry of the Gulf and Peninsular Province of the Californias, *in* Dauphin, J. P., and Simoneit, B.R.T., eds., The Gulf and Peninsular Province of the Californias: American Association of Petroleum Geologists, Memoir 47, p. 21–23.

Dávila-Alcocer, V., and Martínez-Reyes, J., 1987, Una edad cretácica para las rocas basales de la Sierra Guanajuato, Simposio sobre la geología de la región de la Sierra de Guanajuato (Resúmenes), April 1987, p. 19–20.

Davis, G. H., 1979, Laramide faulting and folding in southeastern Arizona: American Journal of Science, v. 279, p. 543–569.

Davis, G. H., Gardulski, A. F., and Anderson, T. H., 1981, Structural and structural-petrologic characteristics of some metamorphic core complex terranes in southern Arizona and northern Sonora, *in* Ortlieb, L., and Roldán-Quintana, J., eds., Geology of northwestern México and southern Arizona: Revista del Instituto de Geología, Universidad Nacional Autónoma de México, Estación Regional del Noreste, p. 323–365.

de Cserna, Z., 1965, Reconocimiento geológico de la Sierra Madre del Sur de México, entre Chilpancingo y Acapulco, Estado de Guerrero: Boletín del Instituto de Geología, Universidad Nacional Autónoma de México, v. 62, p. 1–77.

——, 1971, Precambrian sedimentation, tectonics, and magmatism in México: Geologische Rundschau, v. 60, p. 1488–1513.

——, 1976, Geology of the Fresnillo area, Zacatecas, México: Geological Society of America Bulletin, v. 87, p. 1191–1199.

——, 1981 (1984), Margen continental de colisión activo en la parte suroccidental del Golfo de México: Revista del Instituto de Geología, Universidad Nacional Autónoma de México, v. 5, p. 255–261.

——, 1982 (1983), Hoja Viesca 14Q-g(9), con resumen de la geología de la Hoja Tejupilco, Estados de Guerrero, México y Michoacán: Carta Geológica de México, serie de 1:100,000.

——, 1989, An outline of the geology of México, *in* Bally, A. W., and Palmer, A. R., eds., Decade of North American Geology, Volume A: The geology of North America—An overview: Geological Society of America, p. 233–264.

de Cserna, Z., and Fries, C., Jr., 1981, Hoja Viesca 14Q-h(7), con resumen de la geología de la Hoja Taxco, Estados de Guerrero, México y Morelos: Carta Geológica de México, serie de 1:100,000.

de Cserna, Z., and Ortega-Gutiérrez, F., 1978, Reinterpretation of isotopic age data from the Granjeno Schist, Ciudad Victoria, Tamaulipas; y reinterpretación tectónica del Esquisto Granjeno, Ciudad Victoria, Tamaulipas; Contestación: Revista del Instituto de Geología, Universidad Nacional Autónoma de México, v. 2, p. 212–215.

de Cserna, Z., Schmitter, E., Damon, P. E., Livingston, D. E., and Kulp, J. L., 1962, Edades isotópicas de rocas metamórficas del centro y sur de Guerrero y una monzonita cuarcífera del norte de Sinaloa: Boletín del Instituto de Geología, Universidad Nacional Autónoma de México, v. 64, p. 71–84.

de Cserna, Z., Rincón-Orta, C., Solorio-Munguía, J., and Schmitter-Villada, E., 1968, Una edad radiométrica Pérmica temprana de la región de Placer de Guadelupe, noreste de Chihuahua: Sociedad Geológico Mexicana Boletín, v. 31, p. 158–159.

de Cserna, Z., Fries, C., Jr., Rincón-Orta, C., Westley, A., Solorio-Munguía, J., and Schmitter-Villada, E., 1974, Edad Precámbrica tardía del Esquisto Taxco, Estado de Guerrero: Asociación Mexicanos Geológicos Petroleros Boletín, v. 26, p. 183–193.

de Cserna, Z., Graf, J. L., Jr., and Ortega-Gutiérrez, F., 1977, Alóctono Paleozoico Inferior en la región de Ciudad Victoria, Estado de Tamaulipas: Revista del Instituto de Geología, Universidad Nacional Autónoma de México,

v. 1, p. 33–43.

de Cserna, Z., Ortega-Gutiérrez, F., and Palacios-Nieto, M., 1980, Reconocimiento geológico de la parte central de la cuenca del alto Río Balsas, Estados de Guerrero y Puebla: Libro guía de la excursión geológica a la parte central de la cuenca del alto Río Balsas, Estados de Guerrero y Puebla: Sociedad Geológica Mexicana, p. 1–33.

de Cserna, Z., de la Fuente-Duch, M., Palacios-Nieto, M., Triay, L., Mitre-Salazar, L.-M., and Mota-Palomino, R., 1987 (1988), Estructura geológica, gravimetría, sismicidad y relaciones neotectónicas regionales de la cuenca de México: Boletín del Instituto de Geología, Universidad Nacional Autónoma de México, v. 104, 71 p.

Deaton, B. C., and Burkart, B., 1984a, Time of sinistral slip along the Polochic fault of Guatemala: Tectonophysics, v. 102, p. 297–313.

——, 1984b, K-Ar ages of samples from the Subinal Formation and the Colotenango beds, Guatemala: Isochron/West, v. 39, p. 19–20.

DeJong, K. A., Escarcega-Escarcega, J. A., and Damon, P. E., 1988, Eastward thrusting, southwestward folding, and westward backsliding in the Sierra La Vibora, Sonora, México: Geology, v. 16, p. 904–907.

DeJong, K. A., García y Barragán, J. C., Damon, P. E., Miranda, M., Jacques-Ayala, C., and Almazán-Vázquez, E., 1990, Untangling the tectonic knot of Mesozoic Cordilleran orogeny in northern Sonora, NW México: Geological Society of America Abstracts with Programs, v. 22, A327.

Delevoryas, T., and Srivastava, S. C., 1981, Jurassic plants from the Department of Francisco Morazán, central Honduras: Review of Palaeobotany and Palynology, v. 34, p. 345–357.

Delgado-Argote, L. A., 1988, Geología preliminar de la secuencia volcanosedimentaria y serpentinitas asociadas del jurásico(?) del área de Cuicatlán-Concepción Pápalo, Oaxaca: Revista del Instituto de Geología, Universidad Nacional Autónoma de México, v. 7, p. 127–135.

——, 1989, Regional implications of the Jurassic-Cretaceous volcanosedimentary Cuicateco terrane, Oaxaca, México: Geofísica Internacional, v. 28, p. 939–973.

Delgado-Argote, L. A., and Carballido-Sánchez, E. A., 1990, Análisis tectónico del sistema traspresivo neogénico entre Macuspana, Tabasco, y Puerto Angel, Oaxaca: Revista del Instituto de Geología, Universidad Nacional Autónoma de México, v. 9, p. 21–32.

Delgado-Argote, L. A., Rubinovich-Cogan, R., and Gasca-Duran, A., 1986, Descripción preliminar de la geología y mecánica de emplazamiento del complejo ultrabásico del cretácio de Loma Baya, Guerrero, México: Geofísica Internacional, v. 25, p. 537–558.

Delgado-Argote, L. A., López-Martinez, M., York, D., and Hall, C. M., 1990, Geology and geochronology of ultramafic localities in the Cuicateco and Tierra Caliente Complexes, southern México: Geological Society of America Abstracts with Programs, v. 22, A326.

Delgado-Argote, L. A., Casar-Aldrete, I., González-Caver, E., Morales-Puente, P., and Girón-García, P., 1992a, Geologic controls on the variation of hydrogen isotope ratios of structural water of serpentinites from San Pedro Limon-Palmar Chico, Tierra Caliente terrane, México: Geofísica Internacional, in press.

Delgado-Argote, L. A., López-Martínez, M., York, D., and Hall, C. M., 1992b, Geologic framework and geochronology of ultramafic complexes of southern México, Canadian Journal of Earth Sciences, v. 29, p. 1590–1604.

Dellatre, M. P., 1984, Permian miogeoclinal strata at El Volcan, Baja California, México, in Frizzell, V. A., Jr., ed., Geology of the Baja California Peninsula, Volume 39: Pacific Section, Society of Economic Paleontologists and Mineralogists, p. 23–29.

Demant, A., 1978, Características del Eje Neovolcánico Transmexicano y sus problemas de interpretación: Revista del Instituto de Geología, Universidad Nacional Autónoma de México, v. 2, p. 172–181.

——, 1981, Plio-Quaternary volcanism of the Santa Rosalía area, Baja California, México, in Ortlieb, L., and Roldán-Quintana, J., eds., Geology of northwestern México and southern Arizona: Instituto de Geología, Universidad Nacional Autónoma de México, Estación Regional del Noreste, p. 295–307.

DeMets, C., and Stein, S., 1990, Present-day kinematics of the Rivera plate and implications for tectonics in southwestern México: Journal of Geophysical Research, v. 95, p. 21931–21948.

DeMets, C., Gordon, R. G., Argus, D. F., and Stein, S., 1990, Current plate motions: Geophysical Journal International, v. 101, p. 425–478.

Dengo, C. A., 1986, Comment on "Late Cretaceous allochthons and post-Cretaceous strike-slip displacement along the Cuilco-Chixoy-Polochic fault, Guatemala" by Anderson, T. H., Erdlac, R. J., Jr., and Sandstrom, M. A., Tectonics, v. 5, p. 469–472.

Dengo, G., 1972, Review of Caribbean serpentinites and their tectonic implications: Geological Society of America Memoir 132, p. 303–312.

——, 1975, Paleozoic and Mesozoic tectonic belts in México and Central America, in Nairn, A.E.M., and Stehli, F. G., eds., The ocean basins and margins: Volume 3: The Gulf of Mexico and the Caribbean: New York, Plenum Press, p. 283–323.

——, 1985, Mid-America: tectonic setting for the Pacific margin from southern México to northwestern Colombia, in Nairn, A.E.M., and Stehli, F. G., eds., The ocean basins and margins: Volume 7A: The Pacific Ocean, New York, Plenum Press, p. 123–180.

Denison, R. E., Kenny, G. S., Burke, W. H., Jr., and Hetherington, E. A., Jr., 1969, Isotopic ages of igneous and metamorphic boulders from the Haymond Formation (Pennsylvanian), Marathon basin, Texas, and their significance: Geological Society of America Bulletin, v. 80, p. 245–256.

Denison, R. E., Burke, W. H., Jr., Hetherington, E. A., Jr., and Otto, J. B., 1971, Basement rock framework of parts of Texas, southern New Mexico, and northern México, in The geologic framework of the Chihuahua Tectonic Belt, Midland, Texas: West Texas Geological Society, p. 1–14.

Díaz, T., and Navarro-G., A., 1964, Lithology and stratigraphic correlation of the upper Paleozoic in the region of Palomas, Chihuahua, Geology of Mina Plomosa–Placer de Guadalupe area, Chihuahua, México: West Texas Geological Society, Field Trip Guidebook, Publication 64-50, p. 65–84.

Díaz-García, V., 1980, Las rocas metamórficas de la región de Zucualpán, Estado de México: Revista del Instituto de Geología, Universidad Nacional Autónoma de México, v. 4, p. 1–12.

Dickinson, W. R., 1981, Plate tectonic evolution of the southern Cordillera, in Dickinson, W. R., and Payne, W. D., eds., Relation of tectonics to ore deposits in the southern Cordillera, Volume 14: Arizona Geological Society Digest, p. 113–135.

Dickinson, W. R., and Snyder, W. S., 1979, Geometry of triple junctions related to San Andreas transform: Journal of Geophysical Research, v. 84, p. 561–572.

Dickinson, W. R., Klute, M. A., and Swift, P. N., 1986, The Bisbee Basin and its bearing on late Mesozoic paleogeographic and paleotectonic relations between the Cordilleran and Caribbean regions, in Abbott, P. L., ed., Cretaceous stratigraphy western North America: Pacific Section, Society of Economic Paleontologists and Mineralogists, p. 51–62.

Dilles, J. H., and Wright, J. E., 1988, The chronology of early Mesozoic arc magmatism in the Yerington district of western Nevada and its regional implications: Geological Society of America Bulletin, v. 100, p. 644–652.

Dillon, W. P., Vedder, J. G., and Graf, R. J., 1972, Structural profile of the northwestern Caribbean: Earth and Planetary Science Letters, v. 17, p. 175–180.

Dixon, T. H., González, G., Lichten, S. M., Ness, G. E., Dauphin, J. P., and Tralli, D. M., 1991, Preliminary determination of Pacific–North America relative motion in the southern Gulf of California using the Global Positioning System: Geophysical Research Letters, v. 18, p. 861–864.

Dobson, P. F., and Mahood, G. A., 1985, Volcanic stratigraphy of the Los Azufres geothermal area, México: Journal of Volcanology and Geothermal Research, v. 25, p. 273–287.

Dokka, R. K., 1984, Fission-track geochronologic evidence for Late Cretaceous mylonitization and early Paleocene uplift of the northeastern Peninsular Ranges, California: Geophysical Research Letters, v. 11, p. 46–49.

Dokka, R. K., and Merriam, R. H., 1982, Late Cenozoic extension of northeastern Baja California: Geological Society of America Bulletin, v. 93,

p. 371–378.

Donnelly, T. W., 1989, Geologic history of the Caribbean and Central America, *in* Bally, A. W., and Palmer, A. R., eds., Decade of North American Geology, Volume A: The geology of North America: An overview: Geological Society of America, p. 299–321.

Donnelly, T. W., Melson, W., Kay, R., and Rogers, J.J.W., 1973, Basalts and dolerites of Late Cretaceous age from the central Caribbean, *in* Initial Reports of the Deep Sea Drilling Project, Volume 15: Washington, D.C., U.S. Government Printing Office, p. 989–1012.

Donnelly, T. W., Horne, G. S., Finch, R. C., and López-Ramos, E., 1990a, Northern Central America; The Maya and Chortis blocks, *in* Dengo, G., and Case, J. E., eds., Decade of North American Geology, Volume H: The Caribbean region: Geological Society of America, p. 37–76.

Donnelly, T. W., and 10 others, 1990b, History and tectonic setting of Caribbean magmatism, *in* Dengo, G., and Case, J. E., eds., Decade of North American Geology, Volume H: The Caribbean region: Geological Society of America, p. 339–374.

Dreier, J., 1984, Regional tectonic control of epithermal veins in the western United States and México: Arizona Geological Society Digest, v. 15, p. 28–50.

Dunbar, J. A., and Sawyer, D. S., 1987, Implications of continental crust extension for plate reconstruction: An example from the Gulf of Mexico: Tectonics, v. 6, p. 739–755.

Durham, J. W., Applegate, S. P., and Espinosa-Arrubarrena, L., 1981, Onshore marine Cenozoic along the southwest Pacific Coast of México: Geological Society of America Bulletin, v. 92, p. 384–394.

Dyer, R., 1986, Precambrian and Paleozoic rocks of Sierra el Carrizalillo, Chihuahua, México—A preliminary report: Geological Society of America Abstracts with Programs, v. 18, p. 353.

Dyer, R., Chávez-Quirarte, R., and Guthrie, R. S., 1988, Cordilleran orogenic belt of northern Chihuahua, México: American Association of Petroleum Geologists Bulletin, v. 72, p. 99.

Edwards, J. D., 1955, Studies of some early Tertiary red conglomerates of central México: U.S. Geological Survey Professional Paper 264-H, p. 153–185.

Eissler, H., and McNally, K., 1984, Seismicity and tectonics of the Rivera plate and implications for the 1932 Jalisco, México, earthquake: Journal of Geophysical Research, v. 89, p. 4520–4530.

Elías-Herrera, M., 1987, Metamorphic geology of the Tierra Caliente Complex, Tejupilco region, State of México: Geological Society of America Abstracts with Programs, v. 19, p. 654.

——, 1989, Geología metamórfica del área de San Lucas de Maíz, Estado de México: Boletín del Instituto de Geología, Universidad Nacional Autónoma de México, v. 105, 79 p.

Enciso de la Vega, S., 1968, Hoja Viesca 13R-1(7), con resumen de la geología de la Hoja Cuencame, Estado de Durango: Carta Geológica de México, serie de 1:100,000.

——, 1988, Una nueva localidad pérmica con fusilínidos en Puebla: Revista del Instituto de Geología, Universidad Nacional Autónoma de México, v. 7, p. 28–34.

Engebretson, D. C., Cox, A., and Gordon, R. G., 1985, Relative motions between oceanic and continental plates in the Pacific Basin: Geological Society of America Special Paper, v. 206, 64 p.

Enos, P., 1983, Late Mesozoic paleogeography of México, *in* Reynolds, M. W., and Dolly, E. D., eds., Mesozoic paleogeography of the west-central United States, Rocky Mountain Paleogeography Symposium 2: Rocky Mountain Section, Society of Economic Paleontologists and Mineralogists, p. 133–157.

Erikson, J. P., 1990, The Motagua fault zone of Guatemala and Cenozoic displacement on the Caribbean–North America plate boundary: Geological Society of America Abstracts with Programs, v. 22, A229.

Fakundiny, R. H., 1970, Geology of the El Rosario quadrangle, Honduras, Central America [Ph.D. dissertation]: Austin, University of Texas, 234 p.

Farrar, A., and others, 1970, K-Ar evidence for the post-Paleozoic migration of granitic intrusion foci in the Andes of northern Chile: Earth and Planetary Science Letters, v. 10, p. 60–66.

Feininger, T., 1987, Allochthonous terranes in the Andes of Ecuador and northwestern Peru: Canadian Journal of Earth Sciences, v. 24, p. 266–278.

Fenby, S. S., and Gastil, R. G., 1991, Geologic-tectonic map of the Gulf of California and surrounding areas, *in* Dauphin, J. P., and Simoneit, B.R.T., eds., The Gulf and Peninsular Province of the Californias, American Association of Petroleum Geologists Memoir 47, p. 79–83.

Ferrari, L., Garduño, V. H., Pasquaré, G., and Tibaldi, A., 1991, Geology of Los Azufres Caldera, México, and its relationships with regional tectonics: Journal of Volcanology and Geothermal Research, v. 47, p. 129–147.

Ferrusquía-Villafranca, I., 1976, Estudios geológicos-paleontológicos en la región mixteca, Part 1: Geología del área Tamazulapán-Teposcolula-Yanhuitlán, Mixteca Alta, Estado de Oaxaca, México: Boletín de Instituto de Geología, Universidad Nacional Autónoma de México, v. 97, 160 p.

Ferrusquía-Villafranca, I., and Comas-Rodríguez, O., 1988, Reptiles marinos mesozoicos en el sureste de México y su significación geológico-paleontológica Revista del Instituto de Geología, Universidad Nacional Autónoma de México, v. 7, p. 136–147.

Ferrusquía-Villafranca, I., and McDowell, F. W., 1988, Time constraints on formation of continental Tertiary basins in the state of Oaxaca: Geological Society of America Abstracts with Programs, v. 20, A59.

Fife, D. L., Minch, J. A., and Crampton, P. J., 1967, Late Jurassic age of the Santiago Peak Volcanics, California: Geological Society of America Bulletin, v. 78, p. 299–302.

Filmer, P. E., and Kirschvink, J. L., 1989, A paleomagnetic constraint on the Late Cretaceous paleoposition of northwestern Baja California, México: Journal of Geophysical Research, v. 94, p. 7332–7342.

Finch, R. C., 1981, Mesozoic stratigraphy of central Honduras: American Association of Petroleum Geologists Bulletin, v. 65, p. 1320–1333.

Fix, J. E., 1975, The crust and upper mantle of central México: Geophysical Journal of the Royal Astronomical Society, v. 43, p. 453–499.

Flawn, P. T., and Maxwell, R. A., 1958, Metamorphic rocks in the Sierra del Carmen, Coahuila, México: American Association of Petroleum Geologists Bulletin, v. 42, p. 2245–2249.

Flawn, P. T., Goldstein, A., Jr., King, P. B., and Weaver, C. E., 1961, The Ouachitan System: University of Texas Publication 6120, 401 p.

Flores, T., 1929, Reconocimientos geológicos en la región central del Estado de Sonora: Institúto Geología Mexicano Boletín, v. 49, 267 p.

Flynn, J. J., Cipolletti, R. M., and Novacek, M. J., 1989, Chronology of early Eocene marine and terrestrial strata, Baja California, México: Geological Society of America Bulletin, v. 101, p. 1182–1196.

Forman, J. A., Burke, W. H., Jr., Minch, J. A., and Yeats, R. S., 1971, Age of the basement rocks at Magdalena Bay, Baja California, México: Geological Society of America Abstracts with Programs, v. 3, p. 120.

Freeland, G. L., and Dietz, R. S., 1972, Plate tectonic evolution of the Caribbean–Gulf of Mexico region: Nature, v. 232, p. 20–23.

Frez, J., and González-García, J. J., 1991, Seismotectonics of the border region between southern and northern Baja California: Geological Society of America Abstracts with Programs, v. 23, p. A24.

——, 1991a, Crustal structure and seismotectonics of northern Baja California, *in* Dauphin, J. P., and Simoneit, B.R.T., eds., The Gulf and Peninsular Province of the Californias: American Association of Petroleum Geologists Memoir 48, p. 261–283.

Fries, C., Jr., 1960, Geología del Estado de Morelos y de partes adyacentes de México y Guerrero, región central-meridional de México: Boletín del Instituto de Geología, Universidad Nacional Autónoma de México, v. 60, 236 p.

Fries, C., Jr., and Rincón-Orta, C., 1965, Nuevas aportaciones geocronológicas y tecnicas empleadas en el Laboratorio de Geocronología: Boletín del Instituto de Geología, Universidad Nacional Autónoma de México, v. 73, p. 57–133.

Fries, C., Jr., Schmitter, E., Damon, P. E., Livingston, D. E., and Erikson, R., 1962, Edad de las rocas metamórficas en los Cañones de La Peregrina y de Caballeros, parte centro-occidental de Tamaulipas: Boletín del Instituto de Geología, Universidad Nacional Autónoma de México, v. 64, p. 55–59.

Fries, C., Jr., Schmitter, E., Damon, P. E., and Livingston, D. E., 1962a, Rocas precámbricas de edad grenvilliana de la parte central de Oaxaca en el sur de México: Boletín del Instituto de Geología, Universidad Nacional Autónoma de México, v. 64, p. 45–53.

Fry, J. G., Bottjer, D. J., and Lund, S. P., 1985, Magnetostratigraphy of displaced Upper Cretaceous strata in southern California: Geology, v. 13, p. 648–651.

Fulwider, R. W., 1976, Biostratigraphy of the Tepetate Formation, Baja California del Sur [M.S. thesis]: Los Angeles, University of Southern California, 111 p.

Garrison, J. R., Jr., Ramírez-Ramírez, C., and Long, L. E., 1980, Rb-Sr isotopic study of the ages and provenance of Precambrian granulite and Paleozoic greenschist near Ciudad Victoria, México, *in* Pilger, R. H., Jr., ed., The origin of the Gulf of Mexico and the early opening of the central north Atlantic Ocean: Baton Rouge, Louisiana State University, p. 37–49.

Gastil, R. G., 1983, Mesozoic and Cenozoic granitic rocks of southern California and western México: Geological Society of America Memoir 159, p. 265–275.

—— , 1985, Terranes of peninsular California and adjacent Sonora, *in* Howell, D. G., ed., Tectonostratigraphic terranes of the Circum-Pacific region: Circum-Pacific Council for Energy and Mineral Resources, Earth Science Series, no. 1, p. 273–283.

—— , 1990, The Cordilleran geocline in peninsular California, USA, and México: Geological Society of America Abstracts with Programs, v. 22, p. 24.

—— , 1991, Is there a Oaxaca-California megashear? Conflict between paleomagnetic data and other elements of geology: Geology, v. 19, p. 502–505.

Gastil, R. G., and Jensky, W. A., II, 1973, Evidence for strike-slip displacement beneath the Trans-Mexican Volcanic Belt, *in* Kovach, R. L., and Nur, A., eds., Proceedings of Conference on Tectonic Problems of San Andreas Fault System, Volume 13: Stanford University Publications in Geological Sciences, p. 171–180.

Gastil, R. G., and Krummenacher, D., 1977a, Reconnaissance geological map of coastal Sonora between Puerto Lobos and Bahía Kino: Geological Society of America Map and Chart Series, MC-16, scale 1:150,000.

—— , 1977b, Reconnaissance geology of coastal Sonora between Puerto Lobos and Bahía Kino: Geological Society of America Bulletin, v. 88, p. 189–198.

Gastil, R. G., and Fenby, S. S., 1991, Detachment faulting as a mechanism for tectonically filling the Gulf of California during dilation, *in* Dauphin, J. P., and Simoneit, B.R.T., eds., The Gulf and Peninsular Province of the Californias: American Association of Petroleum Geologists Memoir 47, p. 371–375.

Gastil, R. G., and Miller, R. H., 1983, Pre-batholithic terranes of southern and Peninsular California, U.S.A. and México: Status report, *in* Stevens, C. H., ed., Pre-Jurassic rocks in western North American suspect terranes: Pacific Section, Society of Economic Paleontologists and Mineralogists, p. 49–61.

—— , 1984, Prebatholithic paleogeography of Peninsular California and adjacent México, *in* Frizzell, V. A., Jr., ed., Geology of the Baja California Peninsula, Volume 39: Pacific Section, Society of Economic Paleontologists and Mineralogists, p. 9–16.

Gastil, R. G., Phillips, R. P., and Allison, E. C., 1975, Reconnaissance geology of the state of Baja California: Geological Society of America Memoir 140, 170 p.

Gastil, R. G., Krummenacher, D., and Jensky, W. A., II, 1978, Reconnaissance geology of west-central Nayarit, México: Geological Society of America Map and Chart Series, MC-24, scale 1:200,000, 8 p.

Gastil, R. G., Morgan, G. J., and Krummenacher, D., 1978, Mesozoic history of peninsular California and related areas east of the Gulf of California, *in* Howell, D. G., and McDougall, K. A., eds., Mesozoic paleogeography of the western United States: Pacific Section, Society of Economic Paleontologists and Mineralogists, p. 107–115.

—— , 1981, The tectonic history of peninsular California and adjacent México, *in* Ernst, W. G., ed., The geotectonic development of California, Rubey Volume I: Englewood Cliffs, New Jersey, Prentice-Hall, p. 284–306.

Gastil, R. G., Krummenacher, D., and Minch, J., 1979, The record of Cenozoic volcanism around the Gulf of California: Geological Society of America

Bulletin, v. 90, p. 839–857.

Gastil, R. G., Diamond, J., and Knaack, C., 1986a, The magnetite-ilmenite line in peninsular California: Geological Society of America Abstracts with Programs, v. 18, p. 109.

Gastil, R. G., Miller, R. H., and Campa-Uranga, M., 1986b, The Cretaceous paleogeography of peninsular California and adjacent México, *in* Abbott, P. L., ed., Cretaceous stratigraphy western North America: Pacific Section, Society of Economic Paleontologists and Mineralogists, p. 41–50.

Gastil, R. G., and 10 others, 1991, The relation between the Paleozoic strata on opposite sides of the Gulf of California, *in* Pérez-Segura, E., and Jacques-Ayala, C., eds., Studies of Sonoran geology: Geological Society of America Special Paper 254, p. 7–18.

Gehrels, G. E., and Saleeby, J. B., 1987, Geologic framework, tectonic evolution, and displacement history of the Alexander terrane: Tectonics, v. 8, p. 151–173.

Gilmer, A. L., Clark, K. F., Conde-C., J., Hernández-C., I., Figueroa-S., J. I., and Porter, E. W., 1988, Sierra de Santa María, Velardeña mining district, Durango, México: Economic Geology, v. 83, p. 1802–1829.

Goetz, C. W., Girty, G. H., and Gastil, R. G., 1988, East over west ductile thrusting along a terrane boundary in the Peninsular Ranges: Rancho El Rosarito, Baja California, México: Geological Society of America Abstracts with Programs, v. 20, p. 165.

Goldsmith, R., Marvin, R. F., and Mehnert, H. H., 1971, Radiometric ages in the Santander Massif, Eastern cordillera, Colombia Andes: U.S. Geological Survey Professional Paper 750-D, p. 44–49.

Gomberg, D. N., Banks, P. D., and McBirney, A. R., 1968, Preliminary zircon ages from the Central Cordillera: Science, v. 162, p. 121–122.

Gomberg, J., and Masters, T. G., 1988, Waveform modeling using locked-mode synthetic and differential seismograms: Application to determination of the structure of México: Geophysical Journal of the Royal Astronomical Society, v. 94, p. 193–218.

Gomberg, J., Priestley, K. F., Masters, T. G., and Brune, J. N., 1988, The structure of the crust and upper mantle of northern México: Geophysical Journal of the Royal Astronomical Society, v. 94, p. 1–20.

Gomberg, J., Priestly, K., and Brune, J., 1989, The compressional velocity structure of the crust and upper mantle of northern México and the border region: Bulletin of the Seismological Society of America, v. 79, p. 1496–1519.

González-León, C., 1979, Geology of the Sierra del Alamo, *in* Anderson, T. H., and Roldán-Quintana, J., eds., Geology of northern Sonora, Geological Society of America, Field Trip Guidebook, p. 23–31.

—— , 1980, La Formación Antimonio (Triásico Superior-Jurásico Inferior) en la Sierra del Alamo, Estado de Sonora: Revista del Instituto de Geología, Universidad Nacional Autónoma de México, v. 4, p. 13–18.

—— , 1986, Estratigrafía del Paleozoic de la Sierra del Tule, noreste de Sonora: Revista del Instituto de Geología, Universidad Nacional Autónoma de México, v. 6, p. 117–135.

González-Partida, E., and Martínez-Serrano, R., 1989, Geochronología, termomicrometría e istopoía de azufre y carbono de la brecha cuprífera La Sorpresa, Estado de Jalisco: Revista del Instituto de Geología, Universidad Nacional Autónoma de México, v. 8, p. 202–210.

González-Partida, E., Casar-Aldrete, I., Morales-Puente, P., and Nieto-Obregón, J., 1989, Fechas de Rb-Sr (Maastrichtiano y Oligoceno) de rocas volcánicas e intrusivas de la región de Zihuatanejo, Sierra Madre del Sur de México: Revista del Instituto de Geología, Universidad Nacional Autónoma de México, v. 8, p. 248–249.

González-Ruiz, J. R., and McNally, K. C., 1988, Stress accumulation and release since 1882 in Ometepec, Guerrero, México: Implications for failure mechanisms and risk assessments of a seismic gap: Journal of Geophysical Research, v. 93, p. 6297–6317.

Goodwin, L. B., and Haxel, G. B., 1990, Structural evolution of the southern Baboquivari Mountains, south-central Arizona and north-central Sonora: Tectonics, v. 9, p. 1077–1095.

Goodwin, L. B., and Renne, P. R., 1991, Effects of progressive mylonitization on Ar retention of biotites in the Santa Rosa Mylonite Zone, California, and thermochronologic implications: Contributions to Mineralogy and Petrology, v. 108, p. 283–297.

Gordon, M. B., 1989, Mesozoic igneous rocks on the Chortis block: Implications for Caribbean reconstructions: EOS Transactions of the American Geophysical Union, v. 70, p. 1342.

Gose, W. A., 1985, Paleomagnetic results from Honduras and their bearing on Caribbean tectonics: Tectonics, v. 4, p. 565–585.

Gose, W. A., and Sánchez-Barreda, L. A., 1981, Paleomagnetic results from southern México: Geofísica Internacional, v. 20, p. 163–175.

Gose, W. A., Belcher, R. C., and Scott, G. R., 1982, Paleomagnetic results from northeastern México: Evidence for large Mesozoic rotations: Geology, v. 10, p. 50–54.

Götte, M., 1988, Estudio geológico-estructural de Galeana/N.L. (México) y sus alrededores: Actas de la Facultad de Ciencias de la Tierra, Universidad Autónoma de Nuevo León, Linares, v. 2, p. 61–87.

—— , 1990, Halotektonische Deformationsprozesse in Sulfatgesteinen der Minas Viejas-Formation (ober-Jura) in der Sierra Madre Oriental, nordost-Mexiko [Ph.D. dissertation]: Darmstadt, Germany, Technischen Hochschule Darmstadt, 270 p.

Götte, M., and Michalzik, D., 1991, Stratigraphic relations and facies sequence of an Upper Jurassic evaporitic ramp in the Sierra Madre Oriental (México): Zentralblatt für Geologie und Palaeontologie, Teil I, H. 6, p. 1445–1466.

Grajales-Nishimura, J. M., 1988, Geology, geochronology, geochemistry, and tectonic implications of the Juchatengo green rock sequence, state of Oaxaca, southern México [M.S. thesis]: Tucson, University of Arizona.

Grajales-Nishimura, J. M., and López-Infanzón, M., 1983, Estudio petrogenético de las rocas ígneas y metamórficas en el prospecto Tomatlán-Guerrero-Jalisco: Mexico City, Instituto Mexicano del Petroleo, Internal Report, Proyecto C-1160, 69 p.

Grajales-Nishimura, J. M., Torres, R., and Murillo, G., 1986, Datos isotópicos potasio-argón para rocas ígneas y metamórficas en el Estado de Oaxaca: Sociedad Geológica Mexicana, VIII Convención Nacional, Libro de Resúmenes.

Grajales-Nishimura, J. M., Terrell, D., and Torres-Vargas, R., 1990, Late Cretaceous synorogenic volcanic/sedimentary sequences in eastern Sonora, México: Geological Society of America Abstracts with Programs, v. 22, p. 26.

Grant, G. J., and Ruiz, J., 1988, The Pb-Zb-Cu-Ag deposits of the Granadeña mine, San Francisco del Oro–Santa Barbara district, Chihuahua, México: Economic Geology, v. 83, p. 1683–1702.

Greenwood, E., Kottlowski, F. E., and Thompson, S., III, 1977, Petroleum potential and stratigraphy of Pedregosa Basin: Comparison with Permian and Orogrande basins: American Association of Petroleum Geologists Bulletin, v. 61, p. 1448–1469.

Griffith, R. C., 1987, Geology of the southern Sierra Calamjué area; Structural and stratigraphic evidence for latest Albian compression along a terrane boundary, Baja California, México [M.S. thesis]: San Diego, California, San Diego State University, 115 p.

Griffith, R. C., and Goetz, C. W., 1987, Structural and geochronological evidence for mid-Cretaceous compressional tectonics along a terrane boundary in the Peninsular Ranges: Geological Society of America Abstracts with Programs, v. 19, p. 384.

Grimm, K. A., Ledesma, M., Garrison, R. E., and Fonseca, C., 1991, The Oligo-Miocene San Gregorio Formation of Baja California Sur, México: An early record of coastal upwelling along the eastern Pacific margin: First International Meeting on Geology of the Baja California Peninsula, Universidad Autónoma de Baja California Sur La Paz, B.C.S., p. 35.

Gromet, L. P., and Silver, L. T., 1987, REE variations across the Peninsular Ranges batholith: Implications for batholithic petrogenesis and crustal growth in magmatic arcs: Journal of Petrology, v. 28, p. 75–125.

Guerrero-García, J. C., 1989, Vertical tectonics in southern México and its relation to trench migration: EOS Transactions of the American Geophysical Union, v. 70, p. 1319.

Guerrero-García, J. C., Silver, L. T., and Anderson, T. H., 1978, Estudios geocronológicos en el complejo Xolapa: Sociedad Geológica Mexicana

Boletín, v. 39, p. 22–23.

Guerrero-García, J. C., Herrero-Bervera, E., and Helsley, C. E., 1990, Paleomagnetic evidence for post-Jurassic stability of southeastern México: Maya terrane: Journal of Geophysical Research, v. 95, p. 7091–7100.

Gunderson, R., Cameron, K., and Cameron, M., 1986, Mid-Cenozoic high-K calc-alkalic and alkalic volcanism in eastern Chihuahua, México: Geology and geochemistry of the Benavides-Pozos area: Geological Society of America Bulletin, v. 97, p. 737–753.

Gursky, H.-J., and Michalzik, D., 1989, Lower Permian turbidites in the northern Sierra Madre Oriental, México: Zentralblatt für Geologie und Palaeontologie, Teil I, p. 821–838.

Gursky, H.-J., and Ramírez-Ramírez, C., 1986, Notas preliminares sobre el discubrimiento de volcanicas acidas en el Cañon de Caballeros (Nucleo del Anticlinorio Huizachal-Peregrina, Tamaulipas, México): Actas de la Facultad de Ciencias de la Tierra, Universidad Autónoma de Nuevo León, Linares, v. 1, p. 11–22.

Guzmán-Speziale, M., Pennington, W., and Matumoto, T., 1989, The triple junction of the North America, Cocos, and Caribbean plates: Seismicity and tectonics: Tectonics, v. 8, p. 981–997.

Hagstrum, J. T., and Sedlock, R. L., 1990, Remagnetization and northward translation of Mesozoic red chert from Cedros Island and the San Benito Islands, Baja California, México: Geological Society of America Bulletin, v. 102, p. 983–991.

—— , 1992, Paleomagnetism of Mesozoic red chert from Cedros Island and the San Benito Islands, Baja California, Mexico, revisited: Geophysical Research Letters, v. 19, p. 329–332.

Hagstrum, J. T., McWilliams, M., Howell, D. G., and Gromme, C. S., 1985, Mesozoic paleomagnetism and northward translation of the Baja California peninsula: Geological Society of America Bulletin, v. 96, p. 221–225.

Hagstrum, J. T., Sawlan, M. G., Hausback, B. P., Smith, J. G., and Gromme, C. S., 1987, Miocene paleomagnetism and tectonic setting of the Baja California peninsula, México: Journal of Geophysical Research, v. 92, p. 2627–2640.

Halpern, M., 1978, Geological significance of Rb-Sr isotopic data of northern Chile crystalline rocks of the Andean orogen between latitudes 23° and 27° south: Geological Society of America Bulletin, v. 89, p. 522–532.

Halpern, M., Guerrero-García, J. C., and Ruiz-Castellanos, M., 1974, Rb-Sr dates of igneous and metamorphic rocks from southeastern and central México; A progress report: Unión Geofísica Mexicana, Reunión Annual, Resúmenes, p. 30–31.

Hamilton, W., 1969, Mesozoic California and the underflow of Pacific mantle: Geological Society of America Bulletin, v. 80, p. 2409–2430.

—— , 1987, Mesozoic geology and tectonics of the Big Maria Mountains region, southeastern Arizona, *in* Dickinson, W. R., and Klute, M. A., eds., Mesozoic rocks of southern Arizona and adjacent areas, Volume 18: Arizona Geological Society Digest, p. 33–47.

Handschy, J. W., and Dyer, R., 1987, Polyphase deformation in Sierra del Cuervo, Chihuahua, México: Evidence for ancestral Rocky Mountain tectonics in the Ouachita foreland of northern México: Geological Society of America Bulletin, v. 99, p. 618–632.

Handschy, J. W., Keller, G. R., and Smith, K. J., 1987, The Ouachita system in northern México: Tectonics, v. 6, p. 323–330.

Hardy, L. R., 1981, Geology of the central Sierra de Santa Rosa, Sonora Mexico, *in* Ortlieb, L., and Roldán-Quintana, J., eds., Geology of northwestern México and southern Arizona: Instituto de Geología, Universidad Nacional Autónoma de México, Estación Regional del Noreste, p. 73–98.

Harlow, G. E., 1990, Constraints on the geochemical and P-T evolution of Guatemalan jadeitite and related inclusions in serpentinite: Geological Society of America Abstracts with Programs, v. 22, A349.

Harper, G. D., Saleeby, J. B., and Norman, E.A.S., 1985, Geometry and tectonic setting of sea-floor spreading for the Josephine Ophiolite, and implications for Jurassic accretionary events along the California margin, *in* Howell, D. G., ed., Tectonostratigraphic terranes of the Circum-Pacific region, Volume 1: Circum-Pacific Council for Energy and Mineral Resources,

p. 239–257.

Hatcher, R. D., Jr., 1972, Developmental model for the southern Appalachians: Geological Society of America Bulletin, v. 83, p. 2735–2760.

Hausback, B. P., 1984, Cenozoic volcanic and tectonic evolution of Baja California Sur, México, *in* Frizzell, V. A., Jr., ed., Geology of the Baja California peninsula, Volume 39: Pacific Section, Society of Economic Paleontologists and Mineralogists, p. 219–236.

Havskov, J., Singh, S., and Novelo, D., 1982, Geometry of the Benioff zone in the Tehuantepec area in southern México: Geofísica Internacional, v. 21, p. 325–330.

Hawkins, J. W., 1970, Metamorphosed Late Jurassic andesites and dacites of the Tijuana-Tecate area, California, Pacific slope geology of northern Baja California and adjacent Alta California: Pacific Section, American Association of Petroleum Geologists–Society of Economic Paleontologists and Mineralogists–Society of Exploration Geophysicists Guidebook, p. 25–29.

Haxel, G. B., Tosdal, R. M., May, D. J., and Wright, J. E., 1984, Latest Cretaceous and early Tertiary orogenesis in south-central Arizona: Thrust faulting, regional metamorphism, and granitic plutonism: Geological Society of America Bulletin, v. 95, p. 631–653.

Hayama, Y., Shibata, K., and Takeda, H., 1984, K-Ar ages of the low-grade metamorphic rocks in the Altar massif, northwest Sonora, México: Journal of the Geological Society of Japan, v. 90, p. 589–596.

Hayes, P. T., 1970, Cretaceous paleogeography of southeastern Arizona and adjacent areas: U.S. Geological Survey Professional Paper 658-B, 42 p.

Heim, A., 1922, Notes on the Tertiary of southern Lower California: Geological Magazine, v. 59, p. 529–547.

——, 1940, The front ranges of the Sierra Madre Oriental, México, from Ciudad Victoria to Tamazunchale: Eclogae Geologicae Helvetiae, v. 33, p. 313–352.

Heinrich, W., and Besch, T., 1992, Thermal history of the upper mantle beneath a young back-arc extensional zone: Ultramafic xenoliths from San Luis Potosí, central Mexico: Contributions to Mineralogy and Petrology, v. 111, p. 126–142.

Henry, C. D., 1986, East-northeast–trending structures in western México: Evidence for oblique convergence in the late Mesozoic: Geology, v. 14, p. 314–317.

——, 1989, Late Cenozoic Basin and Range structure in western México adjacent to the Gulf of California: Geological Society of America Bulletin, v. 101, p. 1147–1156.

Henry, C. D., and Aranda-Gómez, J. J., 1992, The real southern Basin and Range: Mid- to late Cenozoic extension in Mexico: Geology, v. 20, p. 701–704.

Henry, C. D., and Fredrikson, G., 1987, Geology of part of southern Sinaloa, México adjacent to the Gulf of California: Geological Society of America Map and Chart Series, MCH063, scale 1:250,000, 14 p.

Henry, C. D., and McDowell, F. W., 1986, Geochronology of magmatism in the Tertiary volcanic field, Trans-Pecos Texas: University of Texas at Austin Bureau of Economic Geology Guidebook, v. 23, p. 91–122.

Henry, C. D., and Price, J. G., 1986, Early Basin and Range development in Trans-Pecos Texas and adjacent Chihuahua: Magmatism and orientation, timing, and style of extension: Journal of Geophysical Research, v. 91, p. 6213–6224.

Henry, C. D., Price, J. G., and James, E. W., 1991, Mid-Cenozoic stress evolution and magmatism in the southern Cordillera, Texas and México: Transition from continental arc to intraplate extension: Journal of Geophysical Research, v. 96, p. 13545–13560.

Henyey, T. L., and Bischoff, J. L., 1973, Tectonic elements of the northern part of the Gulf of California: Geological Society of America Bulletin, v. 84, p. 315–330.

Herrero-Bervera, E., Urrutia-Fucugauchi, J., and Khan, M. A., 1990, A paleomagnetic study of remagnetized Upper Jurassic red beds from Chihuahua, northern México: Physics of Earth and Planetary Interiors, v. 62, p. 307–322.

Herzig, C. T., and Kimbrough, D. L., 1991, Early Cretaceous zircon ages prove a non-accretionary origin for the Santiago Peak volcanics, orthern Santa Ana

Mtns., California: Geological Society of America Abstracts with Programs, v. 23, p. 35.

Hickey, J. J., 1984, Stratigraphy and composition of a Jura-Cretaceous volcanic arc apron, Punta Eugenia, B.C. Sur, México, *in* Frizzell, V. A., Jr., ed., Geology of the Baja California Peninsula, Volume 39, Pacific Section, Society of Economic Paleontologists and Mineralogists, p. 149–160.

Hill, R. I., Silver, L. T., and Taylor, H. P., Jr., 1986, Coupled Sr-O isotope variations as an indicator of source heterogeneity for the northern Peninsular Ranges batholith: Contributions to Mineralogy and Petrology, v. 92, p. 351–361.

Hillhouse, J. W., and Gromme, C. S., 1984, Northward displacement and accretion of Wrangellia: New paleomagnetic evidence from Alaska: Journal of Geophysical Research, v. 89, p. 4461–4477.

Hoffman, P. F., 1989, Precambrian geology and tectonic history of North America, *in* Bally, A. W., and Palmer, A. R., eds., Decade of North American Geology, Volume A: The geology of North America—An overview: Geological Society of America, p. 447–512.

Holguin-Q., N., 1978, Estudio estratigráfico del cretácico inferior en el norte de Sinaloa, México: Revista del Instituto Mexicano del Petróleo, v. 10(1), p. 6–13.

Horne, G. S., Clark, G. S., and Pushkar, P., 1976a, Pre-Cretaceous rocks of northwestern Honduras: Basement terrane in Sierra de Omoa: American Association of Petroleum Geologists Bulletin, v. 60, p. 566–583.

Horne, G. S., Pushkar, P., and Shafiqullah, M., 1976b, Preliminary K-Ar age data from the Laramide Sierras of central Honduras, Informe de la IV Reunión de Geólogos de America Central, Volume 5: Publicaciones Geológicas del ICAITI, Guatemala, p. 92–98.

——, 1976c, Laramide plutons on the landward continuation of Bonacca Ridge, northern Honduras: Transactions, Caribbean Geological Conference, VII, p. 583–588.

Hose, L. D., 1982, The effects of Laramide stress on Cretaceous carbonate sediments in the northwest flanks of the Huizachal-Peregrina anticlinorium, northeastern México, *in* Powers, R. B., ed., Geologic studies of the Cordilleran thrust belt: Denver, Colorado, Rocky Mountain Association of Geologists, p. 869–874.

Howell, D. G., Jones, D. L., and Schermer, E. R., 1985, Tectonostratigraphic terranes of the Circum-Pacific region, *in* Howell, D. G., ed., Tectonostratigraphic terranes of the Circum-Pacific region: Circum-Pacific Council for Energy and Mineral Resources, Earth Science Series, no. 1, p. 3–30.

Howell, D. G., Champion, D. E., and Vedder, J. G., 1987, Terrane accretion, crustal kinematics, and basin evolution, southern California, *in* Ingersoll, R. V., and Ernst, W. G., eds., Cenozoic basin development of coastal California, Rubey Volume VI: Englewood Cliffs, New Jersey, Prentice-Hall, p. 242–258.

Hubberten, H.-W., and Nick, K., 1986, La Sierra de San Carlos, Tamaulipas—Un complejo igneo de la Provincia Alcalina Mexicana Oriental: Actas de la Facultad de Ciencias de la Tierra, Universidad Autónoma de Nuevo León, Linares, v. 1, p. 68–77.

Humphreys, E., and Weldon, R., 1991, Kinematic constraints on the rifting of Baja California, *in* Dauphin, J. P., and Simoneit, B.R.T., eds., The Gulf and Peninsular Province of the Californias, American Association of Petroleum Geologists Memoir 47, p. 217–229.

Imlay, R. W., 1936, Evolution of the Coahuila Peninsula, Part IV, Geology of the western part of the Sierra de Parras: Geological Society of America Bulletin, v. 47, p. 1091–1152.

——, 1938, Studies of the Mexican Geosyncline: Geological Society of America Bulletin, v. 49, p. 1651–1694.

——, 1939, Upper Jurassic ammonites from México: Geological Society of America Bulletin, v. 50, p. 1–77.

——, 1943, Jurassic formations of Gulf region: American Association of Petroleum Geologists Bulletin, v. 27, p. 1407–1533.

——, 1963, Jurassic fossils from southern California: Journal of Paleontology, v. 37, p. 97–107.

——, 1964, Middle and Upper Jurassic fossils from southern California: Journal of Paleontology, v. 33, p. 505–509.

——, 1965, Jurassic marine faunal differentiation in North America: Journal of Paleontology, v. 39, p. 1023–1038.

——, 1980, Jurassic paleobiogeography of the conterminous United States in its continental setting: U.S. Geological Survey Professional Paper 1062, 134 p.

Imlay, R. W., Cepeda, E., Alvarez, M., Jr., and Díaz, T., 1948, Stratigraphic relations of certain Jurassic formations in eastern México: American Association of Petroleum Geologists Bulletin, v. 32, p. 1750–1761.

INEGI, 1980, Carta geológica de México: Instituto Nacional de Estadística Geografía e Informática, scale 1:1,000,000, 8 sheets.

——, 1985, Geología de la República Mexicana: Instituto Nacional de Estadística Geografía e Informática, 2nd ed., 88 p.

Irving, E., Woodsworth, G. J., Wynne, P. J., and Morrison, A., 1985, Paleomagnetic evidence for displacement from the south of the Coast Plutonic Complex, British Columbia: Canadian Journal of Earth Sciences, v. 22, p. 584–598.

Jachens, R. C., Simpson, R. W., Griscom, A., and Mariano, J., 1986, Plutonic belts in southern California defined by gravity and magnetic anomalies: Geological Society of America Abstracts with Programs, v. 18, p. 120.

Jachens, R. C., Todd, V. R., Morton, D. M., and Griscom, A., 1991, Constraints on the structural evolution of the Peninsular Ranges batholith, California, from a new aeromagnetic map: Geological Society of America Abstracts with Programs, v. 23, p. 38.

Jacobo-Arrabán, J., 1986, El basamento del distrito Poza Rica y su implicación en la generación de hidrocarburos: Revista del Instituto Mexicano del Petróleo, v. 18.

Jacques-Ayala, C., 1989, Arroyo Sásabe Formation (Aptian-Albian), northwestern Sonora, México—Marginal marine sedimentation in the Sonora backarc basin: Revista del Instituto de Geología, Universidad Nacional Autonóma de México, v. 8, p. 171–178.

Jacques-Ayala, C., and Potter, P. E., 1987, Stratigraphy and paleogeography of Lower Cretaceous rocks, Sierra El Chanate, northwest Sonora, México, *in* Dickinson, W. R., and Klute, M. A., eds., Mesozoic rocks of southern Arizona and adjacent areas, Volume 18: Arizona Geological Society Digest, p. 203–214.

Jacques-Ayala, C., García y Barragán, J. C., and DeJong, K. A., 1990, Caborca-Altar geology: Cretaceous sedimentation and compression, Tertiary uplift and extension, *in* Gehrels, G. E., and Spencer, J. E., eds., Geologic excursions through the Sonoran Desert region, Arizona and Sonora: Tucson, Arizona Geological Survey Special Paper 7, p. 165–182.

James, E. W., 1989, Southern extension of the Independence dike swarm of eastern California: Geology, v. 17, p. 587–590.

James, E. W., and Henry, C. D., 1991, Compositional changes in Trans-Pecos Texas magmatism coincident with Cenozoic stress realignment: Journal of Geophysical Research, v. 96, p. 13561–13575.

——, 1993, Southeastern extent of the North American craton in Texas and northern Chihuahua as revealed by Pb isotopes: Geological Society of America Bulletin, v. 105, p. 116–126.

James, E. W., and Mattinson, J. M., 1988, Metamorphic history of the Salinian block: An isotopic reconnaissance, *in* Ernst, W. G., ed., Metamorphism and crustal evolution of the western United States, Rubey Volume VII: Englewood Cliffs, New Jersey, Prentice Hall, p. 938–952.

Jarrard, R. D., 1986, Terrane motion by strike-slip faulting of forearc slivers: Geology, v. 14, p. 780–783.

Johnson, C. A., 1987, Regional tectonics in central México: Active rifting and transtension within the Mexican volcanic belt: EOS Transactions of the American Geophysical Union, v. 68, p. 423.

Johnson, C. A., and Harrison, C.G.A., 1987, Tectonics and volcanism in central México: A Landsat Thematic Mapper perspective: Remote Sensing of the Environment, v. 28, p. 273–286.

Johnson, C. A., Barros, J. A., and Harrison, C.G.A., 1988, The Chapala-Oaxaca fault zone: A major trench-parallel fault in southwestern México: EOS Transactions of the American Geophysical Union, v. 69, p. 1451.

Johnson, C. A., Lang, H., Cabral-Cano, E., Harrison, C., and Barros, J. A., 1990, A new Cordilleran fold-thrust model for Laramide deformation in Tierra Caliente, Michoacán and Guerrero states, México: Geological Society of America Abstracts with Programs, v. 22, A186–187.

Johnson, K. R., and Muller, P. D., 1986, Late Cretaceous collisional tectonics in the southern Yucatán (Maya) block of Guatemala: Geological Society of America Abstracts with Programs, v. 18, p. 647.

Jones, D. A., and Miller, R. H., 1982, Jurassic fossils from the Santiago Peak Volcanics, San Diego County, California, *in* Abbott, P. L., ed., Geologic studies in San Diego: San Diego, California, San Diego Association of Geologists, p. 93–103.

Jones, N. W., Dula, R., Long, L. E., and McKee, J. W., 1982, An exposure of a fundamental fault in Permian basement granitoids, Valle San Marcos, Coahuila, México: Geological Society of America Abstracts with Programs, v. 14, p. 523–524.

Jones, N. W., McKee, J. W., Marquez-D., B., Tovar, J., Long, L. E., and Laudon, T. E., 1984, The Mesozoic La Mula island, Coahuila, México: Geological Society of America Bulletin, v. 95, p. 1226–1241.

Jones, N. W., McKee, J. W., and Anderson, T. H., 1986, Pre-Cretaceous igneous rocks in north central México: Geological Society of America Abstracts with Programs, v. 18, p. 649.

Jones, N. W., McKee, J. W., Anderson, T. H., and Silver, L. T., 1990, Nazas Formation: A remnant of the Jurassic arc of western North America in north-central México: Geological Society of America Abstracts with Programs, v. 22, A327.

Karig, D. E., Cardwell, R. K., Moore, G. F., and Moore, D. G., 1978, Late Cenozoic subduction and continental margin truncation along the northern Middle America trench: Geological Society of America Bulletin, v. 89, p. 265–276.

Keller, G. R., Braile, L. W., McMechan, G. A., Thomas, W. A., Harder, S. H., Chang, W.-F., and Jardine, W. G., 1989a, Paleozoic continent-ocean transition in the Ouachita Mountains imaged from PASSCAL wide-angle seismic reflection-refraction data: Geology, v. 17, p. 119–122.

Keller, G. R., Kruger, J. M., Smith, K. J., and Voight, W. M., 1989b, The Ouachita system; A geophysical overview, *in* Hatcher, R. D., Jr., Thomas, W. A., and Viele, G. W., eds., Decade of North American Geology, Volume F-2: The Appalachian and Ouachita orogen in the United States: Geological Society of America, p. 689–694.

Keller, P. C., Bockoven, N. T., and McDowell, F. W., 1982, Tertiary volcanic history of the Sierra del Gallego area, Chihuahua, México: Geological Society of America Bulletin, v. 93, p. 303–314.

Kellum, L. B., Imlay, R. W., and Kane, W. G., 1932, Evolution of the Coahuila Peninsula, México: Geological Society of America Bulletin, v. 47, p. 969–1008.

Kesler, S. E., Josey, W. L., and Collins, E. M., 1970, Basement rocks of western nuclear Central America: The Western Chuacús Group: Geological Society of America Bulletin, v. 81, p. 3307–3322.

Ketner, K. B., 1986, Eureka Quartzite in México?— Tectonic implications: Geology, v. 14, p. 1027–1030.

——, 1990, Stratigraphy, structure, and regional correlatives of tectonically juxtaposed Paleozoic miogeoclinal and eugeoclinal assemblages, Cerro Cobachi, Sonora, México: Geological Society of America Abstracts with Programs, v. 22, p. 34.

Kilmer, F. H., 1979, A geological sketch of Cedros Island, Baja California, México, *in* Abbott, P. L., and Gastil, R. G., eds., Baja California geology: San Diego, Department of Geological Sciences, San Diego State University, p. 11–28.

——, 1984, Geology of Cedros Island, Baja California, México: Arcata, California, Humboldt State University, 69 p.

Kim, J. J., Matumoto, T., and Latham, G. V., 1982, A crustal section of northern Central America as inferred from wide-angle reflections from shallow earthquakes: Bulletin of the Seismological Society of America, v. 72, p. 925–940.

Kimbrough, D. L., 1982, Structure, petrology, and geochronology of Mesozoic paleooceanic terranes on Cedros Island and the Vizcaíno Peninsula, Baja

California Sur, México [Ph.D. dissertation]: Santa Barbara, University of California, 395 p.

——, 1984, Paleogeographic significance of the Middle Jurassic Gran Cañon Formation, Cedros Island, Baja California Sur, *in* Frizzell, V. A., Jr., ed., Geology of the Baja California Peninsula, Volume 39: Pacific Section, Society of Economic Paleontologists and Mineralogists, p. 107–117.

——, 1985, Tectonostratigraphic terranes of the Vizcaíno Peninsula and Cedros and San Benito Islands, Baja California, México, *in* Howell, D. G., ed., Tectonostratigraphic terranes of the Circum-Pacific region: Circum-Pacific Council for Energy and Mineral Resources, Earth Science Series, no. 1, p. 285–298.

Kimbrough, D. L., Hickey, J. J., and Tosdal, R. M., 1987, U-Pb ages of granitoid clasts in upper Mesozoic arc-derived strata of the Vizcaíno Peninsula, Baja California, México: Geology, v. 15, p. 26–29.

Kimbrough, D. L., Anderson, C. L., Glass, S. M., Kenney, M. D., Thomas, A. P., and Vitello, T., 1990, Early Cretaceous zircon U-Pb ages from the Santiago Peak volcanics, western Peninsular Ranges batholith, San Diego County, California: Geological Society of America Abstracts with Programs, v. 22, p. 35.

King, P. B., 1975, The Ouachita and Appalachian orogenic belts, *in* Nairn, A.E.M., and Stehli, F. G., eds., The ocean basins and margins: Volume 3: The Gulf of Mexico and the Caribbean: New York, Plenum Press, p. 201–241.

King, R. E., 1939, Geological reconnaissance in northern Sierra Madre Occidental of México: Geological Society of America Bulletin, v. 50, p. 1625–1722.

King, R. E., Dunbar, C. O., Cloud, P. E., and Miller, A. K., 1944, Geology and paleontology of the Permian area northwest of Las Delicias, southwestern Coahuila, México: Geological Society of America Special Paper 52, 170 p.

Kistler, R. W., 1974, Phanerozoic batholiths in western North America: Annual Review of Earth and Planetary Sciences, v. 2, p. 403–418.

Klein, R. T., Hannula, S. R., Jones, N. W., McKee, J. W., and Anderson, T. H., 1990, Cerro El Pedernal: Block-in-mélange features in northern Zacatecas, México: Geological Society of America Abstracts with Programs, v. 22, p. 35.

Kleist, R., Hall, S. A., and Evans, I., 1984, A paleomagnetic study of the Lower Cretaceous Cupido Formation, northeast México: Evidence for local rotation within the Sierra Madre Oriental: Geological Society of America Bulletin, v. 95, p. 55–60.

Klesse, E., 1968 (1970), Geology of the El Ocotito–Ixcuinatoyac region and of La Dicha stratiform sulfide deposit, State of Guerrero: Sociedad Geológica Mexicana Boletín, v. 31, p. 107–140.

Klitgord, K. D., and Schouten, H., 1987, Plate kinematics of the Central Atlantic, *in* Tucholke, B. E., and Vogt, P. R., eds., Decade of North American Geology, Volume M: The Western Atlantic Region: Geological Society of America, p. 351–378.

Köhler, H., Schaaf, P., Muller-Sohnius, D., Emmermann, R., Negendank, J.F.W., and Tobschall, H. J., 1993, Geochronological and geochemical investigations on plutonic rocks from the complex of Puerto Vallarta, Sierra Madre del Sur (México): Geofísica Internacional (in press).

Kroonenberg, S. B., 1982, A Grenvillian granulite belt in the Colombian Andes and its relation to the Guiana Shield: Geologie en Mijnbouw, v. 61, p. 325–333.

Ladd, J. W., Ibrahim, A. K., McMillen, K. J., Latham, G. V., and von Huene, R. E., 1982, Interpretation of seismic-reflection data of the Middle America trench offshore Guatemala, *in* Initial Reports of the Deep Sea Drilling Project, Volume 67: Washington, D.C., U.S. Government Printing Office, p. 675–689.

Lapierre, H., Ortiz, L. E., Abouchami, W., Monod, O., Coulon, C., and Zimmerman, J.-L., 1992, A crustal section of an intra-oceanic island arc: The Late Jurassic–Early Cretaceous Guanajuato magmatic sequence, central México: Earth and Planetary Science Letters, v. 108, p. 61–77.

Larsen, E. S., Jr., 1948, Batholith and associated rocks of Corona, Elsinore, and San Luis Rey quadrangles, southern California: Geological Society of America Memoir 29, 182 p.

Larson, R. L., 1972, Bathymetry, magnetic anomalies, and plate tectonic history of the mouth of the Gulf of California: Geological Society of America Bulletin, v. 83, p. 3345–3360.

Lawrence, D. P., 1975, Petrology and structural geology of the Sanarate–El Progreso area, Guatemala [Ph.D. dissertation]: Binghamton, State University of New York, 255 p.

——, 1976, Tectonic implications of the geochemistry and petrology of the El Tambor formation: Probable oceanic crust in central Guatemala: Geological Society of America Abstracts with Programs, v. 8, p. 973–974.

Ledezma-Guerrero, O., 1967, Hoja Viesca 13R-1(6), con resumen de la geología de la Hoja Parras, Estados de Coahuila, Durango y Zacatecas: Carta Geológica de México, serie de 1:100,000.

——, 1981, Hoja Viesca 13R-1(12), con resumen de la geología de la Hoja Camacho, Estados de Durango y Zacatecas: Carta Geológica de México, serie de 1:100,000.

LeFevre, L. V., and McNally, K. C., 1985, Stress distribution and subduction of aseismic ridges in the Middle America subduction zone: Journal of Geophysical Research, v. 90, p. 4495–4510.

Legg, M. R., Wong, O., V., and Suárez Vidal, F., 1991, Geologic structure and tectonics of the inner continental borderland of northern Baja California, *in* Dauphin, J. P., and Simoneit, B.R.T., eds., The Gulf and Peninsular Province of the Californias: American Association of Petroleum Geologists Memoir 47, p. 145–177.

Leveille, G., and Frost, E. G., 1984, Deformed upper Paleozoic–lower Mesozoic cratonic strata, El Capitán, Sonora, México: Geological Society of America Abstracts with Programs, v. 16, p. 575.

Lewis, J. F., and Draper, G., 1990, Geology and tectonic evolution of the northern Caribbean margin, *in* Dengo, G., and Case, J. E., eds., Decade of North American Geology, Volume H: The Caribbean region: Geological Society of America, p. 77–140.

Lillie, R. J., and 7 others, 1983, Crustal structure of Ouachita Mountains, Arkansas; A model based on integration of COCORP reflection profiles and regional geophysical data: American Association of Petroleum Geologists Bulletin, v. 67, p. 907–931.

Livingston, D. E., 1973, A plate tectonic hypothesis for the genesis of porphyry copper deposits of the southern Basin and Range province: Earth and Planetary Science Letters, v. 2, p. 171–179.

Livingston, D. E., and Damon, P. E., 1968, The ages of stratified Precambrian rock sequences in central Arizona and northern Sonora: Canadian Journal of Earth Sciences, v. 5, p. 763–772.

Longoria, J. F., 1984, Mesozoic tectonostratigraphic domains in east-central México, *in* Westermann, G.E.G., ed., Jurassic-Cretaceous biochronology and paleogeography of North America: Geological Association of Canada Special Paper 27, p. 65–76.

——, 1988, Late Triassic–Jurassic paleogeography and origin of Gulf of Mexico basin: Discussion: American Association of Petroleum Geologists Bulletin, v. 72, p. 1411–1418.

Lonsdale, P., 1989, Geology and tectonic history of the Gulf of California, *in* Winterer, E. L., Hussong, D. M., and Decker, R. W., eds., Decade of North American Geology, Volume N: The Eastern Pacific region: Geological Society of America, p. 499–521.

——, 1991, Structural patterns of the Pacific floor offshore of peninsular California, *in* Dauphin, J. P., and Simoneit, B.R.T., eds., The Gulf and Peninsular Province of the Californias: American Association of Petroleum Geologists Memoir 47, p. 87–125.

Lonsdale, P., and Becker, K., 1985, Hydrothermal plumes, hot springs, and conductive heat flow in the southern trough of Guaymas Basin: Earth and Planetary Science Letters, v. 73, p. 211–225.

López-Infanzón, M., 1986 (1989), Petrología y radiometría de rocas ígneas y metamórficas de México: Asociación Mexicanos Geológicos Petroleros Boletín, v. 38, p. 59–98.

López-Infanzón, M., and Nelson, S. A., 1990, Geology and K-Ar dating of the Sierra de Chiconquiaco–Palma Sola volcanics, central Veracruz, México: Geological Society of America Abstracts with Programs, v. 22, A165.

López-Ramos, E., 1972, Bosquejo geológico del basamento ígneo y metamórfico de la zonas norte y Poza Rica: Asociación Mexicanos Geológicos Petroleros Boletín, v. 13.

——, 1973, Estudio geológico de la Península de Yucatán: Asociación Mexicanos Geológicos Petroleros Boletín, v. 25, p. 23–76.

——, 1975, Geological summary of the Yucatán peninsula, *in* Nairn, A.E.M., and Stehli, F. G., eds., The ocean basins and margins: Volume 3: The Gulf of Mexico and the Caribbean: New York, Plenum Press, p. 257–282.

——, 1981 (1984), Paleogeografía y tectónica del Mesozoico en Mexico: Revista del Instituto de Geología, Universidad Nacional Autónoma de México, v. 5, p. 158–177.

——, 1983, Geología de México, Tomo III, edición 3a: Mexico, D.F., 454 p.

——, 1985, Geología de México, Tomo II, edición 3a, primera reimpresión: Mexico, D.F., 453 p.

Lothringer, C. J., 1984, Geology of a Lower Ordovician allochthon, Rancho San Marcos, Baja California, México, *in* Frizzell, V. A., Jr., ed., Geology of the Baja California Peninsula, Volume 39: Pacific Section, Society of Economic Paleontologists and Mineralogists, p. 17–22.

Lozáno-Romen, R. F., 1975, Evaluación petrolífera de la Península de Baja California: Asociación Mexicana de Geólogos Petroleros Boletín, v. 27, 329 p.

Luhr, J. F., and Carmichael, I.S.E., 1990, Geology of Volcán de Colima: Boletín del Instituto de Geología, Universidad Nacional Autónoma de México, v. 107, 101 p.

Luhr, J. F., Nelson, S. A., Allan, J. F., and Carmichael, I.S.E., 1985, Active rifting in southwestern México: Manifestations of an incipient eastward spreading-ridge jump: Geology, v. 13, p. 54–57.

Luhr, J. F., Aranda-Gómez, J. J., and Pier, J. G., 1989, Spinel-lherzolite-bearing Quaternary volcanic centers in San Luis Potosí, México: 1. Geology, mineralogy, and petrology: Journal of Geophysical Research, v. 94, p. 7916–7940.

——, 1991, Crustal contamination in early Basin and Range alkalic basalts: Petrology of the Los Encinos volcanic field, México: EOS Transactions of the American Geophysical Union, v. 72, no. 44, p. 560.

Luhr, J. F., Pier, J. G., and Aranda-Gómez, J. J., 1990, Geology and petrology of the late Neogene Los Encinos volcanic field of north-central México: Geological Society of America Abstracts with Programs, v. 22, A165.

Lund, S. P., and Bottjer, D. J., 1991, Paleomagnetic evidence for microplate tectonic development of southern and Baja California, *in* Dauphin, J. P., and Simoneit, B.R.T., eds., The Gulf and Peninsular Province of the Californias: American Association of Petroleum Geologists Memoir 47, p. 231–248.

Lund, S. P., Bottjer, D. J., Whidden, K. J., and Powers, J. E., 1991a, Paleomagnetic evidence for the timing of accretion of the Peninsular Ranges terrane in southern California: Geological Society of America Abstracts with Programs, v. 23, p. 74.

Lund, S. P., Bottjer, D. J., Whidden, K. J., Powers, J. E., and Steele, M. C., 1991b, Paleomagnetic evidence for Paleogene terrane displacements and accretion in southern California, *in* Abbott, P. L., and May, J. A., eds., Eocene geologic history, San Diego region, Volume 68: Pacific Section, Society of Economic Paleontologists and Mineralogists, p. 99–106.

Lyle, M., and Ness, G. E., 1991, The opening of the southern Gulf of California, *in* Dauphin, J. P., and Simoneit, B.R.T., eds., The Gulf and Peninsular Province of the Californias: American Association of Petroleum Geologists Memoir 47, p. 403–423.

Macdonald, A. J., Kreczmer, M. J., and Kesler, S. E., 1986, Vein, manto, and chimney mineralization at the Fresnillo silver-lead-zinc mine, México: Canadian Journal of Earth Sciences, v. 23, p. 1603–1614.

Macdonald, K. C., and Holcombe, T. L., 1978, Inversion of magnetic anomalies and seafloor spreading in the Cayman Trough: Earth and Planetary Science Letters, v. 40, p. 407–414.

Maher, D. J., Jones, N. W., McKee, J. W., and Anderson, T. H., 1991, Volcanic rocks at Sierra de Catorce, San Luis Potosí, México: A new piece for the Jurassic arc puzzle: Geological Society of America Abstracts with Programs, v. 23, A133.

Malfait, B. T., and Dinkelman, M. G., 1972, Circum-Caribbean tectonic and igneous activity and the evolution of the Caribbean plate: Geological Society of America Bulletin, v. 83, p. 251–272.

Malpica, C. R., 1972, Rocas marinas del Paleozoico tardío en el área de San José de Gracia, Sinaloa: Sociedad Geológica Mexicana, II Convención Nacional, Libro de Resúmenes, p. 174–175.

Marshall, R. H., 1984, Petrology of the subsurface Mesozoic rocks of the Yucatán Platform [M.S. thesis]: New Orleans, Louisiana, University of New Orleans, 96 p.

Martin, R. G., 1980, Distribution of salt structures in the Gulf of Mexico: Map and descriptive text: U.S. Geological Survey Miscellaneous Field Studies, Map MF—1213.

Martínez-Reyes, J., and Nieto-Samaniego, A. F., 1990, Efectos geológicos de la tectónica reciente en la parte central de México: Revista del Instituto de Geología, Universidad Nacional Autónoma de México, v. 9, p. 33–50.

Marvin, R. F., Mehnert, H. H., and Zartman, R. E., 1988, U.S. Geological Survey radiometric ages—compilation "C." Part 1, México, and Wisconsin, Michigan, and North and South Carolina: Isochron/West, v. 51, p. 3–4.

Mauger, R. L., McDowell, F. W., and Blount, J. G., 1983, Grenville-age Precambrian rocks of Los Filtros area, near Aldama, Chihuahua, México, *in* Clark, K. F., and Goodell, P. C., eds., Geology and mineral resources of north-central Chihuahua: El Paso Geological Society Guidebook, p. 165–168.

Maurrasse, F.J.-M.R., 1990, Stratigraphic correlation for the circum-Caribbean region, *in* Dengo, G., and Case, J. E., eds., Decade of North American Geology, Volume H: The Caribbean region: Geological Society of America, Plates 4, 5A, B.

May, S. R., and Butler, R. F., 1986, North American Jurassic apparent polar wander: Implications for plate motion, paleogeography, and Cordilleran tectonics: Journal of Geophysical Research, v. 91, p. 11519–11544.

Mayer-Pérez Rul, F. A., 1967, Hoja Viesca 13R-1(5), con resumen de la geología de la Hoja Viesca, Estados de Coahuila y Durango: Carta Geológica de México, serie de 1:100,000.

McBirney, A. R., 1963, Geology of a part of the central Guatemalan Cordillera: University of California Publications in the Geological Sciences, v. 38(4), p. 177–242.

McBirney, A. R., and Bass, M. N., 1969, Structural relations of pre-Mesozoic rocks of northern Central America: American Association of Petroleum Geologists Memoir 11, p. 269–280.

McBirney, A. R., Aoki, K. I., and Bass, M., 1967, Eclogite and jadeite from the Motagua fault zone, Guatemala: American Mineralogist, v. 52, p. 908–918.

McBride, E. F., Weidie, A. E., Jr., Wolleben, J. A., and Laudon, P. C., 1974, Stratigraphy and structure of the Parras and La Popa basins, northeastern México: Geological Society of America Bulletin, v. 85, p. 1603–1622.

McBride, S. L., Caelles, J. C., Clark, A. H., and Farrar, E., 1976, Paleozoic radiometric age provinces in the Andean basement, latitudes 25°-30°S: Earth and Planetary Science Letters, v. 29, p. 373–383.

McCabe, C., Van der Voo, R., and Urrutia-Fucugauchi, J., 1988, Late Paleozoic or early Mesozoic magnetizations in remagnetized Paleozoic rocks, State of Oaxaca, México: Earth and Planetary Science Letters, v. 91, p. 205–213.

McCloy, C., 1984, Stratigraphy and depositional history of the San José del Cabo Trough, Baja California Sur, México, *in* Frizzell, V. A., Jr., ed., Geology of the Baja California Peninsula, Volume 39: Pacific Section, Society of Economic Paleontologists and Mineralogists, p. 267–273.

McDowell, F. W., and Clabaugh, S., 1979, Ignimbrites of the Sierra Madre Occidental and their relation to the tectonic history of western México: Geological Society of America Special Paper 180, p. 113–124.

McDowell, F. W., and Keizer, R. P., 1977, Timing of mid-Tertiary volcanism in the Sierra Madre Occidental between Durango City and Mazatlan, México: Geological Society of America Bulletin, v. 88, p. 1479–1487.

McDowell, F. W., Wark, D. A., and Aguirre-Díaz, 1990, The Tertiary ignimbrite flare-up in western Mexico: Geological Society of America Abstracts with Programs, v. 22, p. 66.

McKee, J. W., Jones, N. W., and Anderson, T. H., 1988, Las Delicias basin: A record of late Paleozoic arc volcanism in northeastern México: Geology, v. 16, p. 37–40.

McKee, J. W., Jones, N. W., and Long, L. E., 1990, Stratigraphy and provenance of strata along the San Marcos fault, central Coahuila, México: Geological Society of America Bulletin, v. 102, p. 593–614.

McLean, H., 1988, Reconnaissance geologic map of the Loreto and part of the San Javier quadrangles, Baja California Sur, México: U.S. Geological Survey, Geologic Map MF-2000, scale 1:50,000.

McLean, H., Hausback, B. P., and Knapp, J. H., 1987, The geology of west-central Baja California Sur, México: U.S. Geological Survey Bulletin 1579, p. 1–16.

McMenamin, M.A.S., Pittenger, S. L., Carson, M. R., and Larrabee, E. M., 1993, Upper Precambrian–Cambrian faunal sequence, Sonora, México and Lower Cambrian fossils from New Jersey, United States, in Landing, E., ed., Festschrift, Studies in stratigraphy and paleontology in honor of Donald W. Fisher, New York State Museum Bulletin 481 (in press).

McWilliams, M. O., and Howell, D. G., 1982, Exotic terranes of western California: Nature, v. 297, p. 215–217.

Mead, R. D., Kesler, S. E., Foland, K. A., and Jones, L. M., 1988, Relationship of Sonoran tungsten mineralization to the metallogenic evolution of México: Economic Geology, v. 83, p. 1943–1965.

Meiburg, P., and 6 others, 1987, El basamento precretácico de Aramberri—estructura clava para comprender et decollement de la cubierta jurásica-cretácica de la Sierra Madre Oriental, México?: Actas de la Facultad de Ciencias de la Tierra, Universidad Autónoma de Nuevo León, Linares, v. 2, p. 15–22.

Mejía-Dautt, O., and many others, 1980, Evaluación geológica petrolera, plataforma Valles–San Luis Potosí: XI Excursión geológica, Petroleros Mexicanos, Superintendencia General de Distritos de Exploración Petrolera, Zona Norte, 123 p.

Mellor, E. I., and Breyer, J. A., 1981, Petrology of late Paleozoic basin-fill sandstones, north-central México: Geological Society of America Bulletin, v. 92, p. 367–373.

Menard, H. W., 1978, Fragmentation of the Farallon plate by pivoting subduction: Journal of Geology, v. 86, p. 99–110.

Meyer, R. P., Steinhart, S. S., and Woollard, G. P., 1961, Central plateau, México, Explosion studies of continental structure: Washington, D.C., Carnegie Institution of Washington Publication 622, p. 199–225.

Michalzik, D., 1991, Facies sequence of Triassic-Jurassic red beds in the Sierra Madre Oriental (NE México) and its relation to the early opening of the Gulf of Mexico: Sedimentary Geology, v. 71, p. 243–259.

Michaud, F., 1988, Apports de la micropaleontologie a la connaissance stratigraphique de la Formation San Ricardo (Callovien-Neocomien) état du Chiapas: Revista del Instituto de Geología, Universidad Nacional Autónoma de México, v. 7, p. 35–40.

Michaud, F., Bourgois, J., Barrier, E., and Fourcade, E., 1989, La série cretacée de Tecoman (État de Colima): Conséquences sur les rapports sturcturaux entre zones internes et zones de l'édifice montagneux mexicain: Comptes Rendus l'Academie de Sciences, v. 309, p. 587–593.

Mickus, K. L., and Keller, G. R., 1992, Lithospheric structure of the south-central United States: Geology, v. 20, p. 335–338.

Miller, R. H., and Dockum, M. S., 1983, Ordovician conodonts from metamorphosed carbonates of the Salton Trough, California: Geology, v. 11, p. 410–412.

Mills, R. A., and Hugh, K. E., 1974, Reconnaissance geologic map of Mosquitia region, Honduras, and Nicaragua Caribbean coasts: American Association of Petroleum Geologists Bulletin, v. 80, p. 189–207.

Mina, F., 1957, Bosquejo geológico del Territorio Sur de la Baja California: Asociación Mexicana de Geólogos Petroleros Boletín, v. 9, p. 139–270.

Minch, J. A., 1969, A depositional contact between the pre-batholithic Jurassic and Cretaceous rocks in Baja California, México: Geological Society of America Abstracts with Programs, v. 42–43.

Minch, J. C., Gastil, R. G., Finch, W., Robinson, J., and James, A. H., 1976, Geology of the Vizcaíno Peninsula, in Howell, D. G., ed., Aspects of the geologic history of the California continental borderland: Pacific Section, Society of Economic Paleontologists and Mineralogists Miscellaneous Publi-

cation 24, p. 136–195.

Mitre-Salazar, L.-M., 1989, La megafalla laramídica de San Tiburcio, Estado de Zacatecas: Revista del Instituto de Geología, Universidad Nacional Autónoma de México, v. 8, p. 47–51.

Mitre-Salazar, L.-M., and 8 others, 1991, H-1: Southern Baja California to Coahuila: Geological Society of America, Centennial Continent/Ocean Transect #13.

Mixon, R. B., Murray, G. E., and Díaz, G., 1959, Age and correlation of Huizachal Group (Mesozoic), state of Tamaulipas, México: American Association of Petroleum Geologists Bulletin, v. 43, p. 757–771.

Molina-Garza, R. S., Van der Voo, R., and Urrutia-Fucugauchi, J., 1992, Paleomagnetism of the Chiapas Massif, southern Mexico: Evidence for rotation of the Maya Block and implications for the opening of the Gulf of Mexico: Geological Society of America Bulletin, v. 104, p. 1156–1168.

Molnar, P., and Sykes, L. R., 1969, Tectonics of the Caribbean and Middle America regions from focal mechanisms and seismicity: Geological Society of America Bulletin, v. 80, p. 1639–1684.

Monger, J.W.H., Price, R. A., and Tempelman-Kluit, D. J., 1982, Tectonic accretion and origin of the two major metamorphic and plutonic welts in the Canadian Cordillera: Geology, v. 10, p. 70–75.

Monod, O., Lapierre, H., Chiodi, M., Martinez, J., Calvet, P., Ortiz, E., and Zimmermann, J.-L., 1990, Reconstitution d'un arc insulaire intra-océanique au Mexique central: La séquence volcano-plutonique de Guanajuato (Crétacé inférieur): Comptes Rendus l'Académie des Sciences, v. 310, p. 45–51.

Montigny, R., Demant, A., Delpretti, P., Piguet, P., and Cocheme, J. J., 1987, Chronologie K/A des sequences volcaniques du Nord de la Sierra Madre Occidental (Mexique): Comptes Rendus l'Academie des Sciences, v. 304, p. 987–992.

Moore, D. G., 1973, Plate-edge deformation and crustal growth, Gulf of California structural province: Geological Society of America Bulletin, v. 84, p. 1883–1906.

Moore, G. M., and Carmichael, I.S.E., 1991, Late Miocene basaltic volcanism in the Guadalajara area, Jalisco, México, and its relationship to the opening of the Gulf of California: Geological Society of America Abstracts with Programs, v. 23, A332.

Moore, T. E., 1985, Stratigraphy and tectonic significance of the Mesozoic tectonostratigraphic terranes of the Vizcaíno Peninsula, Baja California Sur, México, in Howell, D. G., ed., Tectonostratigraphic terranes of the Circum-Pacific region: Circum-Pacific Council for Energy and Mineral Resources, Earth Science Series, no. 1, p. 315–329.

——, 1986, Petrology and tectonic implications of the blueschist-bearing Puerto Nuevo mélange complex, Vizcaíno Peninsula, Baja California Sur, México, in Evans, B. W., and Brown, E. H., eds., Blueschists and Eclogites: Geological Society of America Memoir 164, p. 43–58.

Mooser, F., Nairn, A.E.M., and Negendank, J.F.W., 1974, Palaeomagnetic investigations of the Tertiary and Quaternary rocks; 8, A palaeomagnetic and petrologic study of volcanics of the Valley of México: Geologische Rundschau, v. 63, p. 451–483.

Morán-Zenteno, D. J., 1992, Investigaciones isotopicas de Rb-Sr y Sm-Nd en rocas cristalinas de la región de Tierra Colorada–Acapulco–Cruz Grande, Estado de Guerrero [Ph.D. dissertation]: México City, Universidad Nacional Autónoma de México, 186 p.

Morán-Zenteno, D. J., Urrutia-Fucugauchi, J., Bohnel, H., and González-Torres, E., 1988, Paleomagnetismo de rocas jurásicas del norte de Oaxaca y sus implicaciones tectónicas: Geofísica Internacional, v. 27, p. 485–518.

Morán-Zenteno, D. J., Köhler, H., Von Drach, V., and Schaaf, P., 1990a, The geological evolution of Xolapa terrane, southern México, as inferred from Rb-Sr and Sm-Nd isotopic data: Munich, Ludwig-Maximilians-Universität, München Geowissenschaftliches Lateinamerika-Kolloquium, 21.11.90–23.11.90.

Morán-Zenteno, D. J., Urrutia-Fucugauchi, J., and Köhler, H., 1990b, Nuevos fechamientos de Rb-Sr en rocas cristalinas del Complejo Xolapa en el estado de Guerrero: Sociedad Geológica Mexicana, X Convención Nacional, Libro de Resúmenes, p. 92–93.

Morán-Zenteno, D. J., Urrutia-Fucugauchi, J., and Köhler, H., 1991, Rb-Sr geochronology and Sr-Nd systematics of crystalline rocks from the Xolapa terrane, southern México: Geological Society of America Abstracts with Programs, v. 23, A136.

Mora, C. I., Valley, J. W., and Ortega-Gutiérrez, F., 1986, The temperature and pressure conditions of Grenville-age granulite-facies metamorphism of the Oaxacan complex, southern México: Revista del Instituto de Geología, Universidad Nacional Autónoma de México, v. 6, p. 222–242.

Morales, M., Bartolini, C., Damon, P., and Shafiqullah, M., 1990, K-Ar age dating, stratigraphy and extensional deformation of Sierra Lista Blanca, central Sonora, México: Geological Society of America Abstracts with Programs, v. 22, A364.

Moran, A. I., 1976, Allochthonous carbonate debris in Mesozoic flysch deposits in Santa Ana Mountains, California: American Association of Petroleum Geologists Bulletin, v. 60, p. 2038–2043.

Morris, L. K., Lund, S. P., and Bottjer, D. J., 1986, Paleolatitude drift history of displaced terranes in southern and Baja California: Nature, v. 321, p. 844–847.

Mossman, R. W., and Viniegra-Osorio, F., 1976, Complex fault structures in Veracruz province of México: American Association of Petroleum Geologists Bulletin, v. 60, p. 379–388.

Muehlberger, W. R., and Ritchie, A. W., 1975, Caribbean-American plate boundary in Guatemala and southern México as seen on Skylab IV orbital photography: Geology, v. 3, p. 232–235.

Mueller, K. J., and Rockwell, T. K., 1991, Late Quaternary structural evolution of the western margin of the Sierra Cucapa, northern Baja California, *in* Dauphin, J. P., and Simoneit, B.R.T., eds., The Gulf and Peninsular Province of the Californias: American Association of Petroleum Geologists Memoir 47, p. 249–260.

Mullan, H. S., 1978, Evolution of the Nevadan orogen in northwestern México: Geological Society of America Bulletin, v. 89, p. 1175–1188.

Muller, P. D., 1980, Geology of the Los Amates quadrangle and vicinity, Guatemala [Ph.D. dissertation]: Binghamton, State University of New York, 326 p.

Murchey, B. L., 1990, Radiolarian biostratigraphy of Paleozoic siliceous sedimentary rocks in central Sonora, México: Geological Society of America Abstracts with Programs, v. 22, p. 71.

Murillo-Muñetón, G., 1991, Análisis petrológico y edades K-Ar de las rocas metamórficas e ígneas precenozoicas de la región de La Paz-Los Cabos, B.C.S., México: First International Meeting on Geology of the Baja California Peninsula, p. 55–56.

Murray, G. E., Weidie, A. E., Jr., Boyd, D. R., Forde, R. H., and Lewis, P. D., Jr., 1962, Formational division of Difunta Group, Parras basin, Coahuila and Nuevo León, México: American Association of Petroleum Geologists Bulletin, v. 46, p. 374–383.

Nairn, A.E.M., 1976, A paleomagnetic study of certain Mesozoic formations in northern México: Physics of the Earth and Planetary Interiors, v. 13, p. 47–56.

Nairn, A.E.M., Negendank, J.F.W., Noltimier, H. C., and Schmitt, T. J., 1975, Paleomagnetic investigations of the Tertiary and Quaternary igneous rocks, X: The ignimbrites and lava units west of Durango, México: Jahrbuch für Geologie und Mineralogie, Monatshefte, v. 11, p. 664–678.

Nava, F., and others, 1988, Structure of the Middle America trench in Oaxaca, México: Tectonophysics, v. 154, p. 241–251.

Navarro-G., A., and Tovar-R., J., 1974, Stratigraphy and tectonics of the state of Chihuahua, Geologic field trip guidebook through the state of Chihuahua and Sonora, West Texas Geological Society Publication 74-63, p. 87–91.

Nelson, S. A., 1990, Volcanic hazards in México: A summary: Revista del Instituto de Geología, Universidad Nacional Autónoma de México, v. 9, p. 71–81.

Nelson, S. A., and Livieres, R. A., 1986, Contemporaneous calc-alkaline and alkaline volcanism at Sanganguey Volcano, Nayarit, México: Geological Society of America Bulletin, v. 97, p. 798–808.

Nelson, S. A., González-Caver, E., and Kyser, T. K., 1991, Constraints on the origin of Late Miocene to Recent alkaline and calc-alkaline magmas from the Tuxtla volcanic field, Veracruz, México: Geological Society of America Abstracts with Programs, v. 23, A332.

Ness, G. E., and Lyle, M. W., 1991, A seismo-tectonic map of the Gulf and Peninsular Province of the Californias, *in* Dauphin, J. P., and Simoneit, B.R.T., eds., The Gulf and Peninsular Province of the Californias: American Association of Petroleum Geologists Memoir 47, p. 71–77.

Ness, G. E., Lyle, M. W., and Couch, R. W., 1991, Marine magnetic anomalies and oceanic crustal isochrons of the Gulf and Peninsular Province of the Californias, *in* Dauphin, J. P., and Simoneit, B.R.T., eds., The Gulf and Peninsular Province of the Californias: American Association of Petroleum Geologists Memoir 47, p. 47–69.

Neuhaus, J. R., Cassidy, M., Krummenacher, D., and Gastil, R. G., 1988, Timing of protogulf extension and transtensional rifting through volcanic/sedimentary stratigraphy of SW Isla Tiburón, Gulf of California, Sonora, México: Geological Society of America Abstracts with Programs, v. 20, p. 218.

Nick, K., 1988, Mineralogische, geochemische und petrographische Untersuchungen in der Sierra de San Carlos (Mexiko) [Ph.D. dissertation]: Karlsruhe, Germany, Universität Karlsruhe, 167 p.

Nixon, G. T., 1982, The relationship between Quaternary volcanism in central México and the seismicity and structure of subducted coeanic lithosphere: Geological Society of America Bulletin, v. 93, p. 514–523.

——, 1989, The geology of Iztaccíhuatl volcano and adjacent areas of the Sierra Nevada and Valley of México: Geological Society of America Special Paper 219, 58 p.

Nixon, G. T., Demant, A., Armstrong, R. L., and Harakal, J. E., 1987, K-Ar and geologic data bearing on the age and evolution of the Trans-Mexican Volcanic Belt: Geofísica Internacional, v. 26, p. 109–158.

Normark, W. R., Spencer, J. E., and Ingle, J. C., Jr., 1987, Geology and Neogene history of the Pacific continental margin of Baja California Sur, México, *in* Scholl, D. W., Grantz, A., and Vedder, J. G., eds., Geology and resource potential of the continental margin of western North America and adjacent ocean basins—Beaufort Sea to Baja California, Volume 6: Circum-Pacific Council for Energy and Mineral Resources, Earth Science Series, p. 449–472.

Nourse, J. A., 1990, Tectonostratigraphic development and strain history of the Magdalena metamorphic core complex, northern Sonora, México, *in* Gehrels, G. E., and Spencer, J. E., eds., Geologic excursions through the Sonoran Desert region, Arizona and Sonora: Arizona Geological Survey Special Paper 7, p. 155–164.

Nowicki, M. J., Hall, S. A., and Evans, I., 1990, Paleomagnetic evidence for local and regional post-Eocene rotations in northern México: EOS Transactions of the American Geophysical Union, v. 71, p. 491.

Obregón-Andria, L., and Arriaga-Arredondo, F., 1991, Coal in Sonora, *in* Pérez-Segura, E., and Jacques-Ayala, C., eds., Studies of Sonoran geology: Geological Society of America Special Paper 254, p. 121–130.

Ortega-Guerrero, B., and Urrutia-Fucugauchi, J., 1989, Paleogeography and tectonics of the Mixteca terrane, southern México, during the interval of drifting between North and South America and Gulf of Mexico rifting: EOS Transactions of the American Geophysical Union, v. 70, p. 1314.

Ortega-Gutiérrez, F., 1974 (1975), Nota preliminar sobre las eclogitas de Acatlán, Puebla: Sociedad Geológica Mexicana Boletín, v. 35, p. 1–6.

——, 1978a, Estratigrafía del Complejo Acatlán en la Mixteca Baja, Estados de Puebla y Oaxaca: Revista del Instituto de Geología, Universidad Nacional Autónoma de México, v. 2, p. 112–131.

——, 1978b, Geología del contacto entre la Formación Acatlán Paleozoico y el Complejo Oaxaqueno Precámbrico, al oriente de Acatlán, Estado de Puebla: Sociedad Geológica Mexicana Boletín, v. 39, p. 27–28.

——, 1978c, El Gneis Novillo y rocas metamórficas asociadas en los cañones del Novillo y de La Peregrina, área de Ciudad Victoria, Tamaulipas: Revista del Instituto de Geología, Universidad Nacional Autónoma de México, v. 2, p. 19–30.

——, 1979, The tectonothermic evolution of the Paleozoic Acatlán Complex of southern México: Geological Society of America Abstracts with Programs,

v. 11, p. 490.

——, 1980, Algunas rocas miloníticas de México y su significado tectonico: Sociedad Geológica Mexicana, V Convención Nacional, Libro de Resúmenes, p. 99–100.

——, 1981a, Metamorphic belts of southern México and their tectonic significance: Geofísica Internacional, v. 20, p. 177–202.

——, 1981b (1984), La evolución tectónica premisisípica del sur de México: Revista del Instituto de Geología, Universidad Nacional Autónoma de México, v. 5, p. 140–157.

——, 1982, Evolución magmática y metamórfica del complejo cristalino de La Paz, Baja California Sur: Sociedad Geológica Mexicana, VI Convención Nacional, Libro de Resúmenes, p. 90.

——, 1984a, Evidence of Precambrian evaporites in the Oaxacan granulite complex of southern México: Precambrian Research, v. 23, p. 377–393.

——, 1984b, Relaciones estratigráficas del basamento pre-oxfordiano de la región de Caopas-Rodeo, Zacatecas, y su significado tectónico: Sociedad Geológica Mexicana, VII Convención Nacional, Libro de Resúmenes, p. 56–57.

Ortega-Gutiérrez, F., and González-Arreola, C., 1985, Una edad cretácica de las rocas sedimentárias deformadas de la Sierra de Juárez, Oaxaca: Revista del Instituto de Geología, Universidad Nacional Autónoma de México, v. 6, p. 100–101.

Ortega-Gutiérrez, F., Anderson, T. H., and Silver, L. T., 1977, Lithologies and geochronology of the Precambrian craton of southern México: Geological Society of America Abstracts with Programs, v. 9, p. 1121–1122.

Ortega-Gutiérrez, F., Prieto-Vélez, R., Zúñiga, Y., and Flores, S., 1979, Una secuencia volcano-plutónico-sedimentária cretácica en el norte de Sinaloa: un complejo ofiolítico?: Revista del Instituto de Geológica, Universidad Nacional Autónoma de México, v. 3, p. 1–8.

Ortega-Gutiérrez, F., Mitre-Salazar, L.-M., Roldán-Quintana, J., Sándrez-Rubio, G., and de la Fuente, M., 1990, H-3: Middle America Trench–Oaxaca–Gulf of Mexico: Geological Society of America, Centennial Continent/Ocean Transect #14.

Ortiz, E., Yta, M., Talavera, O., Lapierre, H., Monod, O., and Tardy, M., 1991, Origine intra-oceanique des formations volcano-plutoniques d'arc du Jurassique superieur-Cretace inferieur du Mexique centro-meridional: Paris, Comptes Rendus l'Académie des Sciences, Serie II, v. 312, p. 399–406.

Ortlieb, L., 1991, Quaternary vertical movements along the coasts of Baja California and Sonora, in Dauphin, J. P., and Simoneit, B.R.T., eds., The Gulf and Peninsular Province of the Californias: American Association of Petroleum Geologists Memoir 47, p. 447–480.

Ortlieb, L., Ruegg, J. C., Angelier, J., Colletta, B., Kasser, M., and Lesage, P., 1989, Geodetic and tectonic analyses along an active plate boundary: The central Gulf of California: Tectonics, v. 8, p. 429–442.

Pacheco-G., C., and Barba, M., 1986, El precámbrico de Chiapas, un terrano estratotectónico: Sociedad Geológica Mexicana, VIII Convención Nacional, Libro de Resúmenes.

Pacheco-G., C., Castro-M., R., and A.-Gómez, M., 1984, Confluencia de terrenos estratotectónicos en Santa María del Oro, Durango, México: Revista del Instituto Mexicano del Petróleo, v. 16(1), p. 7–20.

Padilla y Sánchez, R. J., 1985, Las estructuras de la Curvature de Monterrey, Estados de Coahuila, Nuevo León, Zacatecas y San Luis Potosí: Instituto de Geología, Universidad Nacional Autónoma de México, Revista, v. 6, p. 1–20.

——, 1986, Post-Paleozoic tectonics of northeast México and its role in the evolution of the Gulf of Mexico: Geofísica Internacional, v. 25, p. 157–206.

Page, B. M., and Engebretson, D. C., 1984, Correlation between the geologic record and computed plate motions for central California: Tectonics, v. 3, p. 133–155.

Paine, W. R., and Meyerhoff, A. A., 1970, Gulf of Mexico basin: Interactions among tectonics, sedimentation, and hydrocarbon accumulation: Transactions of the Gulf Coast Geological Society, v. 20, p. 5–44.

Palmer, A. R., 1983, compiler, The Decade of North American Geology 1983 geologic time scale: Geology, v. 11, p. 503–504.

Palmer, A. R., DeMis, W. D., Muehlberger, W. R., and Robison, R. A., 1984,

Geologic implications of Middle Cambrian boulders from the Haymond Formation (Pennsylvanian) in the Marathon basin, west Texas: Geology, v. 12, p. 91–94.

Pantoja-Alor, J., 1963, Hoja Viesca 13R-k(3), con resumen de la geología de la Hoja San Pedro del Gallo, Estado de Durango: Carta Geológica de México, serie de 1:100,000.

——, 1983, Geocronometría de magmatismo cretácico-terciario de la Sierra Madre del Sur: Sociedad Geológica Mexicana, XIV Convención Nacional, Libro de Resúmenes, p. 29.

——, 1988, Petrología e implicaciones tectónicas del evento magmatico Balsas (Paleogene) de la Sierra Madre del Sur: Unión Geofísica Mexicana, Reunión Anual, Resúmenes.

Pantoja-Alor, J., and Estrada-Barraza, S., 1986, Estratigrafía de los alrededores de la mima de fierro de El Encino, Jalisco: Sociedad Geológica Mexicana Boletín, v. 47, p. 1–16.

Pantoja-Alor, J., and Robison, R., 1967, Paleozoic sedimentary rocks in Oaxaca, México: Science, v. 157, p. 1033–1035.

Pantoja-Alor, J., Rincón-Orta, C., Fries, C., Jr., Silver, L. T., and Solorio-Munguía, J., 1974, Contribuciones a la geocronología de Chiapas: Asociación Mexicanos Geológicos Petroleros Boletín, v. 26, p. 207–223.

Pasquaré, G., Ferrari, L., Perazzoli, V., Tiberi, M., and Turchetti, F., 1987, Morphological and structural analysis of the central sector of the Transmexican volcanic belt: Geofísica Internacional, v. 26, p. 177–193.

Pasquaré, G., Garduño, V. H., Tibaldi, A., and Ferrari, M., 1988, Stress pattern evolution in the central sector of the Mexican Volcanic Belt: Tectonophysics, v. 146, p. 353–364.

Pasquaré, G., Ferrari, L., Garduño, V. H., Tibaldi, A., and Vezzoli, L., 1991, Geologic map of the central sector of the Mexican Volcanic Belt, states of Guanajuato and Michoacán, México: Geological Society of America Map and Chart Series MCH072.

Patchett, P. J., and Ruiz, J., 1987, Nd isotopic ages of crust formation and metamorphism in the Precambrian of eastern and southern México: Contributions to Mineralogy and Petrology, v. 96, p. 523–528.

Patterson, D. L., 1984, Paleomagnetism of the Valle Formation and the Late Cretaceous paleogeography of the Vizcaíno Basin, Baja California, México, in Frizzell, V. A., Jr., ed., Geology of the Baja California Peninsula, Volume 39: Pacific Section, Society of Economic Paleontologists and Mineralogists, p. 173–182.

Penfield, G. T., and Camargo-Z., A., 1981, Definition of a major igneous zone in the central Yucatán platform with aeromagnetics and gravity (abs.): 51st annual meeting, Los Angeles, Society of Exploration Geophysicists, p. 448–449.

Pérez-Ramos, O., 1978, Estudio biostratigráfico del paleozoico superior del Anticlinorio de Huayacocotla en la Sierra Madre Oriental: Sociedad Geológica Mexicana Boletín, v. 39, p. 126–135.

Pessagno, E. A., Jr., Finch, J. W., and Abbott, P. L., 1979, Upper Triassic Radiolaria from the San Hipolito Formation, Baja California: Micropaleontology, v. 25, p. 160–197.

Phillips, J. R., 1984, "Middle" Cretaceous metasedimentary rocks of La Olvidada, northeastern Baja California, México, in Frizzell, V. A., Jr., ed., Geology of the Baja California Peninsula, Pacific Section, Society of Economic Paleontologists and Mineralogists, p. 37–41.

Pindell, J., 1985, Alleghenian reconstruction and subsequent evolution of the Gulf of Mexico, Bahamas, and proto-Caribbean: Tectonics, v. 4, p. 1–39.

Pindell, J., and Dewey, J. F., 1982, Permo-Triassic reconstruction of western Pangaea and the evolution of the Gulf of Mexico/Caribbean region: Tectonics, v. 1, p. 179–212.

Pindell, J. L., and Barrett, S. F., 1990, Geological evolution of the Caribbean region; A plate-tectonic perspective, in Dengo, G., and Case, J. E., eds., Decade of North American Geology, Volume H: The Caribbean region: Geological Society of America, p. 405–432.

Pindell, J. L., and 6 others, 1988, A plate-kinematic framework for models of Caribbean evolution: Tectonophysics, v. 155, p. 121–138.

Plafker, G., 1976, Tectonic aspects of the 4 February 1976 Guatemala earthquake—A preliminary evaluation: Science, v. 193, p. 1201–1208.

Ponce, B. F., and Clark, K. F., 1988, The Zacatecas mining district: A Tertiary caldera complex associated with precious and base metal mineralization: Economic Geology, v. 83, p. 1668–1682.

Poole, F. G., and Madrid, R. J., 1986, Paleozoic rocks in Sonora (México) and their relation to the southwestern continental margin of North America: Geological Society of America Abstracts with Programs, v. 18, p. 720–721.

——, 1988, Comparison of allochthonous Paleozoic eugeoclinal rocks in the Sonoran, Marathon, and Antler orogens: Geological Society of America Abstracts with Programs, v. 20, p. A267.

Poole, F. G., Murchey, B. L., and Jones, D. L., 1983, Bedded barite deposits of Middle and Late Paleozoic age in central Sonora, México: Geological Society of America Abstracts with Programs, v. 15, p. 299.

Poole, F. G., Madrid, R. J., and Moráles-Ramírez, J. M., 1990, Sonoran orogen in the Barita de Sonora mine area, central Sonora, México: Geological Society of America Abstracts with Programs, v. 22, p. 76.

Priem, H.N.A., Boelrijk, N.A.I.M., Hebeda, E. H., and Kroonenberg, S. B., 1989, Rb-Sr and K-Ar evidence for the presence of a 1.6 Ga basement underlying the 1.2 Ga Garzon-Santa Marta granulite belt in the Colombian Andes: Precambrian Research, v. 42, p. 315–324.

Prol-Ledesma, R. M., and Juárez, M. G., 1986, Geothermal map of México: Journal of Volcanology and Geothermal Research, v. 28, p. 351–361.

Pubellier, M., and Rangin, C., 1987, Mise en évidence d'une phase cénomano-turonienne en Sonora central (Mexique). Conséquences sur les relations structurales entre domaine cordillérain et domaine téthysien: Comptes Rendus l'Académie des Sciences, Serie II, v. 305, p. 1093–1098.

Quintero-Legorreta, O., and Aranda-García, M., 1985, Relaciones estructurales entre el Anticlinorio de Parras y el Anticlinorio de Arteaga (Sierra Madre Oriental), en la región de Agua Nueva, Coahuila: Instituto de Geología, Universidad Nacional Autónoma de México, Revista, v. 6, p. 21–36.

Quintero-Legorreta, O., and Guerrero, J. C., 1985, Una nueva localidad del basamento precámbrico de Chihuahua en el área de Carrizalillo: Revista del Instituto de Geológica, Universidad Nacional Autónoma de México, v. 6, p. 98–99.

Radelli, L., 1989, The ophiolites of Calmalli and the Olvidada nappe of northern Baja California and west-central Sonora, México, *in* Abbott, P. L., ed., Geologic studies in Baja California, Volume 63: Pacific Section, Society of Economic Paleontologists and Mineralogists, p. 79–85.

Radelli, L., and many others, 1987, Allochthonous Paleozoic bodies of central Sonora: Departamento de Geología—UNISON (Universidad de Sonora) Boletín, v. 4(1 & 2), p. 1–15.

Ramírez-M., J. C., and Acevedo-C., F., 1957, Notas sobre la geología de Chihuahua: Asociación Mexicanos Geológicos Petroleros Boletín, v. 9, p. 583–770.

Rangin, C., 1978, Speculative model of Mesozoic geodynamics, central Baja California to northeastern Sonora (México), *in* Howell, D. G., and McDougall, K. A., eds., Mesozoic Paleogeography of the western United States: Pacific Section, Society of Economic Paleontologists and Mineralogists, p. 85–106.

Rangin, C., and Carrillo, M., 1978, Le complexe ophiolitique à affinité de îles Margarita-Magdalena (Baja California méridionale): Une croûte paléo-océanique obductée: Compte Rendus Sommaire de la Société Géologique de France, fasc. 2, p. 55–58.

Rankin, D. W., 1975, The continental margin of eastern North America in the southern Appalachians: The opening and closing of the Proto-Atlantic Ocean: American Journal of Science, v. 275-A, p. 298–336.

Ranson, W. A., Fernández, L. A., Simmons, W. B., Jr., and Enciso de la Vega, S., 1982, Petrology of the metamorphic rocks of Zacatecas, Zac., México: Sociedad Geológica Mexicana Boletín, v. 43(1), p. 37–60.

Ratschbacher, L., Riller, U., Meschede, M., Herrmann, U., and Frisch, W., 1991, Second look at suspect terranes in southern México: Geology, v. 19, p. 1233–1236.

Reed, J. B., 1989, Lower Triassic(?) strata in the Vallecitos–San Marcos area, Baja California, *in* Abbott, P. L., ed., Geologic studies in Baja California, Volume 63: Pacific Section, Society of Economic Paleontologists and Min-

eralogists, p. 87–102.

Renne, P. R., and 6 others, 1989, ^{40}Ar/^{39}Ar and U-Pb evidence for Late Proterozoic (Grenville-age) continental crust in north-central Cuba and regional tectonic implications: Precambrian Research, v. 42, p. 325–341.

Restrepo, J. J., and Toussaint, J. F., 1988, Terranes and continental accretion in the Colombian Andes: Episodes, v. 11, p. 189–193.

Reynolds, J. H., 1980, Late Tertiary volcanic stratigraphy of northern Central America: Bulletin Volcanologique, v. 45, p. 601–608.

Reynolds, S. J., Spencer, J. E., Richard, S. M., and Laubach, S. E., 1986, Mesozoic structures in west-central Arizona: Arizona Geological Society Digest, v. 16, p. 35–51.

Ritchie, A. W., and Finch, R. C., 1985, Widespread Jurassic strata on the Chortis block of the Caribbean plate: Geological Society of America Abstracts with Programs, v. 17, p. 700–701.

Ritchie, A. W., and McDowell, F. W., 1979, K-Ar ages of plutonic and volcanic rocks from the volcanic highlands of Guatemala northwest of Guatemala City: Isochron/West, v. 25, p. 3–4.

Roberts, R. J., Hotz, P. E., Gilluly, J., and Ferguson, H. G., 1958, Paleozoic rocks of north-central Nevada: American Association of Petroleum Geologists Bulletin, v. 42, p. 2813–2857.

Robin, C., 1982, México, *in* Thorpe, R. S., ed., Andesites, orogenic andesites, and related rocks: New York, Wiley, p. 137–147.

Robinson, J. W., 1975, Reconnaissance geology of the northern Vizcaíno Peninsula, Baja California Sur, México [M.S. thesis]: San Diego, California, San Diego State University, 114 p.

Robinson, K. L., 1991, U-Pb zircon geochronology of basement terranes and the tectonic evolution of southwestern mainland México [M.S. thesis]: San Diego, California, San Diego State University, 190 p.

Robinson, K. L., Gastil, R. G., Campa-Uranga, M. F., and Ramirez, J., 1989, Geochronology of basement and metasedimentary rocks in southern México and their relation to metasedimentary rocks in peninsular California: Geological Society of America Abstracts with Programs, v. 21, p. 135.

Robinson, K. L., Gastil, R. G., and Girty, M. S., 1990, Eocene intra-arc transtension: The detachment of the Chortis block from southwestern México: Geological Society of America Abstracts with Programs, v. 22, p. 78.

Robison, R., and Pantoja-Alor, J., 1968, Tremadocian trilobites from the Nochixtlán region, Oaxaca, México: Journal of Paleontology, v. 42, p. 767–800.

Rodríguez-Castañeda, J. L., 1988, Estatigrafía de la región de Tuape, Sonora: Revista del Instituto de Geología, Universidad Nacional Autónoma de México, v. 7, p. 52–66.

——, 1990, Relaciones estructurales en la parte centroseptentrional del Estado de Sonora: Revista del Instituto de Geología, Universidad Nacional Autónoma de México, v. 9, p. 51–61.

Roldán-Quintana, J., 1982, Evolución tectónica del Estado de Sonora: Revista del Instituto de Geología, Universidad Nacional Autónoma de México, v. 5, p. 178–185.

——, 1989, Geología de la Hoja Baviácora, Sonora: Revista del Instituto de Geología, Universidad Nacional Autónoma de México, v. 8, p. 1–14.

——, 1991, Geology and chemical composition of the Jaralito and Aconchi batholiths in east-central Sonora, México, *in* Pérez-Segura, E., and Jacques-Ayala, C., eds., Studies of Sonoran geology: Geological Society of America Special Paper 254, p. 69–80.

Roldán-Quintana, J., and González-León, C., 1979, Brief summary of the geology of the Cretaceous and Tertiary in the State of Sonora, México, *in* Anderson, T. H., and Roldán-Quintana, J., eds., Geology of northern Sonora: Geological Society of America, Field Trip Guidebook, p. 49–58.

Roper, P. J., 1978, Stratigraphy of the Chuacús Group on the south side of the Sierra de las Minas range, Guatemala: Geologie en Mijnbouw, v. 57, p. 309–313.

Rosencrantz, E., 1990, Structure and tectonics of the Yucatán basin, Caribbean Sea, as determined from seismic reflection studies: Tectonics, v. 9, p. 1037–1059.

Rosencrantz, E., and Mann, P., 1991, SeaMarc II mapping of transform faults in the Cayman Trough, Caribbean Sea: Geology, v. 19, p. 690–693.

Rosencrantz, E., and Sclater, J. G., 1986, Depth and age in the Cayman Trough: Earth and Planetary Science Letters, v. 79, p. 133–144.

Rosencrantz, E., Ross, M. I., and Sclater, J. G., 1988, Age and spreading history of the Cayman Trough as determined from depth, heat flow, and magnetic anomalies: Journal of Geophysical Research, v. 93, p. 2141–2157.

Rosenfeld, J. H., 1981, Geology of the western Sierra de Santa Cruz, Guatemala, Central America; An ophiolite sequence [Ph.D. dissertation]: Binghamton, State University of New York, 315 p.

Ross, C. A., 1979, Late Paleozoic collision of North and South America: Geology, v. 7, p. 41–44.

—— , 1986, Paleozoic evolution of southern margin of Permian basin: Geological Society of America Bulletin, v. 97, p. 536–554.

Ross, C. A., and Ross, J.R.P., 1985, Paleozoic tectonics and sedimentation in west Texas, southern New Mexico, and southern Arizona, in Dickerson, P. W., and Muehlberger, W. R., eds., Structure and tectonics of Trans-Pecos Texas: West Texas Geological Society Publication 85-81, p. 221–230.

Rowley, D. B., and Pindell, J. L., 1989, End Paleozoic–Early Mesozoic western Pangean reconstruction and its implications for the distribution of Precambrian and Paleozoic rocks around Meso-America: Precambrian Research, v. 42, p. 411–444.

Rudnick, R. L., and Cameron, K. L., 1991, Age diversity of the deep crust in northern México: Geology, v. 19, p. 1197–1200.

Ruiz, J., Patchett, P. J., and Arculus, R. J., 1988a, Reply to "Comments on Nd-Sr isotopic composition of lower crustal xenoliths—Evidence for the origin of mid-Tertiary felsic volcanics in México; by Cameron, K. L., and Robinson, J. V.: Contributions to Mineralogy and Petrology, v. 104, p. 615–618.

Ruiz, J., Patchett, P. J., and Ortega-Gutiérrez, F., 1988b, Proterozoic and Phanerozoic basement terranes of México from Nd isotopic studies: Geological Society of America Bulletin, v. 100, p. 274–281.

—— , 1990, Proterozoic and Phanerozoic terranes of México based on Nd, Sr and Pb isotopes: Geological Society of America Abstracts with Programs, v. 22, A113–114.

Ruiz-Castellanos, M., 1979, Rubidium-strontium geochronology of the Oaxaca and Acatlán metamorphic areas of southern México [Ph.D. dissertation]: Dallas, University of Texas, 188 p.

Salas, G. P., ed., 1991, Economic geology of México, in Bally, A. W., and Palmer, A. R., eds., Decade of North American Geology, Volume P-3: Economic geology of México: Geological Society of America, 448 p.

Salvador, A., 1987, Late Triassic–Jurassic paleogeography and origin of Gulf of Mexico basin: American Association of Petroleum Geologists Bulletin, v. 71, p. 419–451.

Salvador, A., and Green, A. G., 1980, Opening of the Caribbean Tethys, in Auboin, J., ed., Geology of the Alpine chains born of Tethys, Bureau de Recherches Geologiques et Minieres Memoire 115, p. 224–229.

Santillán, M., and Barrera, T., 1930, Las posibilidades petrolíferas en la costa occidental de la Baja California, entre los paralelos 30 y 32 de latitud norte: Anales del Instituto de Geología, v. 5, p. 1–37.

Savage, J. C., 1983, Strain accumulation in western United States: Annual Reviews of Earth and Planetary Sciences, v. 11, p. 11–43.

Sawlan, M. G., 1991, Magmatic evolution of the Gulf of California rift, in Dauphin, J. P., and Simoneit, B.R.T., eds., The Gulf and Peninsular Province of the Californias: American Association of Petroleum Geologists Memoir 47, p. 301–369.

Sawlan, M. G., and Smith, J. G., 1984, Petrologic characteristics, age and tectonic setting of Neogene volcanic rocks in northern Baja California Sur, México, in Frizzell, V. A., Jr., ed., Geology of the Baja California Peninsula, Volume 39: Pacific Section, Society of Economic Paleontologist and Mineralogists, p. 237–251.

Sawyer, D. S., Buffler, R. T., and Pilger, R. H., Jr., 1991, The crust under the Gulf of Mexico basin, in Salvador, A., ed., Decade of North American Geology, Volume J: The Gulf of Mexico Basin: Geological Society of America, p. 53–72.

Schaaf, P., Köhler, H., Muller-Sohnius, D., von Drach, V., and Frank, M. M., 1991, Nd and Sr isotopic data on W-Mexican granitoids: Evidence for different magma sources and implications for tectogenesis: EOS Transactions of the American Geophysical Union, v. 72, no. 44, p. 560.

Schellhorn, R. W., Aiken, C.L.V., and de la Fuente, M. F., 1991, Bouguer gravity anomalies and crustal structure in northwestern México, in Dauphin, J. P., and Simoneit, B.R.T., eds., The Gulf and Peninsular Province of the Californias: American Association of Petroleum Geologists Memoir 47, p. 197–215.

Scheubel, F. R., Clark, K. F., and Porter, E. W., 1988, Geology, tectonic environment, and structural controls in the San Martín de Bolaños district, Jalisco, México: Economic Geology, v. 1988, p. 1703–1720.

Schlager, W., and others, 1984, Deep Sea Drilling Project, Leg 77, southeastern Gulf of Mexico: Geological Society of America Bulletin, v. 95, p. 226–236.

Schmidt, E. K., 1975, Plate tectonics, volcanic petrology, and ore formations in the Santa Rosalía area, Baja California, México [M.S. thesis]: Tucson, University of Arizona, 197 p.

Schmidt-Effing, R., 1980, The Huayacocotla aulacogen in México (Lower Jurassic) and the origin of the Gulf of Mexico, in Pilger, R. H., Jr., ed., The origin of the Gulf of Mexico and the early opening of the central north Atlantic Ocean: Baton Rouge, Louisiana State University, p. 78–86.

Schwartz, D. P., Cluff, L. S., and Donnelly, T. W., 1979, Quaternary faulting along the Caribbean–North American plate boundary in Central America: Tectonophysics, v. 52, p. 431–445.

Scotese, C. R., and McKerrow, W. S., 1990, Revised world maps and introduction, in McKerrow, W. S., and Scotese, C. R., eds., Palaeozoic palaeogeography and biogeography: Geological Society of London Memoir 12, p. 1–21.

Scotese, C. R., Gahagan, L. M., and Larson, R. L., 1988, Plate tectonic reconstruction of the Cretaceous and Cenozoic ocean basins: Tectonophysics, v. 155, p. 27–48.

Scott, R. W., 1984, Mesozoic biota and depositional systems of the Gulf of Mexico–Caribbean region, in Westermann, G.E.G., ed., Jurassic-Cretaceous biochronology and paleogeography of North America: Geological Association of Canada Special Paper 27, p. 49–64.

Scott, R. W., and González-León, C., 1991, Paleontology and biostratigraphy of Cretaceous rocks, Lampazos area, Sonora, México, in Pérez-Segura, E., and Jacques-Ayala, C., eds., Studies of Sonoran geology, Geological Society of America Special Paper 254, p. 51–67.

Seager, W. R., and Morgan, P., 1979, Rio Grande Rift in southern New Mexico, west Texas, and northern Chihuahua, in Riecker, R. E., ed., Rio Grande rift: Tectonics and magmatism: Washington, D.C., American Geophysical Union, p. 87–106.

Sedlock, R. L., 1988a, Metamorphic petrology of a high-pressure, low-temperature subduction complex in west-central Baja California, México: Journal of Metamorphic Geology, v. 5, p. 205–233.

—— , 1988b, Tectonic setting of blueschist and island-arc terranes of west-central Baja California, México: Geology, v. 16, p. 623–626.

—— , 1988c, Lithology, petrology, structure, and tectonics of blueschists and associated rocks in west-central Baja California, México [Ph.D. dissertation]: Stanford, California, Stanford University, 223 p.

—— , 1992, Mesozoic geology and tectonics of the Cedros–Vizcaíno–San Benito and Magdalena-Margarita regions, Baja California, México: First International Meeting on Geology of the Baja California Peninsular, Memoir, Universidad Autónoma de Baja California Sur, La Paz, B.C.S., p. 55–62.

—— , 1993, Mesozoic geology and tectonics of oceanic terranes in the Cedros–Vizcaíno–San Benito and Magdalena-Margarita regions, Baja California, México, in Dunne, G. C., and McDougall, K. A., eds., Mesozoic paleogeography of the western United States: Pacific Section, Society of Economic Paleontologists and Mineralogists (in press).

Sedlock, R. L., and Hamilton, D. H., 1991, Late Cenozoic tectonic evolution of southwestern California: Journal of Geophysical Research, v. 96, p. 2325–2352.

Sedlock, R. L., and Isozaki, Y., 1990, Lithology and biostratigraphy of Franciscan-like chert and associated rocks, west-central Baja California, México: Geological Society of America Bulletin, v. 102, p. 852–864.

Sengor, A.M.C., and Dewey, J. F., 1990, Terranology: Vice or virtue?: Philosophical Transactions of the Royal Society of London, Series A, v. 331, p. 457–477.

Servais, M., Rojo-Yaíz, R., and Colorado-Liévano, D., 1982, Estudio de las rocas básicas y ultrabásicas de Sinaloa y Guanajuato: Postulación de un paleogulfo de Baja California y de una digitación Tethysiana en México central: Geomimet, v. 115, p. 53–71.

Servais, M., Cuevas-Pérez, E., and Monod, O., 1986, Une section de Sinaloa à San Luis Potosi: nouvelle approche de l'évolution du Mexique nord-occidental: Bulletin de la Société Géologique de France, Serie 8, v. 2, p. 1033–1047.

Severinghaus, J., and Atwater, T. M., 1990, Cenozoic geometry and thermal condition of the subducting slabs beneath western North America: Geological Society of America Memoir 176, p. 1–22.

Shafiqullah, M., Damon, P. E., and Clark, K. F., 1983, K-Ar chronology of Mesozoic-Cenozoic continental magmatic arcs and related mineralization in Chihuahua, *in* Clark, K. F., and Goodell, P. C., eds., Geology and mineral resources of north-central Chichuhua: El Paso Geological Society Guidebook, p. 303–315.

Shagam, R., and 6 others, 1984, Tectonic implications of Cretaceous-Pliocene fission-track ages from rocks of the circum-Maracaibo Basin region of western Venezuela and eastern Colombia: Geological Society of America Memoir 162, p. 385–414.

Sharpton, V. L., Schuraytz, B. C., Ming, D. W., Jones, J. H., Rosencrantz, E., and Weidie, A. E., 1991, Is the Chicxulub structure in N. Yucatan a 200 km diameter impact crater at the K/T boundary? Analysis of drill core samples geophysics, and regional geology: Houston, Texas, 22nd Annual Lunar and Planetary Science Conference, Abstracts.

Shurbet, D. H., and Cebull, S. E., 1984, Tectonic interpetation of the Trans-Mexican volcanic belt: Tectonophysics, v. 101, p. 159–165.

—— , 1987, Tectonic interpretation of the westernmost part of the Ouachita-Marathon (Hercynian) orogenic belt, west Texas–México: Geology, v. 15, p. 458–461.

Sieh, K. E., and Jahns, R., 1984, Holocene activity of the San Andreas fault at Wallace Creek, California: Geological Society of America Bulletin, v. 95, p. 883–896.

Siem, M. E., and Gastil, R. G., 1990, Active development of the Sierra Mayor metamorphic core complex, northeastern Baja California: Geological Society of America Abstracts with Programs, v. 22, A228.

—— , 1991, WNW to NW mid-Miocene to Recent extension associated with the Sierra Mayor metamorphic core complex, NE Baja California, México: Geological Society of America Abstracts with Programs, v. 23, p. A247.

Silva-Pineda, A., 1970, Plantas del Pensilvánico de la región Tehuacán, Puebla: Instituto de Geología, Universidad Nacional Autónoma de México, Paleontología Mexicana, v. 29, 109 p.

Silver, L. T., 1979, Peninsular Ranges batholith: A case study in continental margin magmatic arc setting, characteristics, and evolution: Geological Society of America Abstracts with Programs, v. 11, p. 517.

—— , 1986, Observations on the Peninsular Ranges batholith, southern California and Mexico, in space and time: Geological Society of America Abstracts with Programs, v. 18, p. 184.

Silver, L. T., and Anderson, T. H., 1974, Possible left-lateral early to middle Mesozoic disruption of the southwestern North America craton margin: Geological Society of America Abstracts with Programs, v. 6, p. 956.

—— , 1983, Further evidence and analysis of the role of the Mojave-Sonora Megashear(s) in Mesozoic Cordilleran tectonics: Geological Society of America Abstracts with Programs, v. 15, p. 273.

Silver, L. T., and Chappell, B. W., 1988, The Peninsular Ranges Batholith: An insight into the evolution of the Cordilleran batholiths of southwestern North America: Transactions of the Royal Society of Edinburgh: Earth Sciences, v. 79, p. 105–121.

Silver, L. T., Early, T. O., and Anderson, T. H., 1975, Petrological, geochemical, and geochronological asymmetries of the Peninsular Ranges batholith: Geological Society of America Abstracts with Programs, v. 7, p. 376.

Silver, L. T., Taylor, H. P., Jr., and Chappell, B., 1979, Peninsular Ranges Batholith, San Diego and Imperial Counties, *in* Mesozoic Crystalline Rocks, Guidebook for Geological Society of America meeting, San Diego: San Diego, California, Department of Geological Sciences, San Diego State University, p. 83–110.

Simonson, B. M., 1977, Geology of the El Porvenir quadrangle, Honduras, Central America: Tegucigalpa, Departmento de Geología e Hidrografía, Instituto Geográfico Nacionál de Honduras, Open-File Report, 84 p.

Singh, S. K., and Mortera, F., 1991, Source time functions of large Mexican subduction earthquakes, morphology of the Benioff zone, age of the plate, and their tectonic implications: Journal of Geophysical Research, v. 96, p. 21487–21502.

Smit, J., and many others, 1992, Tektite-bearing, deep-water clastic unit at the Cretaceous-Tertiary boundary in northeastern México: Geology, v. 20, p. 99–103.

Smith, C. I., 1981, Review of the geologic setting, stratigraphy, and facies distribution of the Lower Cretaceous in northern México: West Texas Geological Society Field Trip Guidebook, Publication 81-74, p. 1–27.

Smith, D. L., 1974, Heat flow, radioactive heat generation, and theoretical tectonics for northwestern México: Earth and Planetary Science Letters, v. 23, p. 43–52.

Smith, D. L., Nuckels, C. E., Jones, R. L., and Cook, G. A., 1979, Distribution of heat flow and radioactive heat generation in northern México: Journal of Geophysical Research, v. 84, p. 2371–2379.

Smith, D. P., 1987, Fault-controlled sedimentation in a mid-Cretaceous forearc basin: Valle Formation, Cedros Island, Baja California, México: Geological Society of America Abstracts with Programs, v. 19, p. 452.

Smith, D. P., and Busby-Spera, C. J., 1989, Evidence for the early remagnetization of Cretaceous strata on Cedros Island, Baja California Norte (México): EOS Transactions of the American Geophysical Union, v. 70, p. 1067–1068.

—— , 1991, Shallow magnetic inclinations in Cretaceous Valle Fm, Baja California: remagnetization, compaction, or 18 degrees northward drift compared to North America?: Geological Society of America Abstracts with Programs, v. 23, p. 99.

—— , 1992, Sedimentological evidence for syndepositional faulting in Cretaceous strata of Cedros Island, Baja California: First International Meeting on Geology of the Baja California Peninsula, Memoir, Universidad Autónoma de Baja California Sur, La Paz, B.C.S., p. 33–41.

Smith, J. T., 1984, Miocene and Pliocene marine mollusks and preliminary correlations, Vizcaíno Peninsula to Arroyo La Purísima, northwestern Baja California Sur, México, *in* Frizzell, V. A., Jr., ed., Geology of the Baja California Peninsula, Volume 39: Pacific Section, Society of Economic Paleontologists and Mineralogists, p. 197–217.

—— , 1991, Cenozoic marine mollusks and paleogeography of the Gulf of California, *in* Dauphin, J. P., and Simoneit, B.R.T., eds., The Gulf and Peninsular Province of the Californias: American Association of Petroleum Geologists Memoir 47, p. 637–666.

—— , 1992, The Salada Formation of Baja California Sur, México: First International Meeting on Geology of the Baja California Peninsula, Memoir, Universidad Autónoma de Baja California Sur, La Paz, B.C.S., p. 23–32.

Smith, J. T., and 6 others, 1985, Fossil and K-Ar age constraints on upper middle Miocene conglomerate, southwestern Isla Tiburón, Gulf of California: Geological Society of America Abstracts with Programs, v. 17, p. 409.

Smith, S. A., Sloan, R. F., Pavlis, T. L., and Serpa, L. F., 1989, Reconnaissance study of Cretaceous rocks in the state of Colima, México: A volcanic arc deformed by right-lateral transpression: EOS Transactions of the American Geophysical Union, v. 70, p. 1313.

Snow, J. K., Asmerom, Y., and Lux, D. R., 1991, Permian-Triassic plutonism and tectonics, Death Valley region, California and Nevada: Geology, v. 19, p. 629–632.

Sosson, M., Calmus, T., Tardy, M., and Blanchet, R., 1990, Nouvelles données sur le front tectonique cénomano-turonien dans le nord de l'État de Sonora (Mexique): Comptes Rendus l'Académie des Sciences, v. 310, p. 417–423.

Speed, R. C., 1979, Collided Paleozoic platelet in the western United States:

Journal of Geology, v. 87, p. 279–292.

Spencer, J. E., and Normark, W. R., 1979, Tosco-Abreojos fault zone: A Neogene transform plate boundary within the Pacific margin of southern Baja California, México: Geology, v. 7, p. 554–557.

Spencer, J. E., and Reynolds, S. J., 1986, Some aspects of the Middle Tertiary tectonics of Arizona and southeastern California: Arizona Geological Society Digest, v. 16, p. 102–107.

Squires, R. L., and Demetrion, R., 1989, An early Eocene pharetronid sponge from the Bateque Formation, Baja California Sur, México: Journal of Paleontology, v. 63, p. 440–442.

—— , 1990a, New early Eocene marine gastropods from Baja California Sur, México: Journal of Paleontology, v. 64, p. 99–103.

—— , 1990b, New Eocene marine bivalves from Baja California Sur, México: Journal of Paleontology, v. 64, p. 382–391.

Stanley, G. D., Jr., Sandy, M. R., and González-León, C., 1991, Upper Triassic fossils from the Antimonio Formation, northern Sonora, support the Mojave-Sonora Megashear: Geological Society of America Abstracts with Programs, v. 23, A127.

Steiger, R. H., and Jager, E., 1977, Subcommission of geochronology: Convention on the use of decay constants in geo- and cosmochronology: Earth and Planetary Science Letters, v. 36, p. 359–362.

Stein, S., and 9 others, 1988, A test of alternative Caribbean plate relative motion models: Journal of Geophysical Research, v. 93, p. 3041–3050.

Stevens, C. H., 1982, The Early Permian *Thysanophyllum* coral belt: Another clue to Permian plate-tectonic reconstructions: Geological Society of America Bulletin, v. 93, p. 798–803.

Stevens, C. H., and Stone, P., 1988, Early Permian thrust faults in east-central California: Geological Society of America Bulletin, v. 100, p. 552–562.

Stevens, C. H., Stone, P., and Kistler, R. W., 1992, A speculative reconstruction of the Middle Paleozoic continental margin of southwestern North America: Tectonics, v. 11, p. 405–419.

Stewart, G., Chael, E., and McNally, K., 1981, The November 29, 1978, Oaxaca, México, earthquake: A large, simple event: Journal of Geophysical Research, v. 85, p. 5053–5060.

Stewart, J. H., 1972, Initial deposits in the Cordilleran geosyncline: Evidence of a late Precambrian (<850 m.y.) continental separation: Geological Society of America Bulletin, v. 83, p. 1345–1360.

—— , 1978, Basin-Range structure in western North America—A review: Geological Society of America Special Paper 152, p. 1–31.

—— , 1988, Latest Proterozoic and Paleozoic southern margin of North America and the accretion of México: Geology, v. 16, p. 186–189.

—— , 1990, Position of Paleozoic continental margin in northwestern México: Present knowledge and speculations: Geological Society of America Abstracts with Programs, v. 22, p. 86–87.

Stewart, J. H., and Poole, F. G., 1974, Lower Paleozoic and uppermost Precambrian miogeocline, Great Basin, western United States, *in* Dickinson, W. R., ed., Tectonics and sedimentation: Society of Economic Paleontologists and Mineralogists Special Publication 22, p. 28–57.

Stewart, J. H., and Roldán-Quintana, J., 1991, Upper Triassic Barranca Group: Nonmarine and shallow-marine rift-basin deposits of northwestern México, *in* Jacques, C., and Pérez, E., eds., Studies in Sonoran geology: Geological Society of America Special Paper 254, p. 19–36.

Stewart, J. H., McMenamin, M.A.S., and Moráles-Ramírez, J. M., 1984, Upper Proterozoic and Cambrian rocks in the Caborca region, Sonora, México—Physical stratigraphy, biostratigraphy, paleocurrent studies, and regional relations: U.S. Geological Survey Professional Paper 109, 36 p.

Stewart, J. H., Anderson, T. H., Haxel, G. B., Silver, L. T., and Wright, J. E., 1986, Late Triassic paleogeography of the southern Cordillera: The problem of a source for voluminous volcanic detritus in the Chinle Formation of the Colorado Plateau region: Geology, v. 14, p. 567–570.

Stewart, J. H., Poole, F. G., Ketner, K. B., Madrid, R. J., Roldán-Quintana, J., and Amaya-Martínez, R., 1990, Tectonics and stratigraphy of the Paleozoic and Triassic southern margin of North America, Sonora, México, *in* Gehrels, G. E., and Spencer, J. E., eds., Geologic excursions through the Sonoran

Desert region, Arizona and Sonora: Arizona Geological Survey Special Paper 7, p. 183–202.

Stock, J. M., 1991, Volcanism and extension at the northern margin of the Puertecitos Volcanic Province, NE Baja California, México: Geological Society of America Abstracts with Programs, v. 23, A194.

Stock, J. M., and Hodges, K. V., 1989, Pre-Pliocene extension around the Gulf of California, and the transfer of Baja California to the Pacific plate: Tectonics, v. 8, p. 99–116.

—— , 1990, Miocene to Recent structural development of an extensional accommodation zone, northeastern Baja California, México: Journal of Structural Geology, v. 12, p. 315–328.

Stock, J., and Molnar, P., 1988, Uncertainties and implications of the Late Cretaceous and Tertiary position of North America relative to the Farallon, Kula, and Pacific plates: Tectonics, v. 7, p. 1339–1384.

Stoiber, R. E., and Carr, M. J., 1973, Quaternary volcanic and tectonic segmentation of Central America: Bulletin Volcanologique, v. 37, p. 304–325.

Stone, P., and Stevens, C. H., 1988, Pennsylvanian and Early Permian paleogeography of east-central California: Implications for the shape of the continental margin and the timing of continental truncation: Geology, v. 16, p. 330–333.

Strand, G. H., Wack, J. F., Gastil, R. G., and Elder, W. P., 1991, New age determination for the volcanic-volcaniclastic section at Arroyo San José, Baja California: Geological Society of America Abstracts with Programs, v. 23, A194.

Suárez-Vidal, F., Armijo, R., Morgan, G., Bodin, P., and Gastil, R. G., 1991, Framework of recent and active faulting in northern Baja California, *in* Dauphin, J. P., and Simoneit, B.R.T., eds., The Gulf and Peninsular Province of the Californias: American Association of Petroleum Geologists Memoir 47, p. 285–301.

Sun, S.-S., 1980, Lead isotopic study of young volcanic rocks from mid-ocean ridges, ocean islands and island arcs: Philosophical Transactions of the Royal Society of London, v. A297, p. 409–445.

Suppe, J., and Armstrong, R. L., 1972, Potassium-argon dating of Franciscan metamorphic rocks: American Journal of Science, v. 272, p. 217–233.

Suter, M., 1984, Cordilleran deformation along the eastern edge of the Valles–San Luis Potosí carbonate platform, Sierra Madre Oriental fold-thrust belt, east-central México: Geological Society of America Bulletin, v. 95, p. 1387–1397.

—— , 1987, Structural traverse across the Sierra Madre Oriental fold-thrust belt in east-central México: Geological Society of America Bulletin, v. 98, p. 249–264.

—— , 1991, State of stress and active deformation in Mexico and western Central America, *in* Slemmons, D. B., Engdahl, E. R., Zoback, M. D., and Blackwell, D. D., eds., Neotectonics of North America, Decade Map Volume: Geological Society of America, Boulder, Colorado, p. 401–421.

Suter, M., Quintero-Legorreta, O., and Johnson, C. A., 1992, Active faults and state of stress in the central part of the Trans-Mexican Volcanic Belt, Mexico: 1. The Venta de Bravo fault: Journal of Geophysical Research, v. 97, p. 11983–11993.

Sutter, J. F., 1979, Late Cretaceous collisional tectonics along the Motagua fault zone, Guatemala: Geological Society of America Abstracts with Programs, v. 11, p. 525.

Swanson, E. R., and McDowell, F. W., 1984, Calderas of the Sierra Madre Occidental volcanic field, western México: Journal of Geophysical Research, v. 9, p. 8787–8799.

Swanson, E. R., Keizer, R. P., Lyons, J. I., and Clabaugh, S. E., 1978, Tertiary volcanism and caldera development in the Durango city area, Sierra Madre Occidental, México: Geological Society of America Bulletin, v. 89, p. 1000–1012.

Sykes, L. R., McCann, W. R., and Kafka, A. L., 1982, Motion of the Caribbean plate during the last 7 million years and implications for earlier Cenozoic movements: Journal of Geophysical Research, v. 87, p. 10656–10676.

Taliaferro, N. L., 1933, An occurrence of Upper Cretaceous sediments in northern Sonora, México: Journal of Geology, v. 41, p. 12–37.

Tarduno, J. A., McWilliams, M., Debiche, M. G., Sliter, W. V., and Blake, M. C., Jr., 1985, Franciscan Complex Calera limestones: Accreted remnants of Farallon Plate oceanic plateaus: Nature, v. 317, p. 345–347.

Tarduno, J. A., McWilliams, M., Sliter, W. V., Cook, H. E., Blake, M. C., Jr., and Premoli-Silva, I., 1986, Southern hemisphere origin of the Cretaceous Laytonville limestone of California: Science, v. 321, p. 1425–1428.

Tardy, M., 1975, La nappe de Parras: Un trait essential de la structure laramienne du secteur transverse de la Sierra Madre Oreintale Mexique: Bulletin de la Société Géologique de France, Ser. 7, v. 17, p. 77–87.

Tardy, M., and Maury, R., 1973, Sobre la presencia de elementos de origen volcánico en las areniscas de los flyschs de edad cretácica superior de los Estados de Coahuila y de Zacatecas—México: Sociedad Geológica Mexicana Boletín, v. 34, p. 5–12.

Tardy, M., Carfantan, J.-C., and Rangin, C., 1986, Essai de synthèse sur la structure du Mexique: Bulletin de la Société Géologique de France, v. 6, p. 1025–1031.

Taylor, D. G., Callomon, J. H., Hall, R., Smith, P. L., Tipper, H. W., and Westermann, G.E.G., 1984, Jurassic ammonite biogeography of western North America: The tectonic implications *in* Westermann, G.E.G., ed., Jurassic-Cretaceous biochronology and paleogeography of North America: Geological Association of Canada Special Paper 27, p. 121–141.

Teissere, R. F., and Beck, M. E., Jr., 1973, Divergent Cretaceous paleomagnetic pole position for the southern California batholith, U.S.A.: Earth and Planetary Science Letters, v. 18, p. 296–300.

Thatcher, W., 1979, Horizontal crustal deformation from historic geodetic measurements in southern California: Journal of Geophysical Research, v. 84, p. 2351–2370.

Thomas, W. A., 1977, Evolution of the Appalachian-Ouachita salients and recesses from reentrants and promontories in the continental margin: American Journal of Science, v. 277, p. 1233–1278.

——, 1985, The Appalachian-Ouachitan connection; Paleozoic orogenic belt at the southern margin of North America: Annual Reviews of Earth and Planetary Sciences, v. 13, p. 175–199.

——, 1989, The Appalachian-Ouachitan belt beneath the Gulf Coastal Plain between the outcrops in the Appalachian and Ouachita Mountains, *in* Hatcher, R. D., Jr., Thomas, W. A., and Viele, G. W., eds., Decade of North American Geology, Volume F-2: The Appalachian and Ouachita orogen in the United States, Geological Society of America, p. 537–553.

——, 1991, The Appalachian-Ouachita rifted margin of southeastern North America: Geological Society of America Bulletin, v. 103, p. 415–431.

Thompson, S., III, Tovar-R., J. C., and Conley, J. N., 1978, Oil and gas exploration wells in the Pedregosa basin, *in* Land of Cochise: 29th Field Conference, New Mexico Geological Society, p. 331–342.

Todd, V. R., and Shaw, S. E., 1985, S-type granitoids and an I-S line in the Peninsular Ranges batholith, southern California: Geology, v. 13, p. 231–233.

Todd, V. R., Erskine, B. G., and Morton, D. M., 1988, Metamorphic and tectonic evolution of the northern Peninsular Ranges batholith, southern California, *in* Ernst, W. G., ed., Metamorphism and crustal evolution of the western United States, Rubey Volume VII, Englewood Cliffs, New Jersey, Prentice-Hall, p. 894–937.

Todd, V. R., Shaw, S. E., Girty, G. H., and Jachens, R. C., 1991, A probable Jurassic plutonic arc of continental affinity in the Peninsular Ranges batholith, southern California: Tectonic implications: Geological Society of America Abstracts with Programs, v. 23, p. 104.

Tolson, G., 1990, Structural development and tectonic evolution of the Santa Rosa area, SW state of México, México: Geological Society of America Abstracts with Programs, v. 22, p. A328.

Torres-Roldán, V., and Wilson, J. L., 1986, Tectonics and facies in the late Paleozoic Plomosas Formation of the Pedregosa basin of Chihuahua: Geological Society of America Abstracts with Programs, v. 18, p. 774.

Torres-Vargas, R., Murillo, G., and Grajales, M., 1986, Estudio petrográfico y radiométrico de la porción norte del límite entre los Complejos Acatlán y Oaxaca: Sociedad Geológica Mexicana, VIII Convención Nacional, Libro de Resúmenes.

Tosdal, R. M., Haxel, G. B., and Wright, J. E., 1990a, Jurassic geology of the Sonoran Desert region, southern Arizona, southeastern California, and northermost Sonora, *in* Jenny, J. P., and Reynolds, S. J., eds., Geologic evolution of Arizona, Volume 17: Arizona Geological Society Digest, p. 397–434.

Tosdal, R. M., Haxel, G. B., Anderson, T. H., Connors, C. D., May, D. J., and Wright, J. E., 1990b, Highlights of Jurassic, Cretaceous to early Tertiary, and mid-Tertiary tectonics, south-central Arizona and north-central Sonora, *in* Gehrels, G. E., and Spencer, J. E., eds., Geologic excursions through the Sonoran Desert region, Arizona and Sonora: Arizona Geological Survey Special Paper 7, p. 76–89.

Tovar-R., J. C., 1981, Provincias con possibilidades petroleras en el distrito de Chihuahua: Asociación Mexicanos Geológicos Petroleros Boletín, v. 33, p. 25–51.

Tozer, E. T., 1982, Marine Triassic faunas of North America: Their significance for assessing plate and terrane movements: Geologische Rundschau, v. 71, p. 1077–1104.

Troughton, G. H., 1974, Stratigraphy of the Vizcaíno Peninsula near Asunción Bay, Territorio de Baja California, México [M.S. thesis]: San Diego, California, San Diego State University, 83 p.

Tschanz, C. M., Marvin, R. F., Curuz, B., Mehnert, H. H., and Cebula, G. T., 1974, Geologic evolution of the Sierra Nevada de Santa Marta, northwestern Colombia: Geological Society of America Bulletin, v. 85, p. 273–284.

Tull, J. F., Harris, A. G., Repetski, J. E., McKinney, F. K., Garrett, C. B., and Bearce, D. N., 1988, New paleontologic evidence constraining the age and paleotectonic setting of the Talladega slate belt, southern Appalachians: Geological Society of America Bulletin, v. 100, p. 1291–1299.

Umhoefer, P. J., Dragovich, J., Cary, J., and Engrebretson, D. C., 1989, Refinements of the "Baja British Columbia" plate-tectonic model for northward translation along the margin of western North America: American Geophysical Union Geophysical Monograph 50, p. 101–111.

Umhoefer, P. J., Teyssier, C., and Dorsey, R. J., 1991, Plio-Quaternary dextral-normal faulting, Gulf Extensional Province, Tres Virgenes to Loreto basin, Baja California Sur: Geological Society of America Abstracts with Programs, v. 23, p. A232.

United States Geodynamics Committee, 1989, North American Continent-Ocean Transects Program: Washington, D.C., National Academy Press, 87 p.

Urrutia-Fucugauchi, J., and Böhnel, H. N., 1987, Tectonic interpretation of the Trans-Mexican volcanic belt—Discussion: Tectonophysics, v. 138, p. 319–323.

——, 1988, Tectonics along the Trans-Mexican volcanic belt according to palaeomagnetic data: Physics of Earth and Planetary Interiors, v. 52, p. 320–329.

Urrutia-Fucugauchi, J., and Linares, E., 1981, Dating of hydrothermal alteration, Ixtapán de la Sal, México State, México: Isochron/West, v. 131, p. 15.

Urrutia-Fucugauchi, J., and Valencio, D. A., 1986, Paleomagnetic study of Mesozoic rocks from Ixtapán de la Sal, México: Geofísica Internacional, v. 25, p. 485–502.

Urrutia-Fucugauchi, J., Morán-Zenteno, D. J., and Cabral-Cano, E., 1987, Palaeomagnetism and tectonics of México: Geofísica Internacional, v. 26, p. 429–458.

Valdes, C. M., and others, 1986, Crustal structure of Oaxaca, México, from seismic refraction measurements: Bulletin of the Seismological Society of America, v. 76, p. 547–563.

van Andel, T., 1964, Recent marine sediments of Gulf of California: American Association of Petroleum Geologists Memoir 3, p. 216–230.

Van der Roo, R., Peinado, J., and Scotese, C. R., 1984, A paleomagnetic reevaluation of Pangea reconstructions, *in* Van der Voo, R., Scotese, C. R., and Bonhommet, N., eds., Plate reconstructions from Paleozoic paleomagnetism, Volume 12: American Geophysical Union Geodynamics Series, p. 11–26.

Vega-Vera, F. J., Mitre-Salazar, L.-M., and Martínez-Hernández, E., 1989, Contribución al conocimiento del estratigrafía del Grupo Difunta (cretácico superior-terciario) en le noreste de México: Revista del Instituto de Geología, Universidad Nacional Autónoma de México, v. 8, p. 179–187.

Verma, S. P., 1987, Mexican Volcanic Belt: Present state of knowledge and unsolved problems: Geofisica Internacional, v. 26, p. 309–340.

Verma, S. P., and Nelson, S. A., 1989, Isotopic and trace element constraints on the origin and evolution of alkaline and calc-alkaline magmas in the northwestern Mexican Volcanic Belt: Journal of Geophysical Research, v. 94, p. 4531–4544.

Viele, G. W., 1979a, Geological map and cross section, eastern Ouachita Mountains, Arkansas: Geological Society of America Map and Chart Series, MC-28F.

—— , 1979b, Geological map and cross section, eastern Ouachita Mountains, Arkansas: Map summary: Geological Society of America Bulletin, v. 90, p. 1096–1099.

Viele, G. W., and Thomas, W. A., 1989, Tectonic synthesis of the Ouachita orogenic belt, in Hatcher, R. D., Jr., Thomas, W. A., and Viele, G. W., eds., Decade of North American Geology, Volume F-2: The Appalachian and Ouachita orogen of the United States: Geological Society of America, p. 695–728.

Villaseñor-Martínez, A. B., Martínez-Cortes, A., and Contreras y Montero, B., 1987, Bioestratigrafía del paleozoico superior del Salvador Patlanoaya, Puebla, México: Revista del Sociedad Mexicana Paleontológica Bulletín, v. 1, p. 396–417.

Viniegra-Osorio, F., 1971, Age and evolution of salt basins of southeastern México: American Association of Petroleum Geologists, v. 55, p. 478–494.

—— , 1981, Great carbonate bank of Yucatán, southern México: Journal of Petroleum Geology, v. 3, p. 247–278.

von Herzen, R. P., 1963, Geothermal heat flow in the gulfs of California and Aden: Science, v. 140, p. 1207–1208.

Wadge, G., and Burke, K., 1983, Neogene Caribbean plate rotation and associated Central American tectonic evolution: Tectonics, v. 2, p. 633–643.

Walck, M. C., 1984, The P-wave upper mantle structure beneath an active spreading centre: The Gulf of California: Geophysical Journal of the Royal Astronomical Society, v. 76, p. 697–723.

Walker, J. D., 1988, Permian and Triassic rocks of the Mojave Desert and their implications for timing and mechanisms of continental truncation: Tectonics, v. 7, p. 685–709.

Walker, N. W., 1992, Middle Proterozoic geologic evolution of Llano uplift, Texas: Evidence from U-Pb zircon geochronometry crust in the Llano: Geological Society of America Bulletin, v. 104, p. 494–504.

Wallace, P., Carmichael, I.S.E., Righter, K., and Becker, T. A., 1992, Volcanism and tectonism in western Mexico: A contrast of style and substance: Geology, v. 20, p. 625–628.

Walper, J. L., 1980, Tectonic evolution of the Gulf of Mexico, in Pilger, R. H., Jr., ed., The origin of the Gulf of Mexico and the early opening of the central north Atlantic Ocean: Baton Rouge, Louisiana State University, p. 87–98.

Walper, J. L., and Rowett, C. L., 1972, Plate tectonics and the origin of the Caribbean and the Gulf of Mexico: Transactions of the Gulf Coast Geological Society, v. 22, p. 105–116.

Ward, P. L., 1991, On plate tectonics and the geologic evolution of southwestern North America: Journal of Geophysical Research, v. 96, p. 12479–12496.

Wardlaw, B., Furnish, W. M., and Nesiell, M. K., 1979, Geology and paleontology of the Permian beds near Las Delicias, Coahuila, México: Geological Society of America Bulletin, v. 90, part I, p. 111–116.

Wark, D. A., 1991, Oliogocene ash flow volcanism, northern Sierra Madre Occidental: Role of mafic and intermediate-composition magmas in rhyolite genesis: Journal of Geophysical Research, v. 96, p. 13389–13411.

Wark, D. A., Kempter, K. A., and McDowell, F. W., 1990, Evolution of waning, subduction-related magmatism, northern Sierra Madre Occidental, México: Geological Society of America Bulletin, v. 102, p. 1555–1564.

Watkins, J. S., McMillen, K. J., Bachman, S. B., Shipley, T. H., Moore, J. C., and Angevine, C., 1981, Tectonic synthesis, Leg 66: Transect and vicinity and geodynamics, in Initial Reports of the Deep Sea Drilling Project, Volume 66: Washington, D.C., U.S. Government Printing Office, p. 837–849.

Weber, R., Centeno-García, E., and Magallón-Puebla, S., 1987, La formación Matzitzi (Estado de Puebla) tiene edad PermoCarbonífera: Segundo Simposio de Geología Regional de México, p. 57–58.

Weidie, A. E., Jr., Ward, W. C., and Marshall, R. H., 1979, Geology of Yucatán platform, Geology of Cancun, Quintana Roo, México, volume 79-72: West Texas Geological Society Field Trip Guidebook, p. 50–68.

Westermann, G.E.G., Corona, R., and Carrasco, R., 1984, The Andean Mid-Jurassic Neuqueniceras ammonite assemblage of Cualac, México, in Westermann, G.E.G., ed., Jurassic-Cretaceous biochronology and paleogeography of North America: Geological Association of Canada Special Paper 27, p. 99–112.

Weyl, R., 1980, Geology of Central America: Berlin, Gebruder Borntraeger, 371 p.

White, E. H., and Guiza, R., 1949, Antimony deposits of El Antimonio district, Sonora, México: U.S. Geological Survey Bulletin 962-B, p. 81–116.

Whittington, H. B., and Hughes, C. P., 1972, Ordovician geography and faunal provinces deduced from trilobite distribution: Philosophical Transactions of the Royal Society of London, Series B, v. 263, p. 235–278.

Williams, H., McBirney, A. R., and Dengo, G., 1964, Geologic reconnaissance of southeastern Guatemala: University of California Publications in Geological Sciences, v. 50, p. 1–56.

Wilson, H. H., 1974, Cretaceous sedimentation and orogeny in nuclear Central America: American Association of Petroleum Geologists Bulletin, v. 58, p. 1348–1396.

Wilson, I. F., and Rocha, V. S., 1949, Coal deposits of the Santa Clara district near Tónichi, Sonora, México: U.S. Geological Survey Bulletin 962-A, p. 1–80.

Wilson, J. L., 1990, Basement structural controls on Mesozoic carbonate facies in northeastern México—A review: Special Publications of the International Association of Sedimentologists, v. 9, p. 235–255.

Windh, J., Griffith, R. C., and Girty, G. H., 1989, Tectonic significance of kink bands in Arroyo Calamajué, Baja California, México, in Abbott, P. L., ed., Geologic studies in Baja California, Volume 63: Pacific Section, Society of Economic Paleontologists and Mineralogists, p. 75–78.

Winker, C. D., and Buffler, R. T., 1988, Paleogeographic evolution of early deep-water Gulf of Mexico and margins, Jurassic to Middle Cretaceous (Comanchean): American Association of Petroleum Geologists Bulletin, v. 72, p. 318–346.

Wolf, M. B., and Saleeby, J. B., 1992, Jurassic Cordilleran dike swarm-shear zones: Implications for the Nevadan orogeny and North American plate motion: Geology, v. 20, p. 745–748.

Woods, M. T., and Davies, G. F., 1982, Late Cretaceous genesis of the Kula plate: Earth and Planetary Science Letters, v. 58, p. 161–166.

Wollard, G. P., and Monges-Caldera, J., 1956, Gravedad, geología regional y estructura cortical en México: Anales del Instituto de Geofísica, Universidad Nacional Autónoma de México, v. 2, p. 60–112.

Yañez, P., Ruiz, J., Patchett, P. J., Ortega-Gutiérrez, F., and Gehrels, G., 1991, Isotopic studies of the Acatlán Complex, southern México: Implications for Paleozoic North American tectonics: Geological Society of America Bulletin, v. 103, p. 817–828.

Young, K., 1983, Mexico, in the Phanerozoic geology of the world, II: The Mesozoic, B: Amsterdam, Elsevier, p. 61–88.

Zaldívar-Ruiz, J., and Garduño-M., V. H., 1984, Estudio estratigráfico y estructural de las rocas del paleozoico superior de Santa María del Oro, Durango y sus implicaciones tectónicas: Sociedad Geológica Mexicana, VII Convención Nacional, Memoria, p. 28–39.

Ziagos, J. P., Blackwell, D. D., and Mooser, F., 1985, Heat flow in southern México and the thermal effects of subduction: Journal of Geophysical Research, v. 90, p. 5410–5420.

Zoback, M. L., Anderson, R. E., and Thompson, G. A., 1981, Cenozoic evolution of the state of stress and style of tectonism in the western United States: Philosophical Transactions of the Royal Society of London, Series A, v. 300, p. 407–434.

MANUSCRIPT ACCEPTED BY THE SOCIETY AUGUST 4, 1992

Index

[Italic page numbers indicate major references]

A

Acadian orogenesis, 35, 86
Acapulco, 11, 73
Acapulco Batholith, 11
Acapulco earthquake (1907), 5
Acapulco earthquake (1957), 5
Acapulco intrusive suite, 11
Acapulco Trench, 5, 7
Acateco Subgroup, 34
Acatlán Complex, 23, *34*, 35, 37, 72, 85, 86, 89, 93
accretion, *107*
Acuitlapán Formation, 40
adamellite, 13
aeromagnetic data, Maya terrane, 33
agglomerate, Yuma composite terrane, 63
Agua Blanca fault, 5, 63, 65, 107, 118
Aguán fault, 13
Aibo granite, 49
Alaska, southeastern, 112
Aldama, 42
Alexander terrane, 112
Alisitos Formation, 63, 64, 66, 70, 105
Alisitos–Santiago Peak–Peninsular Range batholith, 111
Alisitos terrane, 63
alkali basalts, Tepehuano terrane, 62
alkalic magma, Nahuatl terrane, 42
alkalic rocks, Nahuatl terrane, 41
alkalic volcanic rocks, 8
Alleghany orogen, 91
alluvial fans, Tepehuano terrane, 62
Altar, 47
Altiplano. *See* Mesa Central
Altiplano geomorphic province, 58
ammonia, 10
ammonites
   Cuicateco terrane, 23
   Serí terrane, 51
   Tepehuano terrane, 59
   Yuma composite terrane, 65, 81
ammonoids, Serí terrane, 52
amphibole, Cochimí terrane, 22
amphibolite
   Chatino terrane, 11, 13
   Cochimi terrane, 18, 22
   garnetiferous, 19
   Guachichil terrane, 25
   Maya terrane, 28, 33
   Mixteco terrane, 34
   Nahuatl terrane, 37, 38
   Pericú terrane, 47
   Serí terrane, 49
   Tahué terrane, 54
   Tepehuano terrane, 59
   Yuma composite terrane, 65
amphibolite dikes, North America terrane, 42

amphibolite gneiss, North America terrane, 42
andalusite, Pericú terrane, 47
andesite
   Alisitos Formation, 70
   Chatino terrane, 14
   Chortis Formation, 15
   Chortis terrane, 15
   Cuicateco terrane, 24
   Maya terrane, 32, 33, 34
   Nahuatl terrane, 40, 41, 71
   North America terrane, 43, 46
   Serí terrane, 52
   Tahué terrane, 55
   Tepehuano terrane, 62
   Yuma composite terrane, 64, 65, 66
   Zapoteco terrane, 68
andesitic conglomerate, Nahuatl terrane, 41
andesitic dikes, Pericú terrane, 47
andesitic flows
   Nahuatl terrane, 41
   Tahué terrane, 54, 55
   Yuma composite terrane, 63
andesitic hypabyssal rocks, Mixteco terrane, 37
andesitic lavas
   Mixteco terrane, 37
   Nahuatl terrane, 40
   Serí terrane, 52
   Tepehuano terrane, 62
andesitic tuffs, Tahué terrane, 55
andesitic volcanic rocks, 7
anhydrite, 72
anomaly, gravity, 15, 57
anorogenic granites, North America terrane, 42
anorogenic granitoids, North America terrane, 42
anticlinoria, Guachichil terrane, 25, 27
antimony, 10
Antler orogeny, 78, 89
apatite, Cochimí terrane, 22
Appalachian orogen, 76
aragonite, Cochimí terrane, 22
Aramberri, Nuevo León, 27
arc assemblages, Nahuatl terrane, 40
arc magmatism, 1, *98*
arc rocks
   Cochimi terrane, 19
   Cuicateco terrane, 23
   Maya terrane, 34
   Nahuatl terrane, 42
Arcelia, 38
arcs, 1
   continental, 5, 7
Argentina, 86
argillite
   Cochimí terrane, 22
   Serí terrane, 49
   Tahué terrane, 54

Arizona, 42
   southeastern, 43
   southern, 42, 47, 85, 99
   southwestern, 79, 108
Arperos Formation, 61, 62
Arroyo Calamajué, 64, 65
Arteaga terrane, 41
Artesa sequence, 43
asbestos, 10
Ayotusco Formation, 37, 38, 40

B

Bacurato Formation, 54, 55
Bahía de Campeche, 32
Bahía Sebastián Vizcaíno, 66, 70
Baja block, 103, 111
   defined, 80
Baja California, 3
   northeastern, 52, 54
   northern, 52
Baja California Norte, 49, 50, 65
Baja California Sur, 52, 66, 115
Balsas Group, 41
barite, 10
   Serí terrane, 49
Barranca Group, 52, 70
basalt
   Chortis Formation, 15
   Coahuiltecano terrane, 17
   Cuicateco terrane, 25
   Maya terrane, 33
   Nahuatl terrane, 41
   ocean-floor, 18
   Tahué terrane, 55
   Tepehuano terrane, 62
   Yuma composite terrane, 63, 64, 66
basaltic andesites, North America terrane, 46
basaltic flows, Tahué terrane, 54
basaltic volcanic rocks, 8
basanites, Tepehuano terrane, 62
basement rocks
   Chortis terrane, *13*
   Serí terrane, 49
Basin and Range extension, 2, 4, 5, 47, 52, 54, 62, 114, 115, 117, 118, 119
Basin and Range faulting, 52
Basin and Range province, 2, 3, 4, 8, 68, 117, 118
basin subsidence, Cochimi terrane, 18
basinal allochthon, Serí terrane, 50
basinal assemblages, Nahuatl terrane, 40
basinal rocks, Serí terrane, *49*
bathymetry, 4
Baucarit Formation, 46, 52
Bedford Canyon Formation, 65
belemnites, Yuma composite terrane, 63

Typeset by WESType Publishing Services, Inc., Boulder, Colorado
Printed in U.S.A. by Malloy Lithographing, Inc., Ann Arbor, Michigan